Ruppert's Contemporary Bangerpedia 2000 - 2010

James Ruppert

ISBN: 978-1-9160522-3-9

RUPPERT'S BANGERPEDIA 2000 - 2010

BANGERDEDICATION

To everyone who drives on the motoring margins, or aspires to.

EXCUSES EXCUSES

The Publisher and Author cannot guarantee the pin point accuracy of every single stat in this Bangerpedia, so can't be responsible for any decisions the reader makes based on the information contained inside.
Any questions? By all means email the Bangernomics organisation for friendly free advice. Or just to stay hello.

ABOUT THE COVER

I changed it. The ones on the original were mostly car park spots. I quite liked the ropey Micra, althought it was probably more of a '90s survivor. The Vauxhall Monaro belonged to my local garage, but it wasn't posed , I just happen to take a picture of it, The Mercedes Estate needed some colourisation to cheer it up, so let's change things. There has to be a MINI because it is one of the most important cars of the decade that gradually grew into its Banger status. Meanwhile the Golf based Audi TT proved that it was something of a high mileage star and always affordable. Finally and it was final, the rather handsome Rover 75 in Touring format.

WHAT'S IN THE BANGERPEDIA?

THANKS VERY MUCH

Dee & Olivia
(their names are above the title as well)
Car manufacturers who took most of the pictures and then expected rather better reproduction and a more prestigious publication for them to feature in.
All the writers and sub editors I have worked with over the years, on various publications including Parker's Price Guide, Used Car Buyer, Car Week, Car, Practical Classics, Classic Cars, Fast Ford, 911 and Porsche World, Golf Xtreme, Mini World, Mini Magazine, The Independent, The Times, The Evening Standard, Performance Car, Autocar, Free Car Mag, Bangernomics Mag and loads I've forgotten.

READ THIS FIRST

They don't make Bangers anymore. You know, cars that eventually become cheap enough to buy with some modest savings or disposable income. They should be easy to fix, possibly cosmetically compromised, but they will do a job. That means getting you from A to B most of the time.

Motor manufacturers now build four wheeled software programmes that could 'crash' at any moment. Increasingly they hum, have a couple of engines and sometimes a boot full of batteries which is no way to construct a practical, easy to fix and live with vehicle. That is not their game anyway, the industry wants you to move onto the next wonderful thing. It is important to remember that car manufacturers don't care about your bank balance, metal health or general well being. They just want your money. Try not to give it to them, unless you are buying some OE parts which they may not have made anyway.

Meanwhile, proper Bangers are also fading away, so before they all do, I would like to draw your attention to what might, or might not, just about pass for one. What am I talking about here?

Let's remember that there are the Seven Ages of Car. First off it is showroom fresh and Brand New. For the next few months and maybe a year, so long as the mileage is low it will be regarded as Nearly New. After that, the motor will qualify as a Used Approved under a manufacturer scheme. After that, it will be Out of Warranty. That's when they fetch up in the classified ads and on used car lots. Beyond that as they become cheap and unloved motors slip into the banger class. Several years down the line two things can happen: either the car gets scrapped, or it survives and starts to be regarded by enthusiasts as a Classic. All of a sudden Bangers go up in value and will become a classic. Maybe there will be a Classic Bangerpedia eventually if I can be bothered.

For the moment the Contemporary Bangerpedia seeks to identify the cars which are Out of Warranty and will be mostly 10+ years old. I would argue that it takes a decade to find out whether a model is any good or not. That's why we should be highly suspicious of all those European Car of the Year Awards. Often decided upon and given the gong before any consumer has actually stumped up their own cash to actually own them.

What the Bangerpedia contains is most of the models offered for sale in Britain between 2000 and 2010 with one Twingo shaped exception. I avoided everything launched in 2010 as that is a new era, with one Jaguar shaped exception. That means there will be some '90s classics and also relative recent gear from the 10s as well as the noughties. However, some models, whatever happens in the world are unlikely to ever be considered a Banger, or at least affordable. Astons, Lamborghinis, Ferraris and the high end Porsches are mostly outside of the remit. Anything else I've just forgotten. The point is that products from some of the posh manufacturers might seem very tempting, what with massive depreciation and all that, so effectively there will be some warnings to stay away. Actually most late models are so complicated, a word I use far too much, that any problem requires a four figure solution.

If I have missed anything, by all means let me know, I might care enough to possibly include it in the future, with a credit to you. This does though have my name on it, so this is my angle on the whole encyclopedia thing and what is and isn't a Banger. If you don't like it go and write your own. There is a lot for anoraks to argue about, so be my guest.

This Bangerpedia was a difficult bit of literature to assemble and I may not bother again. I tried to use printed sources, you know, books and magazines, as I have a lot of that real stuff going back many years. Some of it I actually wrote. Indeed, I was part of the team that assembled the Parker's Chooser magazine twenty years ago. This Bangerpedia is in many ways a less comprehensive version of that publication. For reference I could not use printed manufacturer brochures all the time, so did stray into various magazines, but very few websites. There certainly were some inconsistencies which explained why publications disagreed about the length of a Ford Focus in 2005, but they did.

What I did when it came to many of the variable figures was put a spread of relevant data, be it official combined MPG, or CO2. That's because when people search for used cars they need to know what the best and worst case scenario is. Not only that, the information varies slightly as a 4×4 buyer might want to know the maximum braked towing weight rather than the top speed or acceleration. The way buying used cars actually works it is that you often end up with something you quite possibly had not even considered. So don't use this as a definitive reference work. It is imperfect and this Bangerpedia will give experts like Keith WR Jones at Parker's some databased nightmares, but not sleepless nights, that there is a rival brand to worry about, (especially as I might have deleted the number of doors or seats, when it is bloody obvious, so that the entry will fit the format).

The pictures aren't perfect. Reproduction is not brilliant when you are relying on Amazon to do it for you. So I decided to make the black and white images impressionistic. It is still a recognition guide of sorts.

It was tempting to list a whole slew of 'what to look fors'. However, I have kept those to minimum on the basis that, 'they don't all do that'. Some might, others might not. Every used car is different. That's because it may not have been serviced, ever, could have been in an accident, or done loads of short stop/start journeys that may be slowly killing the engine. When it comes to mechanicals, I'll only take advice from Mark, Ian, Ted, Rob and Kenny because they are actually qualified to tell me off. If your best mate's girlfriend's mother had a Renault Scenic that did two million miles on its original parts, I don't actually care.

I'm sure you have got the message that this is my Bangerpedia. My rules and mistakes. Rather like I believe it is possible for a convertible or a sports car to be a Banger. This is all about affordable cars and they can be any shape, size or make.

Finally, if you think this is a fairly useless book best left in the smallest room in your house, it may have another purpose. You could play a game of Bangerpedia. All it requires you to do is tick off or highlight the models you have actually seen in real life. Maybe for all time, or perhaps from the moment you have the book delivered. Do it as a competition with a mate. Maybe you need to take a picture to prove it. I would certainly be impressed if you ticked all the models off. It is hard to believe just how many vehicle derivatives were offered by manufacturers over such a short period of time.

I suppose if you could prove you had essentially I-Spyed all the motors in this Bangerpedia I'll put your name in future issues and maybe think of some prize or something, Don't hold me to this though. If all goes well I'll try and develop 'Bangerpedia the Board Game'. Or something.

I hope the Bangerpedia proves to a bit useful rather than being a definitive guide to all the vehicles produced over a ten year period, because it isn't. Where it could come in handy is as a choc on one of the wheels when you inevitably have to jack your Banger up. Mostly though I think it will just start arguments between used car clever dicks. Provided they have bought a copy I don't mind. I do feel the need to restate that I accept no responsibility for any decision you make based on the contents of this Bangerpedia.

With that in mind, always remember that buying any used car, whether or not you have a Bangerpedia stuffed up your jumper, or not, is much a matter of luck as well as judgement. That's why I always remind used car buyers to always beware of the dog.

James Ruppert...at various points in 2020 and 2021 (revised in 2025 to right all the sloppy wrongs) in Bangerfolkshire, The Great Banger Isles

Bangerpedia Decoded

2000 - 2009 are the build years although sometimes there is a disparity between the on sale, or launch, production and registration years. Overall I have triangulated the data to arrive at the best possible beginning and end dates based on the available facts. When there are some open brackets ( ) then we don't know the end date, because the model is still in production. So by all means pencil that in yourself.

Engines: Engine sizes for petrol and a D means diesel, H means Hybrid (electric and at one point F meant Flexifuel in the Volvo section **CO2:** Spread of g/km for taxation purposes **MPG**: Best to worst official combined figures **Gearbox:** Man/ Auto **Boot:** Litres for where relevant seats up and folded **Seats**: Number of, sometimes 2+2 when there is barely enough room **Doors:** Number of, unless it was obvious. **Towing:** Maximum braked in kg **Top Speed**: Fastest and Slowest, not always the turbo-nutter one **0-60:** Best and worst in Seconds, it could actually be 0-62 because of kph element, but thought that the 60 is neater.

Almost forgot, then there is the rating system. This involves Slog the Bangernomic Dog who has been on this journey for some time. It used to be beware of the dog, but in the interests of clarity it has to be the more Slogs the better. Readers would associate it with a star system anyway.

Just the one Slog means you should definitely avoid. That seems obvious. This is probably a triple whammy of potential unreliability, compromised fixability and the overall expense of doing so. The truth is that many upmarket late model cars should be avoided, but might have other saving graces, like being nice to drive. Or just does what it is designed to do, like go off road.

Two Slogs is a fairly questionable Banger. Quite likely to have an average reputation for not breaking down. It might have some added element of overall

averageness, be it in the practicality, driving or repair departments.

Three Slogs is most Bangers. It averages out as being good value for money, easy to own, does what it originally said in the manufacturers brochure. This is the default rating. Every Banger starts with a three and we work up or down from there.

Four Slogs is jolly good going for any Banger. If you've got any standards then this is the one to aim for, even though as we know buying a Banger is not a precise science. You end up with what you can afford or what is available. If it is a four star Banger then great.,This will do a job with some style.

Five Slogs is the Banger Gold standard. These are going to be the real legends which are unarguably brilliant Bangers. Most of these are going to be Japanese, but they will have a genuine reputation for being very good indeed at being a Banger.

No Slogs? I think there might be one entry which scores a big fat Slog shaped zero. Best to run a mile from that one.
Anyway, here they are. Bangers grouped together according to their dimensions and/or their claimed abilities and attributes.

Town Centre Tots - Inner City Bangers

These are really town centre assault vehicles. The minimal amount of motoring you can get away with. A couple of doors and just the right number of wheels and you can do the shopping, commute or the school run.

This is almost the last gasp for the genuinely small car. The '90s actually came up with some properly game changing concepts, and no we are not referring to the Peugeot 1007 and its enormous sliding doors. No one lkked or bought that, except very old people who found it very convenient. Even the more up market Toyota IQ didn't get anywhere despite being turned into an Aston Martin

In theory two seats was all a City dwelling/visiting motorist needed. Consequently the MCC Smart survived long enough to be taken under Mercedes protective wing. The SmartForTwo was still too expensive to succeed and the reality is that you can use bigger motors, like an S-Class with few real downsides. After all, you are still in the same traffic jam as a tewo seat tiddler, so you might as well be comfy.

The reality was that buyers wanted cheap and relatively cheerful which explains the Citroen C1 or Peugeot 107. If the badge matters then there is the Aygo of course. What buyers did respond well too was a bit of high fashion. You could argue that the MINI should be here, but it is simply too big. We think. The Fiat 500 was though the right small car at the right time. Buyers loved it. Meanwhile the all new identical under the skin Ford Ka got on with the business of anonymously doing it's own thing without the same profitable fanfare.

Back to the cheapish and not very cheerfulish are the rather crude Nissan Pixo and related Suzuki Alto plus the surviving Peroduas.

As usual the Koreans had been paying attention and the tiny four door Picanto was a wonderful way to bounce around with a nice big fat warranty as back up. Ditto the Hyundai i10. The automotive bread and butter was starting to shift Eastwards and from the bottom of the small city used car market upwards.

Chevrolet Matiz 2005 - 2009

Like it says under D for Daewoo, but with added dullness to the styling and a badge usually found on something exciting, like a Corvette. Some say it is better built than the plastic sports car but instead of a V8 there is a four cylinder 1.0 litre engine Likely to be less knackered than the Daewoo one. Also buzzy 3-cylinder engine. Reliable.

Engines: 800cc • 1.0 **CO2:** 119-137.7 g/km **MPG**: 49.6-54.3

Gearbox: Man/Auto **Boot:** 170/845 **Seats**: 4 **Doors:** 5

Citroen C1 2005 - 2009

Costs buttons to buy and run, first time drivers don't mind it. Can't get much in the back or the titchy boot. The 1.0 Vibe is all you need, the 1.4 diesel is pointless. Is reliable enough, seems to get through tyres pretty quick. Electrical issues. Also found with Toyota and Peugeot badges. Sometimes a money pit. Controversial two Slog rating.

Engines: 1.0 • 1.4D **CO2:** 109 g/km **MPG**: 61.4 -68.9

Gearbox: Man/AutoEGS **Boot:** 130/712 **Seats**: 4 **Doors:** 3/5

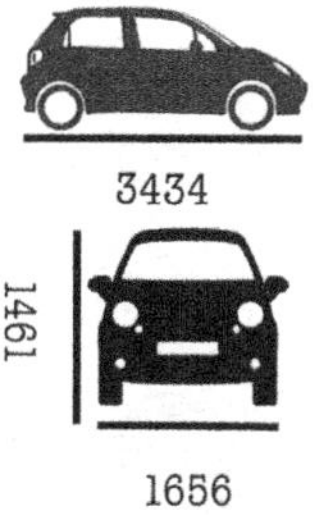

Citroen C1 2009 - 2014

Facelift, just about seats four, but ideally two adults and kids. It also has revised equipment packages and the same rear hatch which has a big lip, when the rear seats are folded there is a ton of space. There was an EGS robot manual gearbox which saves the driver the bother of shifting from around 2010. Stronger clutch, but starter motor can go pop.

Engines: 1.0 • 1.4D **CO2:** 106-109g/km **MPG**: 62.8-68.9

Gearbox: Man/Auto **Boot:** 130/712 **Seats**: 4 **Doors:** 3/5

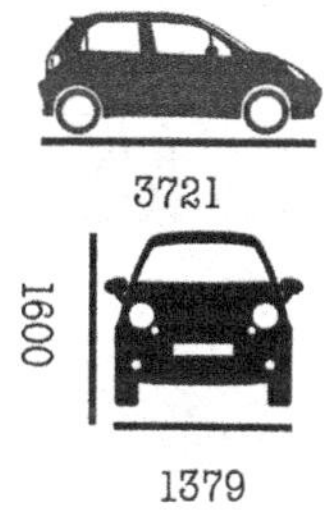

Citroen Saxo 1996 - 2003

Those after the full zero spec, one wheel trim missing Francophile experience (pedals set to the left) will find it in a sexy Saxo. Huge fun to drive, 1.1 First has less than nothing. VTR for when you stray out of town. Often modified very badly, clutches go bang and neglect blows head gaskets. Better built from 2000.

Engines: 1.1 • 1.4 • 1.6 **CO2:** 145-159 g/km **MPG**: 38.7-46.3

Gearbox: Man/Auto **Boot:** 280/953 **Seats**: 4 **Doors:** 3/5

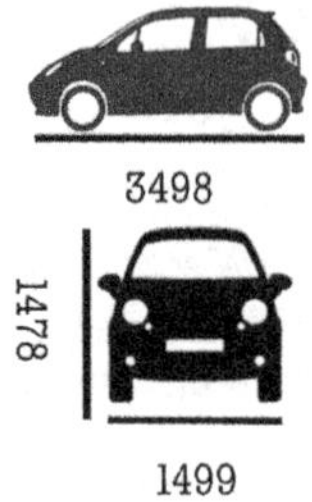

Daewoo Matiz 2001 - 2005

Interesting styling (Giugiaro who did the Golf) which resembles a funky delivery van. Actually quite nippy and easy to live with. The three cylinder engine makes a quite a roarty noise. Seems to have niggly engine issues once it is a few decades old. SE+ model does not add much to the mix. Then became a Chevy see back. Can be neglected and tired.

Engines: 800cc **CO2:** 149 g/km **MPG**: 46.3 **Gearbox:** Man/Auto **Boot:** 167/624 **Seats**: 4 **Doors:** 5

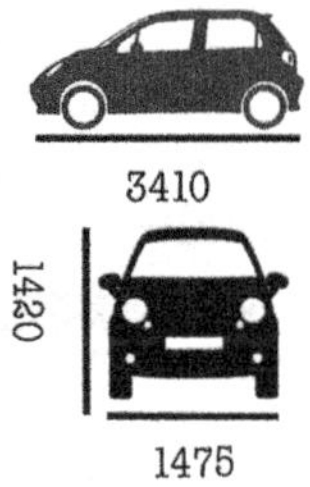

Daihatsu Cuore 1997 - 2003

Not great to look at or drive, but it will last pretty much forever, if not longer. Another three cylinder thrasher that is truly at home in the City, and a bit all at sea elsewhere. Rust is the biggest killer now and finding parts is an ever bigger issue which starts to make insurance and just keeping it running more of a challenge.

Engines: 1.0 **CO2:** 127 g/km **MPG**: 53.3 **Gearbox:** Man/Auto **Boot:** 155/428 **Seats**: 4 **Doors:** 5

Fiat 500 2007 - 2014

Here is the best small Fiat in several generations, that may be because it isn't actually built in Italy anymore. 500 related to the Fiesta, but it is much cuter. Reliability is generally patchy, only problem has been that it costs a lot more to buy because it is pretty and popular. Lots of niggly electrical stuff and rust underneath. Need to be good at DIY or have a decent garage on hand. Two and a half Slogs?

Engines: 1.2 • 1.4 • 1.3D **CO2:** 111-149 g/km **MPG**: 44.8 - 67.3 **Gearbox:** Man/Auto **Boot:** 185/550 **Seats**: 4 **Doors:** 3

Fiat Seicento 1998 - 2004

Fiat are good at tiny cars and although this seems quite ancient it does what a city car should do and soak up pot holes, provided the suspension is OK. Won't notice the lack of power steering. It is a Fiat so expect random failures. High spec SX likely to have a lot less working (windows, boot release). Quite an old piece of kit now. There will be rust.

Engines: 1.1 **CO2:** 143-155g/km **MPG**: 43.5-47.1

Gearbox: Man **Boot:** 170/810 **Seats**: 4 **Doors:** 3

Ford Ka 1996 - 2008

Well this is a legendary little thing which drives like a demon but isn't fast enough because of the ancient iron engine to cause any real damage. Rust is the big killer and the fading plastic bumper just looks sad, but makes it even more lovable. Lots on sale, noisy and under the bonnet a bit tight. Despite rust issues what a Banger should be. Cheap to fix.

Engines: 1.3 **CO2:** 163 g/km **MPG**: 42.2 **Gearbox:** Man **Boot:** 186/724 **Seats**: 4 **Doors:** 3

Ford Ka 2008 - 2014

That's right, this is a Fiat 500, but in very generic Japaneseurohatchstylee. Better value and more reliable. It isn't meant to be pretty the Ka is ready to work. Great to drive and very cheap to run. Don't bother with a diesel, the 1.2 petrol is fine. Starts as basic Studio and anything else is a bonus with Edge, Zetec and Titanium. Water leaks, rusty suspension top mounts, electric power steering failure.

Engines: 1.2 • 1.3D **CO2:** 112-119 g/km **MPG**: 55.4-67.3 **Gearbox:** Man/Auto **Boot:** 224/747 **Seats**: 5 **Doors:** 3

Honda Logo 2000 - 2001

Before the Jazz, there was this, very briefly. The great news is the logo on the Logo is Honda. That means utter reliability, if you can find one. Out of date when it arrived, it nonetheless was a practical little thing which did everything an undemanding owner asked. Basic L even had air con back then, let's hope it still works. Likely to be a bit rusty.

Engines: 1.3 **CO2:** 150g/km **MPG**: 44 **Gearbox:** Man/CVT Auto **Boot:** 228 **Seats**: 4 **Doors:** 3

Hyundai Amica 2000 - 2003

Stumpy and unnecessarily ugly little thing. Previously the entertainingly named AtoZ, with sophisticated one litre engine that was fairly refined. Inside it was quite spacious for a tiny car, with lots of clever touches, yet smaller than many 3 door rivals. Wins any localised specification face off. Actually quite accomplished for a tiddler. Bit of rust.

Engines: 1.0 **CO2:** 151g/km **MPG**: 44.8 **Gearbox:** Man/Auto **Boot:** 342 **Seats**: 5 **Doors:** 5

Hyundai Amica 2006 - 2008

Powered by a new 1.1 litre engine. Dirt cheap motoring when new, even dirtier used. Couldn't be easier to drive in town, small enough to fit in a dog kennel. Had a good level of specification as a GSi had electric windows and the CDX super posh leather and air just like a grown up car. Just wear and tear issues, maybe ABS sensors.

Engines: 1.1 **CO2:** 128g/km **MPG**: 52.3 **Gearbox:** Man/Auto

Boot: 342 **Seats**: 5 **Doors:** 5

Hyundai i10 2008 - 2015

Like an Amica but a zillion times better.. Still had the 1.1 engine in the 00s, but body higher and wider than the average tot, so comfier to be in. From the Classic upwards you've got air and a CD, there is a Comfort with more stuff and a Style with heated seats. Pot holes can break the suspension and wreck the tyres.

Engines: 1.2 **CO2:** 119g/km **MPG**: 56.5 **Gearbox:** Man/Auto

Boot: 258/910 **Seats**: 5 **Doors:** 5

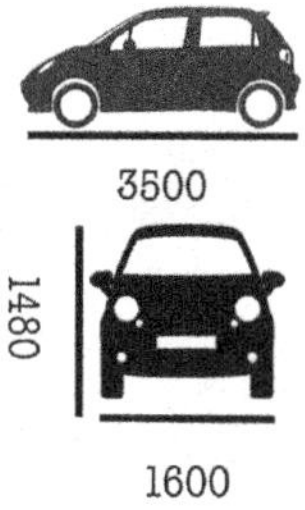

Kia Picanto 2004 - 2011

A micro five door with lots of charm, equipment and best of all, reliability, Slight price spike '08 with a new frontage. Crash rating average, but that's not a deal breaker. Also not much in it between 1.0 and 1.1 engines. Confusing specs LX and SE have most of what you want. Oil leaks, crankshaft retaining bolts most serious issue if not serviced.

Engines: 1.0 • 1.1 **CO2:** 117-126 g/km **MPG**: 56.5-57.6

Gearbox: Man/Auto **Boot:** 220/882 **Seats**: 5 **Doors:** 5

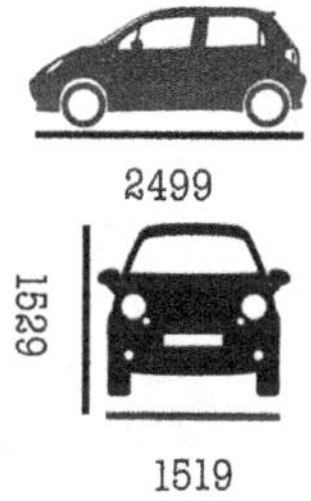

MCC Smart City Coupe 1998 - 2003

In the beginning there was the MCC one. A Swatch on four wheels and it will always look like it comes from a utopian future. The originals came to the UK as grey imports and the left hooker layout meant you stepped onto the pavement and that makes it the best City car ever. Plastic panels and scaffolding grade safety cage. ECU failure kills the car.

Engines: 698cc **CO2:** 118g/km **MPG**: 57.6 **Gearbox:** SemiAuto **Boot:** 150 **Seats**: 2 **Doors:** 2

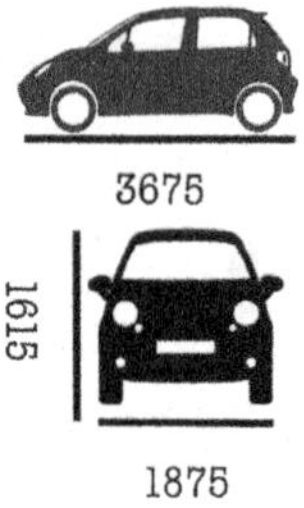

Mitsubishi | 2007 - 2012

No one has ever seen one, but they do exist. Like a Smart with a couple of extra doors. Interesting layout with a tiny turbocharged 660cc in the back. Not that economical considering it is so titchy. Being titchy is an issue, so you can get four inside but marginal luggage. Reliable but you would struggle to get body parts. Indeed, any parts.

Engines: 660cc **CO2:** 114g/km **MPG**: 54.6 **Gearbox:** Auto

Boot: 246 **Seats**: 4 **Doors:** 5

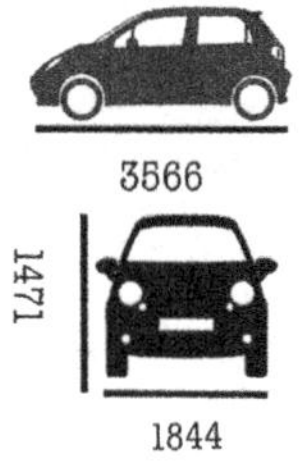

Nissan Pixo 2009 - 2013

There's a Suzuki Alto version of this, which does not make it right. Built in India to be cheap and nasty really. OK, so it is cheap to run and will be reliable, but it does feel cynically thrown together. 1.0 engine and huge mpg returns well into the 60s. Buy if there is no alternative and you get change from a fiver. Airbag warning sensor failure, water leaks.

Engines: 1.0 **CO2:** 103-122g/km **MPG**: 64 **Gearbox:** Man/ Auto **Boot:** 129 **Seats**: 4 **Doors:** 5

Perodua Kelisa 2002 - 2008

It's only claim to any real fame was as the poster car for the cheapest motor on sale in the UK, That's not really a recommendation. Doesn't have ABS, but that's OK as no one knows how to use it. As a brain dead runabout with buzzy 3-cylinder engine and almost zero parts availability, it is perfectly suited to the utterly bankrupt Bangernomicist..

Engines: 1.0 **CO2:** 121 g/km **MPG**: 55.4

Gearbox: 5spd/Auto **Boot:** 260/1220 **Seats**: 4 **Doors:** 5

Perodua Kenari 2000 - 2009

Like most Peroduas, it started life as an old Daihatsu, in this case the very odd micro MPV Move. The Daihastu connection is a good thing as it is Toyota's very own micro brand. A box on wheels, available in basic EX trim, slightly more plush GX and an auto EZ, which makes some sort of sense in the gridlocked City. Head gaskets, fluid leaks.

Engines: 1.0 **CO2:** 136g/km **MPG**: 50.4 **Gearbox:** Man/Auto

Boot: 260/1220 **Seats**: 4 **Doors:** 5

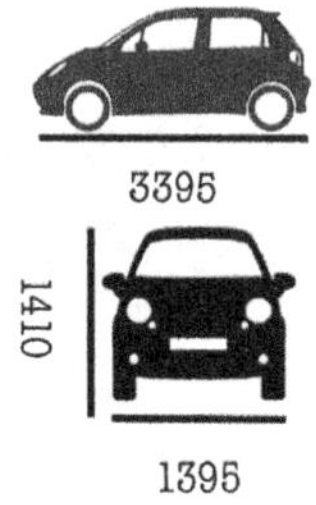

Perodua Nippa 1997 - 2002

In a previous life a Daihatsu Mira. A basic town centre assault vehicle. Basic model was the EX and the GX delivered little more than some flash paint that was metallic and body coloured, which was no big deal. The car was no big deal new or used. It was OK. Has a great reputation for not breaking down. Just a handful left. Body damage and rust.

Engines: 850cc **CO2:** 129g/km **MPG**: 51.9

Gearbox: Man/Auto **Boot:** 190 **Seats**: 4 **Doors:** 5

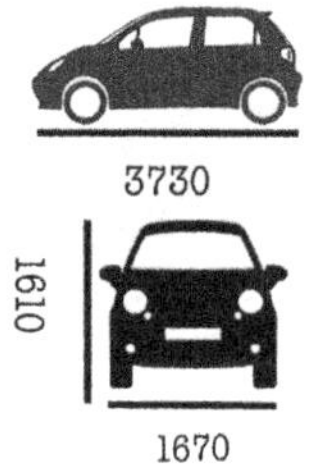

Peugeot 1007 2005 - 2009

Incredibly not a Bond car, unless that is, he makes it to retirement. By then he will find the electric sliding doors a boon, or broken, Yes, the unintended core market was OAPs. Go for the semi automatic for the full immersive retirement experience. Heavy little bus, so smaller engines have a harder time. Avoid the diesels, failing timing belts.

Engines: 1.4 • 1.6 • 1.4D **CO2:** 124-156g/km **MPG**: 44.1-64.1

Gearbox: 5spd/Auto **Boot:** 246/1192 **Seats**: 4 **Doors:** 3

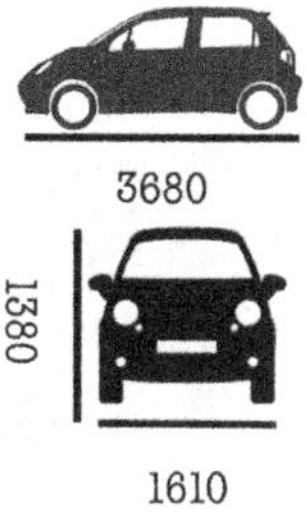

Peugeot 106 1991 - 2003

Also known as the Citroen Saxo, but with a Lion on the front. The littlest engines like the 1.0 deliver economy around town, otherwise slow, unless you go for the 1.6 GTi of course. Noisy and unsophisticated by modern standards, but who cares about modern standards? Steer clear of oil leaks and smoky diesels. Lots to like about these.

Engines: 1.1 • 1.6 • 1.5D **CO2:** 138-194 g/km **MPG**: 34.9-53.3
Gearbox: Man/Auto **Boot:** 215/953 **Seats**: 5 **Doors:** 3/5

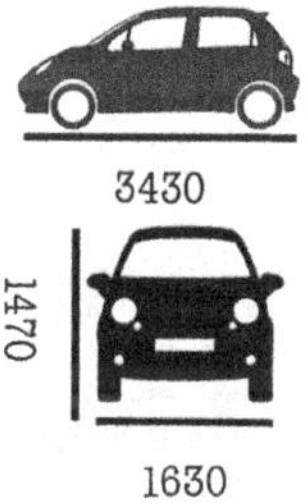

Peugeot 107 2005 - 2014

A relative of the Toyota Aygo and Citroen C1. Petrol 1.0 engine only. Revamp in 2009 with a more efficient engine ? Economy, so 60+ mpg and there is a semi automatic option. It is pretty much the perfect modern small car with loads of room and good to drive. Failing clutches, uprated from 2009 and dodgy water pumps. Water leaks into cabin.

Engines: 1.0 **CO2:** 109g/km **MPG**: 61.4 **Gearbox:** Man/Auto
Boot: 139 **Seats**: 3/5 **Doors:** 5

Proton Savvy 2005 - 2012

Proton owned Lotus at the time, sadly this is a deathly dull, breeze block with rounded edges. Another silly small car name, with few frills and thrills. Makes sense if give away cheap and better than Bangers you've actually heard of. Find the air conned Style model which will probably need a re-gas and so cost a bit more. Parts can take a while to find.

Engines: 1.2 **CO2:** 134g/km **MPG**: 49.6 **Gearbox:** Man/Auto **Boot:** 207/909 **Seats**: 4 **Doors:** 5

Renault Twingo 1993 - 2007

Euro only, but we love it. Drivers will be sitting on the wrong side which makes it cooler. 1.2 engines did the hard work and there was the choice of a 8V (60hp) or a 16V (75hp). Has all the basic charm of a Renault 4. Got styling makeovers in '98, '00 and '04. Worth rescuing from abject neglect on the continent. Timing belt changes needed.

Engines: 1.2 **CO2:** 138 g/km **MPG**: 43.5 **Gearbox:** Man **Boot:** 168/261 **Seats**: 4 **Doors:** 3

Renault Twingo 2007 - 2014

Not as lovable as the original bug eyed, left hand drive one. The 1.2 engine is averagely all right and if you want excitement then there is a GT. Dynamique delivers air con and rear seats that do their own slide and fold thing. Faulty coil packs and wrong spark plugs cause rough running. Water leaks into cabin, cockpit electrics.

Engines: 1.2 • 1.6 **CO2:** 135-140g/km **MPG**: 47.8-49.5

Gearbox: 5spd **Boot:** 165/959 **Seats**: 4 **Doors:** 3

SEAT Arosa 1997 - 2004

Down market and duller VW Lupo. Don't put anyone other than toddlers in the back. S and SE trim will give you things like folding rear seats, ABS brakes and fairly pointless central locking. Diesels are an option, 1.0 is fine for buzzing around 1.4 offers more, especially as a Sport. Facelift 2001. Failing oil pumps, seized brake calipers.

Engines: 1.0 • 1.4 • 1.4D **CO2:** 119-180g/km **MPG**: 37.7- 64.2

Gearbox: 5spd/Auto **Boot:** 130/790 **Seats**: 4 **Doors:** 3

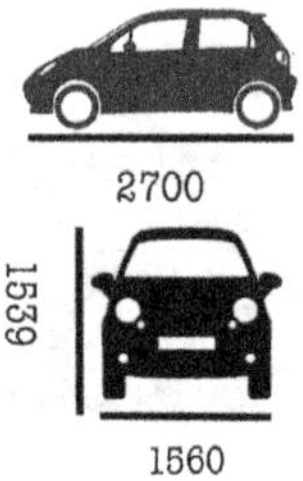

SmartForTwo 2003 - 2007

Good time to bring up the Smart again, now with full Benz backing and new name which described what it did. ForTwo City Coupe with the main trims still being Pulse, Pure and Passion. The big has been the gearbox which is not the best and can be jerky and broken. Middle order Pulse the pick of the bunch. As ever, ECU failure kills the car.

Engines: 698cc **CO2:** 116-127g/km **MPG**: 53.3-60.1

Gearbox: Semi-Auto **Boot:** 150/550 **Seats**: 2 **Doors:** 2

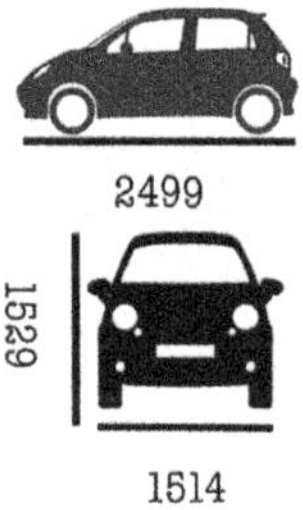

SmartForTwo 2007 - 2012

Next generation which is better built a bit bigger (to make it more pedestrian crash friendly) and better to drive, with more room inside but still bangs and crashes around pot holed towns. Standard 1000cc power now and a turbocharger at the upper end. The diesel engine finally came to the UK in 2008. Warning: ECU failure kills the car.

Engines: 1.0 • 800D **CO2:** 103-116g/km **MPG**: 57.6-65.7

Gearbox: SemiAuto **Boot:** **220**/340 **Seats**: 2 **Doors:** 2

Suzuki Alto 2003 - 2005

Yet another Indian built model that was always marginal when it came to cost and anything resembling satisfaction. It just started and stopped, but still has more charm than the latest Nissan Pixo related model we have already dismissed. ABS, airbag, electric front windows, central locking, so just enough spec. Not much goes wrong or wears out.

Engines: 1.1 **CO2:** 119g/km **MPG**: 57.6 **Gearbox:** Man/Auto

Boot: 432/725 **Seats**: 5 **Doors:** 5

Suzuki Alto 2009 - 2014

It's another 1.0 litre three cylinder, made in India job with pop out rear windows. It'll do 60mpg, which is the biggest plus point for most buyers, but real life return can be less. There's also a four speed automatic for those who can't drive. 2012 facelift also led to a cleaner engine. Reliability record very good indeed, few problems reported

Engines: 1.0 **CO2:** 103g/km **MPG**: 64.2 **Gearbox:** Man/Auto

Boot: 129/367 **Seats**: 4 **Doors:** 5

Suzuki Wagon R+ 2000 - 2003

One of those micro vans with windows. There was an adequate 1.3 replaced an almost as average 1.2 in '05. Became less boxy, slightly bigger and remained just as practical. Suzuki reliability remains the best reason for buying, likely to be cheaper than the Vauxhall version. Clutch and brakes need attention, just the usuals then.

Engines: 1.3 **CO2:** 145g/km **MPG**: 46.3 **Gearbox:** Man/Auto
Boot: 300/1250 **Seats**: 5 **Doors:** 5

Toyota Aygo 2005 - 2014

You might slightly recognise these with a Peugeot or Citroen frock on and it is probably the worst Toyota in the range as a result and not exactly built to their high standards. 1.0 petrol is perfect for what it is rather than the 1.4 diesel which was around for 5 minutes. The French ones likely to be cheaper. Clutch, cabin water leaks, corroded brake discs.

Engines: 1.0 **CO2:** 109g/km **MPG**: 62.8 **Gearbox:** Man/Auto
Boot: 139 **Seats**: 4 **Doors:** 3/5

Toyota IQ 2009 - 2014
A rather adorable thing which is like a less silly Smart as you can actually get a couple of bodies in the back (one big and one small) or rather more shopping. The 1.0 litre three cylinder gets over 60mpg, in theory. This is a properly put together Toyota which is the perfect package for knocking around the City. Might be pricey, but perfect microBanger.

Engines: 1.0 • 1.33 **CO2:** 99-113 g/km **MPG**: 58.9-65.7
Gearbox: CVT Auto **Boot:** 32/242 **Seats**: 4 **Doors:** 3

Vauxhall Agila 2000 - 2007
Suzukiesque, but with a Griffin on the grille + 1.0 + 1.2 litre engines, as if that makes a huge difference. Equipment levels not as generous as the Wagon. There is a Comfort pack with electric windows and central locking. It is possible to get a sliding roof and air conditioning if the original buyer bothered. Pretty reliable, few problems. Just wear and tear.

Engines: 1.0 • 1.2 **CO2:** 139-144g/km **MPG**: 47.1-48.7
Gearbox: Man/Auto **Boot:** 240/1250 **Seats**: 4 **Doors:** 5

Vauxhall Agila 2008 - 2013

As before but much less right angled. A teeny tiny people carrier. It is related to the Suzuki Splash which is elsewhere in the Bangerpedia. The interior has bags of room, but isn't the cleverest when it comes to utilising that space, seats four easily. Three engines and the 1.2 petrol is the best all rounder. No problems reported seems cheap and cheerful

Engines: 1.0 • 1.2 • 1.3D **CO2:** 120-131g/km **MPG**: 51.4-62.8

Gearbox: Man/Auto **Boot:** 225/1050 **Seats**: 5 **Doors:** 5

Volkswagen Fox 2006 - 2012

Instead of relying on Skoda to do the cheap city car shtick, VW went it alone, but it wasn't half as charming as the Lupo. Probably better though because of the bigger boot and better interior. There is 1.2 and 1.4 power in Basic and Urban trim, but there isn't a lot to make a fuss about. On the whole a lazy urban fox. Not that well built. Better to buy a Polo.

Engines: 1.2 • 1.4 **CO2:** 144-159g/km **MPG**: 23.1/42.2-46.3

Gearbox: Man **Boot:** 270/1016 **Seats**: 4 **Doors:** 3

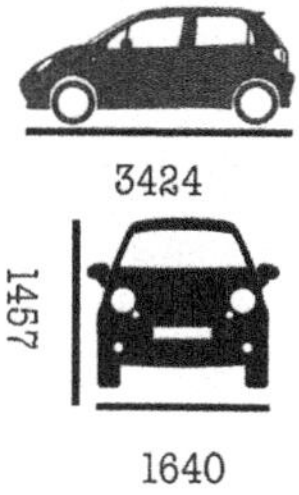

Volkswagen Lupo 1999 - 2005

A high quality small car, a VW Phaeton but with less doors and a lot less bother to own. The diesel does a million miles to the gallon, the Eco 3L (78mpg) and the 1.4 is shockingly perky, GTI though is one of those appreciating classics. This is a very classy little thing which only just about takes a bag of shopping in the boot, if that matters. Apart from some electrics. Basic SDI diesel excellent.

Engines: 1.0 • 1.4 • 1.4D • 1.7D **CO2:** 119-168g/km

MPG: 40.4-64.2 **Gearbox:** Man **Boot:** 130/830 **Doors:** 3

Happy Shoppers - Bangerminis for Errands

Arguably the hardest working bunch of models in Bangerdom. You have everything here, stopgap cars, first cars, second cars, station run cars, they are required to do so much with so little. For many these might also be only cars a well, they quite literally have to do everything. As well as the shopping.

Which brings us very quickly to the MINI which remains the cleverest marketing operations of all time as a little legend was successfully reinvented for the millennium. Not that big, just very clever. What was so very clever is Audi's A2. An aluminum design icon which was too costly and complicated, but the market for a shopping Audi at the time wasn't there when there were Polos, Fiestas and Jazzers around.

Mercedes A Class was a rather more successful assault on this marketplace, but the fact it did not make it too far out of the 2010s proves that the demand for cleverly packaged, premium badged runabouts was limited when new and could be troublesome when used.

Getting back to the models you could always rely on the Honda Jazz remains the very best of it's type. Unremarkable and boring, but oh so reliable and practical. It comes into it's own once used. Same thing with the Toyota Yaris. Dependability is the key just as it is with all those wonderful variations on a Polo theme from Volkswagen. Skoda Fabia and SEAT Ibiza are the great value used buys.

There is always safety in numbers which takes us to the staple offerings from Ford and Vauxhall. Hard to go wrong when there is such a wide choice at the usual outlets whether it is a Corsa, or Fiesta. Often overlooked though are the Mazda 2s which were Ford related and great value. The Nissan Micra got an overdue upgrade, more practical but didn't seem as indestructible as the old one.

The Rover 25 was certainly dated, but as a Banger dependable enough. For a bit of style the restyled Suzuki Swift won a lot of buyer support as sort alternative MINI which could rock an offset racing stripe or two.

Alfa Romeo MiTo 2009 - 2018

Intended as an upmarket handy hatchback thing despite the pig nosed appearance. It has been a fairly disappointing package as the two cylinder TwinAir is underwhelming and only the MultiJet diesel delivers some proper economy. Not great to drive and the fit and finish only average. Oil leaks, some power steering failures, gearbox can get notchy, overall not the prestige pig you'd expect.

Engines: 0.9 • 1.4 • 1.3D **CO2:** 89 -124g/km **MPG**:52.3-83.1

Gearbox: Man/Auto **Boot: 270/ Seats**: 4 **Doors:** 3

Audi A2 1999 - 2005

It looked like the future or at the very least like a very posh van. It was incredibly sophisticated with the ultra lightweight Audi space frame. Using aluminum made it very light at under a ton, but pricey to fix if bent. Old factory radios rubbish and diesel turbo could be blown.1.4 petrol or diesel is pretty economical. Quite possibly the classiest small Banger ever built.

Engines: 1.4 • 1.6 • 1.4D **CO2:** 116-144g/km **MPG**: 46.3-64.2

Gearbox: Man/Auto **Boot:** 390/1085 **Seats**: 4/5 **Doors:** 5

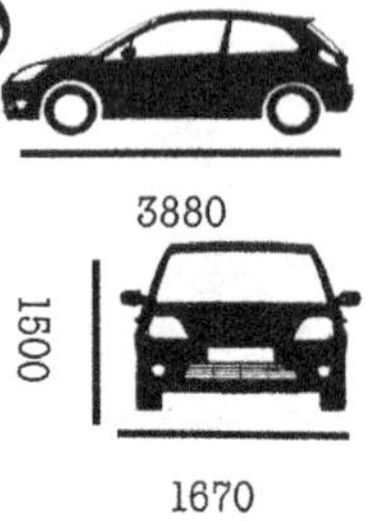

Chevrolet Kalos 2005 - 2009

In the beginning it was a Daewoo, incredibly it is a Giorgetto Giugiaro, not his finest. Worth it? Just grim on so many levels of 1.2 and 1.4 engined dreariness. Those can suffer bore wear and become smokey. Feels cheap and fairly nasty, but at least we were spared the saloon, the Kalos never went away due to being depressingly reliable.

Engines: 1.2 • 1.4 **CO2:** 153-165g/km **MPG**: 42.1-44.1

Gearbox: Man/Auto **Boot:** 220/980 **Seats**: 5 **Doors:** 3/5

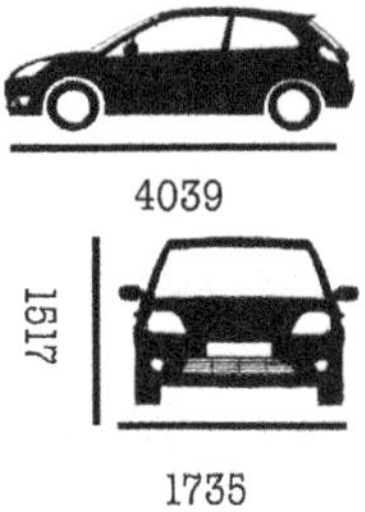

Chevrolet Aveo 2008 - 2012

The world never asked for a third Kalos reincarnation. Big in Eastern Europe. It was the same but slightly different with a new face and repurposed interior, plus a lot of standard kit, so you get a CD that you can hook an MP3 up to and electric windows up front. Basic transportation, slow and just needs basic servicing and brake overhauls.

Engines: 1.2 • 1.4 **CO2:** 132-140g/km **MPG**: 47.9-51.4

Gearbox: Man/Auto **Boot:** 220/980 **Seats**: 5 **Doors:** 5

Citroen C2 2003 - 2009

Dull successor to lively Saxo, effectively a stunted C3 which really does look like a minor household appliance. Had big car tech that can go expensively wrong: stop/start & flappy steering wheel paddles SensoDrive gearbox. 1.1 petrol is fine, 1.4 faster and 1.4HDi delivers 60mpg. Electricals can malfunction. ECU, and catalytic converter failures.

Engines: 1.1 • 1.4 • 1.6 • 1.4D **CO2:** 108-163g/km **MPG**: 47.9-68.9 **Gearbox:** Man/Auto **Boot:** 220/980 **Doors:** 3

Daihatsu Charade 2003 - 2007

Inoffensive and anonymous little shopping cart with a 1.0 three-cylinder engine. Almost extinct, but there was a minor revamp in 2006 that firmed up the suspension. Impressive specification included air conditioning and a CD player, which is all anyone needs to keep them cool, calm collected and entertained. Bit of rust is all you need to worry about.

Engines: 1.0 **CO2:** 114g/km **MPG**: 58.9 **Gearbox:** Man/Auto **Boot:** 157 **Seats**: 5 **Doors:** 3/5

Daihatsu Sirion 1998 - 2004

In Japanese market form, looked like an over chromed grey import (Doria/Boon & Toyota Duet/Passo), a posh Charade and the 1.3 is ex Yaris, so reliability never an issue. 3 cylinder 1.0 otherwise with a decent spec beyond the basic E, otherwise you have air con and eventually side airbags. Owner neglect damages suspension and coolant leaks.

Engines: 1.0 • 1.3 **CO2:** 129-134g/km **MPG**: 49.6-51.4

Gearbox: Man/Automatic **Boot:** 235/890 **Seats**: 5 **Doors:** 5

Daihatsu Sirion 2005 - 2008

Then they tuned the tiddler for Europe and dumped the chrome. Even the S spec has air con and all the electric windows, central locking and state of the art CD. SE has alloys and reversing sensors for the hard of driving. Room for people not luggage. Impressive levels of reliability, will eventually cause problems if not serviced.

Engines: 1.0 • 1.3 **CO2:** 118-137g/km **MPG**: 48.7-56.5

Gearbox: Man/Auto **Boot:** 225/630 **Seats**: 5 **Doors:** 5

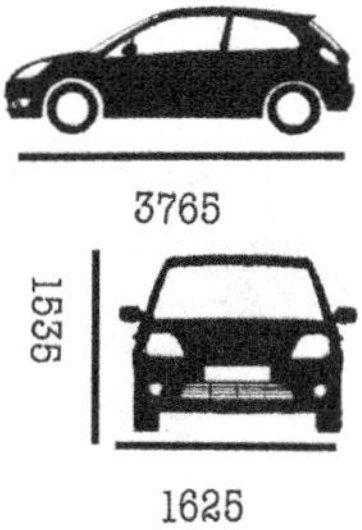

Daihatsu YRV 2001 - 2005

It's a micro MPV with a sliding rear seat and not much else to recommend it apart from decent reliability. That 1.3 Toyota Yaris engine helps. Turbo and 4Trak models rather pointless if amusing fun. You have a YRV because it does 47mpg. There is a Radical spec, whereas Premium ticked all boxes. Can be worn out, just check the brakes and rust.

Engines: 1.3 **CO2:** 145-165g/km **MPG**: 40.4-47.1 **Gearbox:** Man/Auto **Boot:** 240/1980 **Seats**: 5 **Doors:** 5

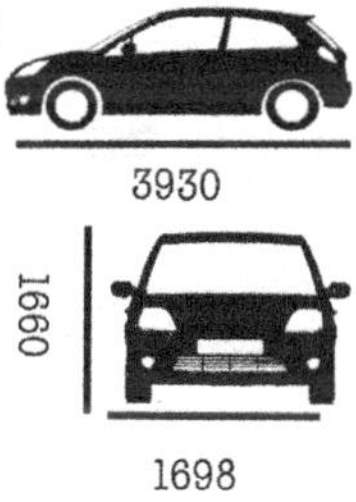

Fiat Idea 2004 - 2007

Short lived A-Class wannabe, but more alt-Meriva, without the performing seats. Well actually the back seat reclines and tumbles forward, but that's it. Meanwhile the hatchback lip is a bit high. Unremarkable 1.4 petrol power or more lively 62mpg 1.3 diesel. Eleganza has side airbag safety kit. Head gaskets blow, poor paint and battery earths rot.

Engines: 1.4 • 1.3D **CO2:** 135-157g/km **MPG**: 42.8-55.4 **Gearbox:** Man/Auto **Boot:** 320/1420 **Seats**: 5 **Doors:** 5

Fiat Panda 1986 - 2002

Still going strong a few decades ago. This Mark 2 ran from 1986 onwards and there was a ton of models to choose from. Charming, but often in a badly finished, falling apart sort of way. 1000 is the pick of engine bunch, CVT might not work anymore and the 4×4 is sadly very rare indeed. Rust, gear selection, bore wear blue smoke.

Engines: 800 • 900 • 1.0 • 1.1 **CO2:** 152-160g/km **MPG**: 40.9-48.6 **Gearbox:** 5spd **Boot:** 269 **Seats**: 4 **Doors:** 3

Fiat Panda 2004 - 2011

Decades overdue. Good engines, the 1.2 petrol is good and the 1.3 MultiJet diesel is 65mpg great. Active had a cassette, Dynamics had a CD or even aircon. Buy a Skydome just because of the name, never mind the hole in the roof. Split and slide rear seats. Leaky suspension dampers, failing clutch master cylinders and electric power steering.

Engines: 1.1 • 1.2 • 1.4 • 1.3D **CO2:** 114-135g/km **MPG**: 49.6-65.7 **Gearbox:** Man/Auto **Boot:** 206/860 **Doors:** 5

Fiat Punto 1999 - 2003

The snazziest looking hatch on the block. Can be flakey when it comes to using as a daily. 1.2 petrol is the basic choice, either 8V or 16V and there are diesels. Big old boot, HLX is the air con spec and ELX has a sunroof. Academic if it doesn't start. Original has bags of character and possibly rust. Head gasket failure.

Engines: 1.2 • 1.8 • 1.9D **CO_2:** 136-197g/km **MPG**: 34-57.6
Gearbox: Man/Auto **Boot:** 264/1080 **Seats**: 5 **Doors:** 3/5

Fiat Punto 2003-2006

From snazzy to a normie via a dreary restyle which didn't hurt sales, just added dullness. Boringly better with a 1.4 petrol, although actual daily users prefer the workaday Multijet diesels which can do an economical job and you should be able to get 60mpg plus. Failing ECUs, fragile electrics and parts, never cheap to fix.

Engines: 1.2 • 1.4 • 1.3D • 1.9D **CO_2:** 119-145g/km **MPG**: 46.3-62.8 **Gearbox:** Man/Auto **Boot:** 264/1080 **Doors:** 3/5

Fiat Grande Punto 2006 - 2010

Is it really grander than a Punto? Certainly prettier and it is bigger with an OK boot and rear accommodation. Has the amusing 'City' button lightens steering when parking in a tight spot. You can make do with a 1.4 petrol, otherwise it is the Multijet diesels are perfect. ECU software may need updating, worn shock absorbers, lots of warning lights.

Engines: 1.4 • 1.3D • 1.9D **CO2:** 119-139g/km **MPG**: 47.9-62.8 **Gearbox:** Man/Auto **Boot:** 275/1030 **Doors:** 3/5

Ford Fiesta 1999 - 2002

The post fish face one is the Mark 5. The 1.25 Zetec still does the hard work and will keep going forever but now there is the luxurious 1.6. Ghia and Zetec-S. Good to drive, not that spacious, but it is cheap and fixable, which is all the committed Bangerthusiast needs. Worn suspension bushes, noisy tappets, broken gearbox. Just over tired by now.

Engines: 1.25 • 1.4 • 1.6 • 1.4D • 1.6D **CO2:** 144-171g/km **MPG**: 38.7-53.3 **Gearbox:** Man/A **Boot:** 250/929 **Doors:** 3/5

Ford Fiesta 2002 - 2008

Big headlights and the rebranding of engines as Duratec. Still a groovy drive and there was a '05 spruce-up inside and out which made little difference, Baffling number of specs, Zetec does it for most. 1.4 probably the best petrol and reasonable economy. Otherwise 1.4TDCi delivers 60 mpg+. Oil leaks especially diesels, ECU failure, broken heaters.

Engines: 1.25 • 1.3 • 1.6 • 1.8D **CO2:** 119-153g/km **MPG**: 42.8-64.2 **Gearbox:** Man/Auto **Boot:** 284/947 **Doors:** 3/5

Ford Fiesta 2008 - 2012

You could argue that this is peak Fiesta. It has been revamped and Aston grilled since. The later ECOnetic was bang on green trend back then, but then the 1.25 and 1.4 TDCi don't need to brag just deliver decent mpg. All the usual reasons to buy include being a relatively easy fix and fun drive. Although steering rack wear. Door seals failure.

Engines: 1.25 • 1.4 • 1.6 • 1.4D • 1.8D **CO2:** 98-133g/km **MPG**: 49.6-76.3 **Gearbox:** Man/Auto **Boot:** 295/979 **Doors:** 3/5

Ford Fusion 2002 - 2012

A Fiesta on stilts and old model at that. Perfect for hip replacement OAPs. Those were the one that bought, so good to get one from them. Reasonable performance and economy from petrols and TDCi diesels. Often in better nick than an equivalent year Fiesta. Diesel timing belts, Durashift gearbox can fail.

Engines: 1.4 • 1.6 • 1.4D • 1.6D **CO2:** 122-160g/km **MPG**: 42.2-64.2 **Gearbox:** 5spd/Auto **Boot:** 337/1175 **Doors:** 5

Honda Jazz 2002 - 2008

Cat food run special, so well cared for dented ones around. Sit high and drive slow with the good little 1.2 and 1.4 petrol engines. Interior space is good with a decent boot, rear seats fold flat or cushion upright if you need more depth for the tins. S model has most things and an SE adds climate control. Suspension wear and tear, brakes. That's it.

Engines: 1.4 **CO2:** 134g/km **MPG**: 49.6 **Gearbox:** Man/Auto **Boot:** 354/1314 **Seats**: 5 **Doors:** 5

Honda Jazz 2008 - 2015

Clever seats, starting to get complicated with hybrid engines. Revamp in 2011. Early iShift gearboxes not brilliant, Go for the CVT. 1.4 is mildly more exciting, but really this is all about not breaking and having enough room for shopping and OAPs. Same old Jazz then. No spare wheel. Otherwise just old age and neglect niggles.

Engines: 1.2 • 1.4 **CO2:** 120-124g/km **MPG**: 52.3-53.3

Gearbox: Man/Auto **Boot:** 366/883 **Seats**: 5 **Doors:** 5

Hyundai Getz 2002 - 2009

Nothing remotely clever going on apart from rear seats that fold forward for a van like rear. 1.4 engine is fine and the 1.5CRTD delivers more mpg, but increasingly an ULEZ ban. Reliable and every model has air con. CDX all the pointless fancy alloy wheel etc trimmings. No issues, maybe brake discs need replacing. All wear and tear related failures.

Engines: 1.1 • 1.4 • 1.5D **CO2:** 118-141g/km **MPG**: 47.1-62.8.0

Gearbox: Man/Auto **Boot:** 289/988 **Seats**: 5 **Doors:** 5

Hyundai i20 2009 - 2014

This just sneaks into the Bangerpedia as one of the newest generation of shopping hatches. 5 year warranty factors now irrelevant, one family owner more important. As for diesels, always worth avoiding low mile ones which just store up trouble. Good specs obvs, top Style has a reversing camera..Clutch wear, surface rust. That's it.

Engines: 1.2 • 1.4 • 1.4D **CO2:** 124-133g/km **MPG**: 54.3-60.1
Gearbox: Man/Auto **Boot:** 295/1060 **Seats**: 5 **Doors:** 5

Mazda 2 2003 - 2007

Funny bubble roofed Fiesta with hollowed out interior so surprising amount of Luton van like room. Comfier than the Ford. Clone engines and in these times either a 1.25 petrol or the really frugal 60mpg 1.4Ds. Equipment levels average. Antares and Capella are the higher spec ones. Simply wear and tear issues, not underlying problems. Wonderful.

Engines: 1.4 • 1.6 • 1.4D **CO2:** 119-168g/km **MPG**: 39.8-62.8
Gearbox: Man/Auto **Boot:** 267/1004 **Seats**: 5 **Doors:** 3/5

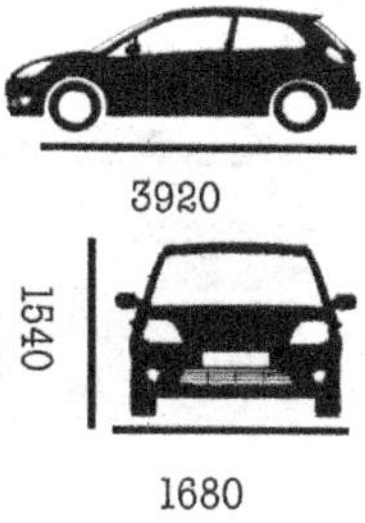

Mazda 2 2007 - 2015

Not as good, not as pretty as a Fiesta. This is massively average which might help to lower the asking price. The 1.3 and 1.5 petrol engines are average enough. SE models have everything you need including outdated sat nav and anything the daft Tamura and Sport Venture delivers is a bonus. Wheel bearing and shock absorber wear.

Engines: 1.3 • 1.5 • 1.4D **CO2:** 129-140g/km **MPG**: 47.9-52.3

Gearbox: Man/Auto **Boot:** 250/787 **Seats**: 5 **Doors:** 3/5

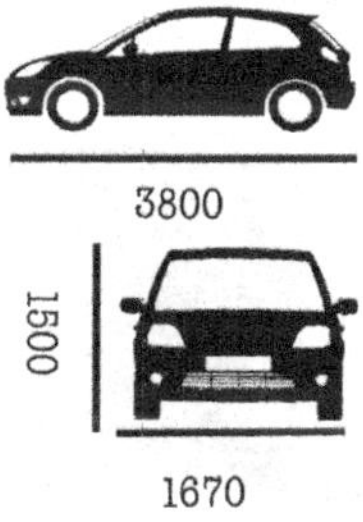

Mazda Demio 1998 - 2003

More of a maxi hatch in more ways than one. Replaced the long forgotten 121 as a roof railed micro MPV. The party trick though was reclining the folding seats into a cheeky double, or a flat bed van that could swallow almost anything. Enthusiastic 1.3 engine. No weak points, just the fact these can be old and tired. So just check brakes, exhaust etc.

Engines: 1.3 • 1.5 **CO2:** 169-176g/km **MPG**: 38.7-40.3

Gearbox: Man/Auto **Boot:** 280/950 **Seats**: 5 **Doors:** 5

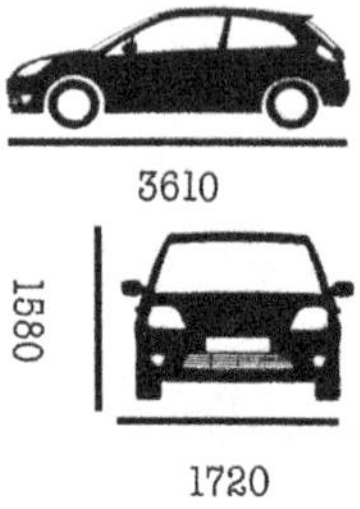

Mercedes A Class 1998 - 2004

A modern people/luggage packaging masterclass, that's a doddle to park. Sliding and removable seats, so it's a van too. Good petrols, quick 1.9, 2.1, but 1.4 is fine just for basic bombing about on the rock hard suspension. Stretched 'L 'ones from 2001.Diesel glow plugs trouble fix, water leaks, suspension bushes, power steering failure.

Engines: 1.4 • 1.6 • 1.9 • 1.7D **CO2:** 131-184g/km **MPG**: 36.9-57.9 **Gearbox:** Man/Auto **Boot:** 350/1190 **Seats**: 5 **Doors:** 5

Mercedes A Class 2005 - 2012

A2 still outclasses it, for style. Three door is just showing off, better buy a CL. Two level floor is clever, big old boot area. All the airbaggery safety kit. 2008 revamp led to better fuel economy, BlueEFFICIENCY will do over 60mpg. 1.5 petrol is adequately effective. Fuel pump and ECU failures. Brake discs and some localised rust.

Engines: 1.5 • 1.7 • 2.0D **CO2:** 137-148g/km **MPG**: 45.6-56.5 **Gearbox:** Man/Auto **Boot:** 435/1995 **Seats**: 5 **Doors:** 3/5

MG ZR 2001 - 2005

Rover 25 with performance knobs on. That means uprated suspension and steering and extra body kittage. Three and five doors as a plain ZR or with a + which added electric windows, sunroof and the fancy 160 had aircon and big alloys. Late 2004 with even more suspension tweaks and restyled bodywork. Head gasket failure. Steering arm bushes, ABS sensor and ECU, check all electricals.

Engines: 1.8 **CO2:** 173-179g/km **MPG**: 37.6-38.9 **Gearbox:** Man **Boot:** 448/1375 **Seats**: 5 **Doors:** 3/5

MINI One + Cooper 2001 - 2006

1.6 engine average. Cooper just has a chip and S added a supercharger. MINI is mostly marketing. MINI D with Toyota power from 2003. MINIs were all about the options, few were zero spec and Salt had rev counters and Peppers added alloys. Shop wisely. Cockpit electrical failures, ABS pump, rear wishbone wear, gearboxes, electric steering.

Engines: 1.6 • 1.4D **CO2:** 129-166g/km **MPG**: 40.9-58.9 **Gearbox:** Man/Auto **Boot:** 150/605 **Seats**: 4 **Doors:** 3

MINI One + Cooper 2006 - 2015

New and improved apparently. Cooper D diesel the most significant arrival. Specs became even more complicated and there are millions of terrible, twee special editions so agreeing a price can be fun. 1.4 One until 2009 when it went back to 1.6. and in came a 1.4. First as the basic, basic One. Clutch and Gearbox need checking as do electrics.

Engines: 1.4 • 1.6 • 1.6D **CO2:** 118-139g/km **MPG**: 48.7-64.2

Gearbox: Man/Auto **Boot:** 160/680 **Seats**: 4 **Doors:** 3

Mitsubishi Colt 1992 - 2004

Ancient from the '80s. In '92 it was refined and restyled. There's a load of engines from the shopping 1.3, mid range 1.5 and hottish 1.6 and GTi 1.8. Blimey there were updates every five minutes. There was a clever automatic gearbox. The only model you remember is the sporty Mirage. Old Bangers now. Rust and neglect, brakes and exhaust.

Engines: 1.3 • 1.6 **CO2:** 156-171g/km **MPG**: 39.2-42.8

Gearbox: Man/Auto **Boot:** 240/830 **Seats**: 4 **Doors:** 3

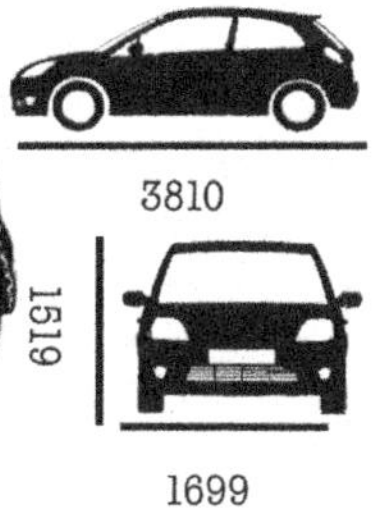

Mitsubishi Colt 2004 - 2013

It has space on its side and a line up of little petrol engines. All equally uninteresting and delivering 40mpg. The diesel is the ecochoice. Mostly though it won't break down. Elegance is the top trim and there was a lukewarm 1.5 Turbo, Classic had most things, they rebranded it but no one really cares. Just niggly day to day things, but overall just wear and tear.

Engines: 1.1 • 1.3 **CO2:** 119-138g/km **MPG**: 48.7-56.5

Gearbox: Man/Auto **Boot:** 155/760 **Seats**: 5 **Doors:** 3/5

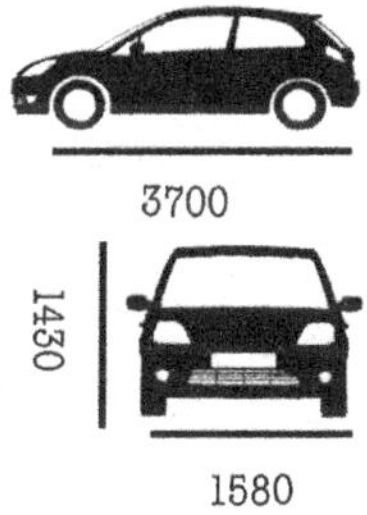

Nissan Micra 1992 - 2003

Driving school legend. Arguably the best and easiest to live with incarnation, but the older ones are getting rusty. Great little 1.3 engine. CVT gearboxes making driving a doddle except when it malfunctions. Like these 'Noddy' cars a lot. Rattly timing chains. Rusty bodywork and suspension. Mass airflow sensors and starting issues.

Engines: 1.0 • 1.4 **CO2:** 147-149g/km **MPG**: 46.3-47.1

Gearbox: Man/Auto **Boot:** 206/960 **Seats**: 5 **Doors:** 3/5

Nissan Micra 2003 - 2010

These are not half bad as it continued where the old one left off with a lot more safety like airbags and comfort features, still very easy to own and drive. There's a dCi which will do 60mpg. Still driving school and OAP specials. Sliding rear seats make small the boot bigger and lots of clever cubbies, best reason to buy. Has Renault related parts. Electrics not great or the diesel. Poor compared to previous one.

Engines: 1.2 • 1.4 • 1.5D **CO2:** 125-154g/km **MPG**: 44.8-60.1

Gearbox: Man/Auto **Boot:** 155/760 **Seats**: 5 **Doors:** 3/5

Nissan Note 2006 - 2013

Shrunk in the wash MPV cheaply made in India. Underneath it is a Clio, which is not always a good thing. Inside is why you would buy, tons of room, so maybe it is an MPV, especially if you like cup holders. The petrols are fine and the 1.5 diesel delivers 60mpg. Steering rack failure, water leaks from air con, electrics not great, eg electric windows.

Engines: 1.4 • 1.6 • 1.5D **CO2:** 119-155g/km **MPG**: 42.8-62.8

Gearbox: Man/Auto **Boot:** 280/1663 **Seats**: 5 **Doors:** 5

Perodua Myvi 2006 - 2015

Better than you would ever imagine with the added bonus of failing to break down, probably because it is also related to the Daihatsu Sirion and Subaru Justy. This makes it a transportation device to get you places, if you can find one. Well built, roomy and cubby hole heavy. Lolly name is great. Meanwhile just wear and tear items to check.

Engines: 1.3 **CO_2:** 137g/km **MPG**: 48.7 **Gearbox:** Man/Auto **Boot:** 255/630 **Seats**: 5 **Doors:** 5

Peugeot 206 1998 - 2009

Fun to drive, best appreciated with a missing wheel trim, it Feels cheap, only average amount of room inside, drive before buying as it can be oddly uncomfortable. The engines are pretty strong, overall cheap to run and economical. Perfect little Pug really. Worn lower ball joints, broken engine mounts, rattly sunroofs, ECU auto gearbox problems.

Engines: 1.1 • 1.4 • 1.6 • 1.4D • 1.6D • 2.0D **CO_2:** 113-166g/km **MPG**: 45.6-65.7 **Gearbox:** Man/Auto **Boot:** 245/1130

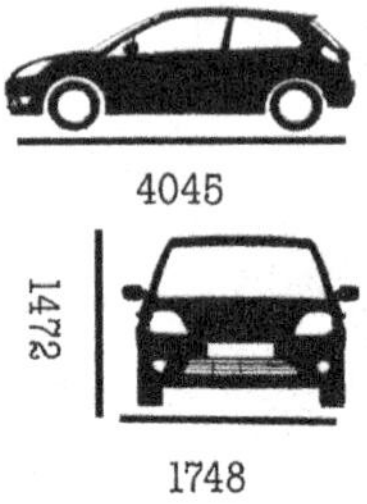

Peugeot 207 2006 - 2014

Next generation 206 without much of the basic original's charm. 1.6 petrol is the one that powers along enthusiastically. Restyled in 2009 and apparently better quality plus it does the Bluetoothy thing as standard. Urbans and Access have nowt and for many that is the pure, trouble free buy. ABS ECU failure. Dashboard goes dark. Gearbox.

Engines: 1.4 • 1.6 • 1.4D • 1.6D **CO2:** 126-150g/km **MPG**: 44.8-60.1 **Gearbox:** Man/Auto **Boot:** 310/1195 **Doors:** 3/5

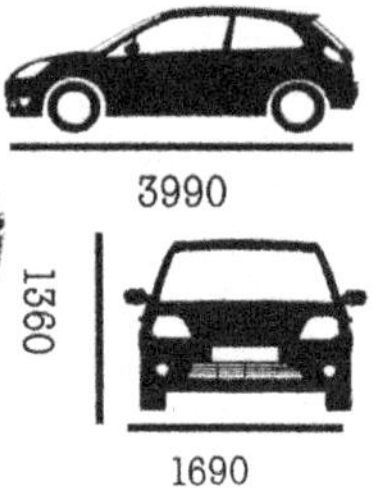

Proton Satria 2000 - 2003

The good news is that underneath it is an old Mitsubishi Colt. Badge is meaningless, average at just about everything. Not that economical and the case for the GTi that Lotus played with, is overplayed. It is reliable and that's why you buy. Can't think of any other reason. Reliable, maybe check for head gasket, oil leaks and just worn out suspension parts.

Engines: 1.6 • 1.8 **CO2:** 206-211g/km **MPG**: 32.7-35.0 **Gearbox:** 5spd **Boot:** 240/574 **Seats**: 5 **Doors:** 3/4/5

Proton Satira Neo 2007 - 2012

Behind the times using old tech and a slightly fresher body, so our sort of Banger then. They went full Lotus connection, (as well as cook Delia Smith) to little real effect apart from a boot badge. Proton died in 2012 and the Satira along with it. That makes it the perfect Banger. No future. No past. No one cares., just wear and tear, suspension, brakes etc

Engines: 1.6 **CO2:** 157g/km **MPG**: 42.8 **Gearbox:** Man/Auto

Boot: 286/615 **Seats**: 5 **Doors:** 3

Renault Clio 1998 - 2001

Bigger so better? Certainly felt more solid, which is good, but maybe less fun than before. Renault building their reputation for safe cars, so lots bought for first time teens. Cheap to run when not breaking down. Owners took long service intervals literally. Diesel EGR valves faulty, electric window failure, ignition coils on the blink, sunroof rattles.

Engines: 1.2 • 1.4 • 1.6 • 1.9D **CO2:** 146-179g/km **MPG**: 37.2-47.1 **Gearbox:** Man/Auto **Boot:** 255/1037 **Doors:** 3/5

Renault Clio 2001 - 2005

Looks the part now with fashionably huge headlamps. Almost impossible to unravel the various specification levels. This is the really safe one with four airbags and stuff. Split/ fold rear seat is a standard fitting for flexibility. First frugal thought is the dCi which delivers comfortably over 60mpg. Diesel EGR valves, ignition coils. Water pumps.

Engines: 1.2 • 1.4 • 1.6 • 1.5D **CO2:** 115-175g/km **MPG**: 47.1-65.7 **Gearbox:** Man/Auto **Boot:** 252/595 **Doors:** 3/5

Renault Clio 2005 - 2013

These are quite solidly built and almost classy inside. There is a 1.2 Turbo from '07 which is great, but pricey when it blows. 2009 specs/styling were revised there's your possible price point and cut off date for shopping if you are all about the budget. Diesel EGR valves and turbos, air con faulty, electricals such as central locking and ignition issues.

Engines: 1.2 • 1.6 • 2.0 • 1.5D **CO2:** 123-179g/km **MPG**: 47.9-70.6 **Gearbox:** Man/Auto **Boot:** 288/1038 **Doors:** 3/5

Renault Modus 2004 - 2008

Micro Espace in looks and execution, very spacious and has the sort of clever seats that should put it in the MPV class. The rear bench seats slides for legroom or luggage and it can seat two or three. They also fold down so that you can have a little flat floored van. Yes, there's a lot to like and there's six airbags. Ignition coils, rusty suspension, blown bulbs.

Engines: 1.2 • 1.6 • 1.5D **CO2:** 120-140g/km **MPG**: 47.8-62.8
Gearbox: Man/Auto **Boot:** 305/1454 **Seats**: 5 **Doors:** 5

Rover CityRover 2003 - 2005

A Tata Indica with a Union flag on the boot. Undistinguished little thing. A future micro collectible. Terrible in most ways, but reasonably spacious. Mechanicals are fairly robust. Sensible Bangernomicists buy a Yaris, others with a sense of irony bag a CityRover. Oil leaks, air con failure, water leaks into cabin, faulty door locks.

Engines: 1.4 **CO2:** 167g/km **MPG**: 37.9 **Gearbox:** Man/Auto
Boot: 220/610 **Seats**: 5 **Doors:** 5

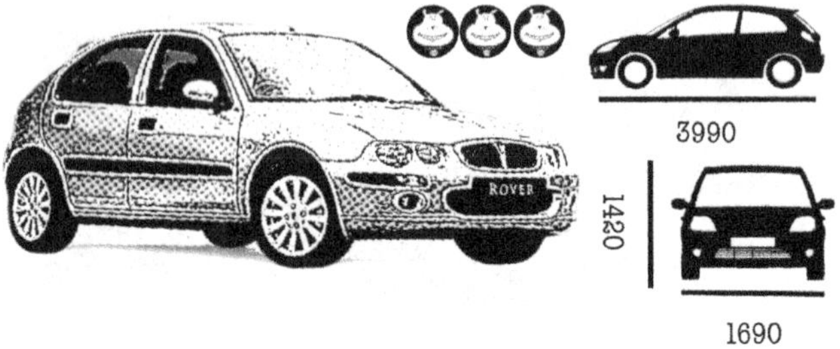

Rover 25 1999 - 2005

Rover 200, but with a smaller number. It hasn't got any bigger, simply rebadged. Notorious head gasket nonsense on the petrols might put you off and there is a gruff old diesel. Build quality can be questionable, so we are running out of reasons to buy. Three door is titchy really. Buy for old times sake only. Head gaskets, suspension wear, ABS +ECU issues.

Engines: 1.1 • 1.4 **CO2:** 160-164g/km **MPG**: 42.0-42.8

Gearbox: Man/Auto **Boot:** 304/1086 **Seats**: 5 **Doors:** 3/5

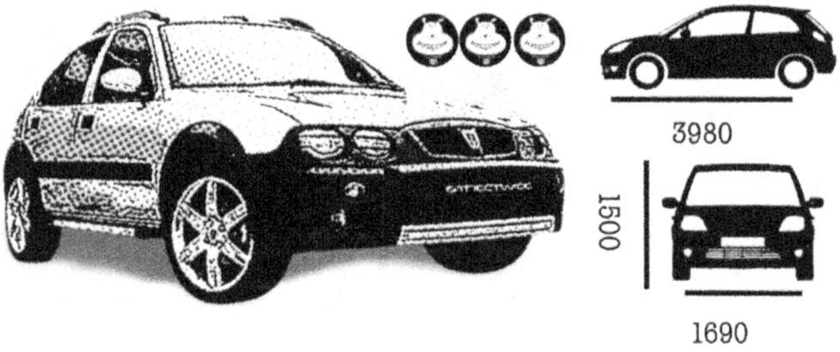

Rover Streetwise 2003 - 2005

Rover 25 rather brilliantly reimagined as a lifestyle crossover. We almost forgot to include it and in many way s this is just a daft derivation of a pretty ordinary small hatch. It is rather endearing and may eventually become some sort of collectible. Head gaskets, suspension wear, ABS+ECU issues, but at least it is all the SUV you will never need.

Engines: 1.4 **CO2:** 164g/km **MPG**: 38.0 **Gearbox:** Man/Auto

Boot: 304/1086 **Seats**: 5 **Doors:** 3/5

SEAT Ibiza 1999-2002

The Spanish Polo. Which means it is well built, has a range of strong engines and they won't easily breakdown, unless previously owned by a cheapskate, which most are. TDI for maximum economy. Three door is on the small side at the back, but there are less common five doors. These are great. Worn suspension, timing chain tensioners, central locking.

Engines: 1.4 • 1.6 • 1.8 • 1.9D **CO2:** 138-187g/km **MPG**: 36.2-56.5 **Gearbox:** Man/Auto **Boot:** 250/782 **Seats**: 5 **Doors:** 3/5

SEAT Ibiza 2002 - 2009

Styling is a bit wonky now, but never mind, the alt-Polo is still a great buy. 1.4 Ecomotive was the, er Eco one, low tax and high mpg. The 1.4 TDI also delivers a strong 60mpg+ Not bad size boot. Specs better from '04 and most have air con. There are the sporty Cupra ones if you must. Plastic timing belt tensioners, air conditioning, worn suspension.

Engines: 1.2 • 1.4 • 2.0 • 1.4D • 1.9D **CO2:** 124-187g/km **MPG**: 36.2-61.4 **Gearbox:** Man/Auto **Boot:** 267/1030 **Doors:** 3/5

Skoda Fabia 1999-2007

More VW magic with a Czech twist. Proper unpretentious little box and you can enjoy it even more with a 1.9 SDI which won't ever let you down. If you fancied one as a saloon, you are in luck. 2004 was a revamp for specs and bodywork. Coolant and oil leaks, timing belt tensioner, head gasket failure.

Engines: 1.2 • 1.4 • 2.0 • 1.4D • 1.9D **CO2:** 124-187g/km **MPG**: 36.2-61.4 **Gearbox:** Man/Auto **Boot:** 260/1016 **Doors:** 4/5

Skoda Fabia 2007 - 2014

Mark 2 is available in just one hatchback flavour. 1.4TDI GreenLine was the super eco one. Big revamp in 2010 with even more efficient engines and most irrespective of fuel deliver nearly or well over 50mpg. Reliable so long as they are properly serviced, need regular service fluid changes to stay current. No real problems, just fit and finish.

Engines: 1.2 • 1.4 • 1.6 • 1.4D • 1.9D **CO2:** 109-180g/km **MPG**: 36.2-61.4 **Gearbox:** Man/Auto **Boot:** 260/1016 **Doors:** 5

Subaru Justy 1996 - 2002

This is also a Suzuki Swift and built in Hungary. The Justy was more complicated as it had four wheel drive. 'Just' the one engine 1.3 and spec GX available with three or five doors. Proper oddity. No one really needs an ugly 4 x 4 mini hatch. Cracked cylinder heads, rust, broken transmission on 4 x 4 kills it.

Engines: 1.3 **CO2:** 165g/km **MPG**: 40.9 **Gearbox:** Man/Auto
Boot: 250/900 **Seats**: 4 **Doors:** 3/5

Subaru Justy 2007 - 2009

Nobody cared that it came back into the nation's showrooms, but at least it was based on an '06 Toyota Yaris and came with just the one 1.0 engine. Seats split/fold and the rear seat bases flip up as well. Comprehensive equipment list, but no one cared so not many to choose from. Toyota parts mean that it is only wear and tear items when buying.

Engines: 1.0 **CO2:** 118g/km **MPG**: 56.5 **Gearbox:** Man/Auto
Boot: 220/340 **Seats**: 5 **Doors:** 5

Suzuki Ignis 2000 - 2008

Here's one of those micro MPVs which isn't that useful, (boot very average) which is able to off road with all wheel drive 4Grip. Being a Suzuki means that it will be reliable. Great for fancy and top hat wearers. '03 series 2 models with split/fold/recline rear seats. Yaris engine is good news. Wear and tear items like brakes and tyres etc.

Engines: 1.3 **CO2:** 150g/km **MPG**: 44.1 **Gearbox:** Man/Auto **Boot:** 181/419 **Seats**: 4 **Doors:** 5

Suzuki Splash 2008 - 2014

Designed for youngsters, it was just another OAP special. A less boxy Wagon R which is well built and equipped. There's a diesel that will do over 60mpg, but you will want one because everything folds flat in the back and you get a van to play with. Excellent. Only affected by crappy British road surfaces, so check suspension and the odd electrical issue.

Engines: 1.2 • 1.3 **CO2:** 120-131g/km **MPG**: 51.4-62.8 **Gearbox:** Man/Auto **Boot:** 178/537 **Seats**: 4 **Doors:** 5

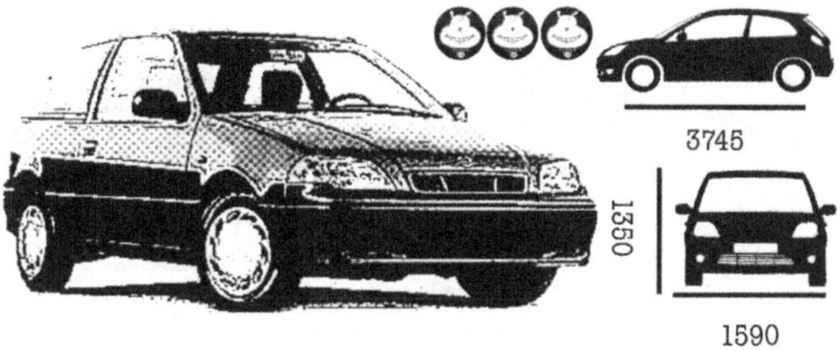

Suzuki Swift 1992 - 2003

Back to basics Japanese delivering reliability and fuel economy. What more do you want? Split fold rear seats standard. 1.0 and 1.3 engines plus a fun filled GTi. Running costs are reasonable enough. Cracked cylinder heads on early cars, don't worry they are all dead. So simply wear and tear and how many have survived abuse/neglect?

Engines: 1.0 • 1.3 **CO2:** 130g/km **MPG**: 51.4 **Gearbox:** Man/Auto **Boot:** 249/579 **Seats**: 4 **Doors:** 3/5

Suzuki Swift 2005 - 2011

Reborn and reimagined as actually a pretty decent supermini with some MINIsh style. Well built and nice to drive and always going to be cheaper and more useful than any bloody MINI. Sport one is interesting, the others just get on with the job of not breaking down. Apart from possible clutch and handbrake wear and tear.

Engines: 1.3 • 1.5 • 1.3D **CO2:** 122-159g/km **MPG**: 43.5-61.4 **Gearbox:** Man/Auto **Boot:** 213/562 **Seats**: 4 **Doors:** 3/5

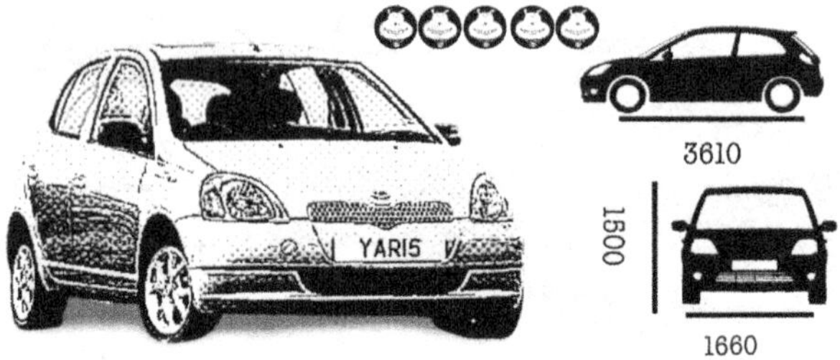

Toyota Yaris 1999 - 2006

Here's the no brainer buy with reliable 1.0 and 1.3 engines Was revamped in 2003. Looks like a tiny conventional hatch from the outside, but actually very roomy inside and a ton of cubbyholes. Decent equipment levels from the S upwards. Automatic gearbox can fail, early model head gaskets can fail. Overall takes a lot of abuse.

Engines: 1.0 • 1.3 • 1.5 • 1.4D **CO2:** 117-134g/km **MPG**: 50.4-64.2 **Gearbox:** Man/Auto **Boot:** 205/950 **Seats**: 5 **Doors:** 3/5

Toyota Yaris 2006 - 2011

Pumped up model was bigger in almost every way, but still stubby and practical. Compared to older generations it was more complicated and that meant more reliability niggles, which was a surprise. Usually worth checking the recall history. Water pumps break, clutch wear, dual mass flywheel on diesel.

Engines: 1.0 • 1.33 • 1.8 • 1.4D **CO2:** 118-119g/km **MPG**: 56.5-68.9 **Gearbox:** Man/Auto **Boot:** 272/1086

Vauxhall Corsa 2000 - 2006

Basic supermini motoring with 1.0 and 1.2 petrols are crudely adequate. 1.4 better if the commute is more dual carriageway. Decent economy from the small petrols and maximum mpg from the diesel of course. Another driving school favourite. Avoid those ones. Steering racks, wheel bearings, water pumps and timing belt servicing.

Engines: 1.0 • 1.2 • 1.4 • 1.3D **CO2:** 122-139g/km **MPG**: 48.7-62.8 **Gearbox:** Man/Auto **Boot:** 260/1060 **Doors:** 3/5

Vauxhall Corsa 2006 - 2014

Third version is bigger and better as usual. Tiny engines just economical which is fine. If you must have a 1.6 then you might as well buy an Astra. There's two versions or 1.3 CDTi. Owners can go on about the clutch and the radiators can leak, issues sorted on 2011 cars that are outside of this Bangerpedia entry. Also, gearbox, suspension top mounts.

Engines: 1.0 • 1.2 • 1.4 • 1.3D • 1.7D **CO2:** 124-149g/km **MPG**: 45.6-61.4 **Gearbox:** Man/Auto **Boot:** 285/1100 **Doors:** 3/5

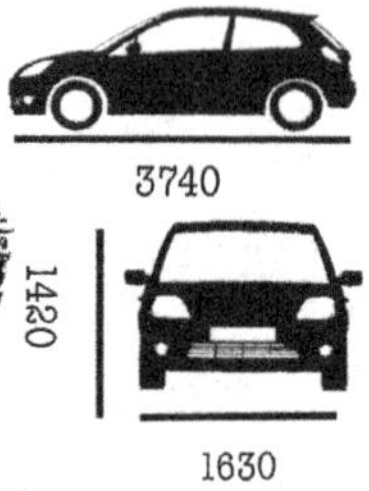

Volkswagen Polo 1999 - 2001

Beefed up Polo re-engineered to be a surrogate Golf . Incredibly there is still a saloon to enjoy. Lots of engines to choose from and big car features. On the whole these are tough and reliable runabouts and get rarer by the day. Difficult to kill. Can chew through tyres and brakes, mostly wear items to look at and rust. Gearbox failure.

Engines: 1.0 • 1.4 • 1.6 • 1.4D • 1.9D **CO2:** 120-173g/km **MPG**: 40.4-62.8 **Gearbox:** Man/Auto **Boot:** 245/975 **Doors:** 3/4/5

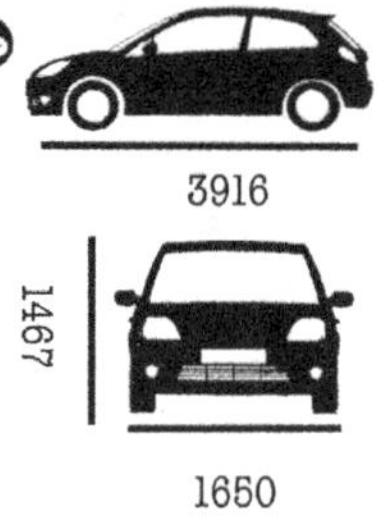

Volkswagen Polo 2002 - 2009

Peak Polo. Comfortable, reliable and practical. Steer away from the overworked three cylinder engines. There is a decent amount of safety kit as well as room inside, whilst the Bluemotion tapped into the economy zeitgeist with 70+mpg. Just get a 1.4 petrol or diesel and be done with it. Plastic timing belt tensioners, gearbox, water leaks into cabin.

Engines:1.2 • 1.4 • 1.6 • 1.4D • 1.9D **CO2:** 124-144g/km **MPG**: 47.1-61.4 **Gearbox:** Man/Auto **Boot:** 270/1030 **Doors:** 3/5

Family Bangers - CommuteShopRepeat

I might have mentioned that those Happy Shoppers were the hardest working Bangers of all time when actually it might well be the classic small family Hatch and odd saloon.

The rest of world stuck by the saloon format in the mass market. Some of those made it to the UK in small numbers so that apprentice traveling reps could lock their samples away from prying eyes. A very odd saloon in particular was the Chrysler Neon which became a Banger pretty quick as did the Daewoo related Cheverolet Lacetti. Volvo could be relied upon with their S40s which became affordable fast because it wasn't an estate. Meanwhile the C30 was a genuinely stylish hatch that rivaled the VW as the premium option.

Overall his is fundamentally the Golf Class that every mass market manufacturer needed to be in, simply to make a decent living. Hence you got brilliant rivals like the Focus and competent Astra. There were proper little weirdos including the boxy Daihatsu Materia and giant glassy hatched Skoda Roomster.

Getting back to the Golf it had started to return to form during this this decade. Maturing from the Mark 4 and after that onwards to the 6. For some reason they Plussed it up with the fat version that no one needed but it kept going in Mark 2 form. There was still room for a Golf with a boot, although the market was shrinking fast, hence the Bora and Jetta before retiring then telling people to transfer to the Passat.

Hot ones? As in Hot Hatches, well yes there are part of the line ups, but we don't make a fuss of them. Just as you would be super careful when looking at one of those, it is the same with the standard cooking models. These can lead really hard lives, or have been on hire fleets. Ideally you want to deal with the careful, mature and considerate family owners who are moving onto the next one.

When there is child shifting, school running, shopping and shift work to be done, it is usually hatches which get things done. After all that's what Golf sized hatches do.

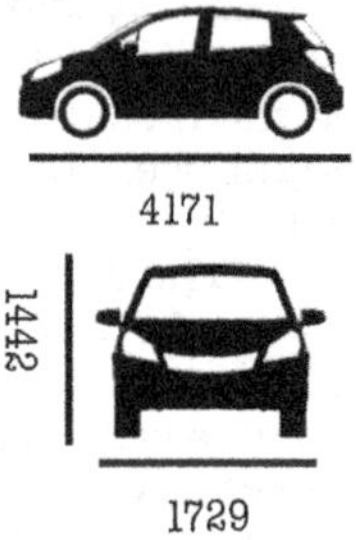

Alfa Romeo 147 2001 - 2009

It was the best pocket sized Alfa for decades with great styling, engines and specs. Choose three doors for looks and five-door for practicality. 2.0 really is Prosecco fizzy. The 3.2 V6 has a whacking 250bhp. There are diesels. It's an Alfa, so expect it to break down, a fact that will upset some. Suspension bushes, Selespeed Auto unreliable.

Engines: 1.6 • 2.0 • 1.9D **CO2:** 157-211g/km **MPG**: 31.7-47.9

Gearbox: Man/Auto **Boot:** 292/1042 **Seats**: 5 **Doors:** 3/5

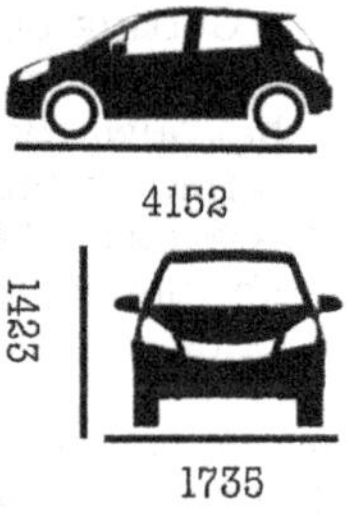

Audi A3 1996 - 2003

Rebirth of the posh hatchback (Audi 50), a Golf with knobs on, not forgetting the three door Sportback. If you want to get places though there is a 2.0 turbo, although the 3.2 V6 is sensationally quick. Five doors from '99, never as much fun as a BMW 3 series. Diesel 1.9 just a Golf TD. Air flow sensors, rusty brake lines, engine management issues.

Engines: 1.6 • 1.8 • 1.9D **CO2:** 135-226g/km **MPG**: 30.1-57.7

Gearbox: Man/Auto **Boot:** 350/1100 **Seats**: 5 **Doors:** 3/5

Audi A3 2003 - 2012

Yes, it is a premium priced Golf, so you really have to want one and be prepared to spend more than average to fix one. Check the recall history as there is a serious flywheel one. The simpler petrol ones make the most sense and any diesel without an EGR valve. Revamped in 2009. Clutch wear, steering racks, radiators, ECU and related sensor problems.

Engines: 1.4 • 1.6 • 1.8 • 2.0 • 1.6D • 2.0D **CO2:** 143-154g/km **MPG**: 43.5-52.3 **Gearbox:** Man/Auto **Boot:** 350/1100

BMW 3-Series Compact 1994 - 2001

Frankenstein hatchback part '90s 3 (E36) and '80s E30, almost like BMW forgot they made the Touring. It's really a 2 plus a very occasional two. Small petrol and diesel engines are fine. Minor revamp in late '96. A budget Bimmer, but not without some real charm. Rust, especially underneath, ABS sensors, new water pump as a precaution.

Engines: 1.8 • 1.9 • 1.9D **CO2:** 183-211g/km **MPG**: 37.2-47.9 **Gearbox:** Man/Auto **Boot:** 306/1030 **Seats**: 5 **Doors:** 3

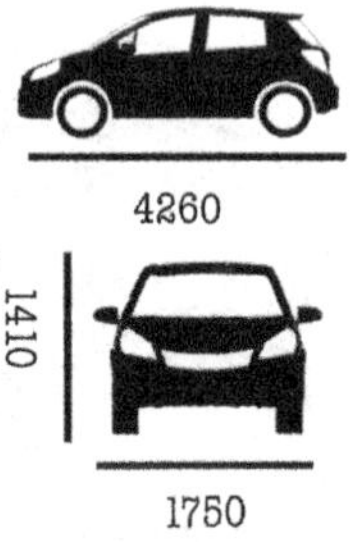

BMW 3 Series Compact 2001 - 2004

Not pretty, there's a bit more room in the back of the E46 version. Added bonus of the 325ti to enjoy if you want a stumpy performance German that isn't an A3. Pretty reliable even if they are starting to get a bit flakey (that means rusty) by now. Gets better with age. Suspension, cooling/oil leaks, some rust now as well.

Engines: 1.8 • 2.0 • 2.5 • 1.9D • 2.0D **CO2:** 148-215g/km **MPG**: 31.7-51.4 **Gearbox:** Man/Auto **Boot:** 310/1100 **Doors:** 3

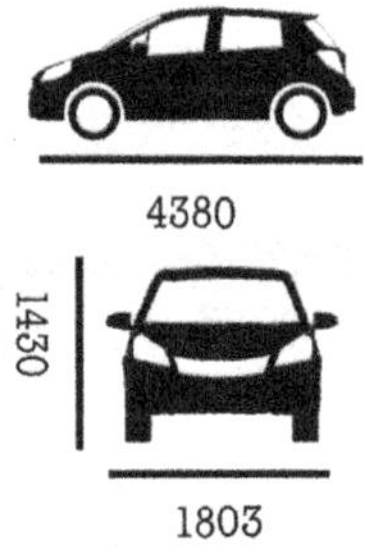

BMW 1 Series 2004 - 2011

Return of the Compact 3, but with extra doors, consequently a bit tight for space. 2007 styling, arrival of the three-door 1-Series. Mostly diesel in the marketplace Alloy wheels & air-con on ES, SE has climate control and parking sensors. 2.0 diesels need caution major failures. Dual mass flywheels. Turbo smoke. ECU failures.

Engines: 1.6 • 2.0 • 3.0 • 2.0D • 3.0D **CO2:** 119-155g/km **MPG**: 40-83.1 **MPH:** 127-138 **0-60:** 8.0-9.6 **Boot:** 330/1150

Chevrolet Cruze 2009 - 2015

Worth considering because it will always be depreciatingly cheap. We'll stick to the hatch and obviously there is an absolute ton of standard equipment. This is a very ordinary package of parts powered by OK petrol and diesels engines. That's it. Electrical issues the most common other than the usual, worn brakes and suspension.

Engines: 1.6 **CO2:** 159g/km **MPG**: 41.5 **Gearbox:** Man/Auto **Boot:** 413/883 **Seats**: 5 **Doors:** 4/5

Chevrolet Lacetti 2005 - 2009

Here's a basic, saloon/hatch that starts to make less sense when it gets older and there are better options. Not a huge boot, but pretty unburstable old engines and the obsolete DNA of dead Daewoos. You buy one because it is cheap, or fairly free. Not a lot of issues just the usual wear and tear ones for brakes and suspension. Electricals worth checking.

Engines: 1.4 • 1.6 **CO2:** 171-178g/km **MPG**: 39.2-39.7 **Gearbox:** Man/Auto **Boot:** 275/1045 **Seats**: 5 **Doors:** 4/5

Chrysler Neon 1999 - 2003

Misunderstood from day one, even when revamped. SE a low spec pointless exercise, better go for the leather lined LX if you really must have a 2.0 litre hatch with a decent amount of space for people and their luggage. A curiously dated American. Head gaskets can go, air con failure, tired suspension parts.

Engines: 2.0 **CO2:** 185-196g/km **MPG**: 25.2-35.3 **Gearbox:** Man/Auto **Boot:** 371 **Seats**: 5 **Doors:** 4

Citroen C3 2002 - 2010

One of the new breed of high seat hatchbacks. Most petrol engines that deliver 40+mpg and the diesels over 60mpg, so that's why you buy. You won't buy because of the reliability though as the Sensodrive semi autobox can fail and the injector pump for 1.6 HDi isn't the best. Leaking air con system, electricals fail such as locks and windows. Not keen.

Engines: 1.4 • 1.6 • 1.4D **CO2:** 109-148g/km **MPG**: 45.6-67.3 **Gearbox:** Man/Auto **Boot:** 282/1155 **Seats**: 5 **Doors:** 5

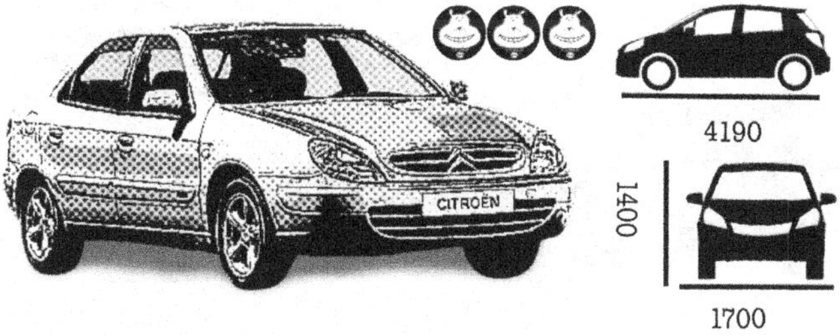

Citroen Xsara 2000 - 2004

Proper old school basic Frenchie. Quite old by this point and it isn't trying to be clever, just reliable and pretty frugal on the old fuel whether you go for a 1.4 or 2.0 HDi. Does not seem to have many reliability downsides provided it has not been run into the ground. ECU issues, some late '90s diesel engine issues can kill the engine. Diesel clutches, also rust.

Engines: 1.4 • 1.6 • 2.0 • 1.9D • 2.0D **CO2:** 138-184g/km **MPG**: 40.9-54.3 **Gearbox:** Man/Auto **Boot:** 408/1190 **Doors:** 5

Citroen C4 2004 - 2010

Not Banger material. Rather complicated as the minimalist dashboard is hiding far too much and the electronic gearbox is another cause for concern. The 1.6 Diesel is probably the sensible choice, but there's a lot to worry about. Dual mass flywheel failure. Air con system, electrics are patchy dashboard and electric windows.

Engines: 1.4 • 1.6 • 1.6D **CO2:** 128-159g/km **MPG**: 42.2-60.1 **Gearbox:** Man/Auto **Boot:** 348 **Seats**: 5 **Doors:** 3/5

Daewoo Lanos 1997 - 2002

Hard to be enthusiastic about these blobby hatches and saloons. They do exist and largely don't break down, but they do suffer from neglect because no one cares so long as they work. Parts can cost more than you would ever believe. A last resort purchase. Not much goes wrong, just old age issues like brakes and suspension.

Engines: 1.4 • 1.6 **CO_2:** 184-192g/km **MPG**: 34.4-35.6
Gearbox: Man/Auto **Boot:** 250/886 **Seats**: 5 **Doors:** 3/4

Daewoo Nubira 2003 - 2005

Saloon no one needed, the recently deceased Daewoo, kept on life support by owner General Motors. Unrefined 1.6 engine, but it is reliable, fixtures and fittings with SX include air con and alloys. Quite likely to be neglected and worn out, but still working. Known problems include wheel bearings and also minor electrics like the windows. Reliable.

Engines: 1.6 • 2.0 **CO_2:** 198-214g/km **MPG**: 32.8-36.2
Gearbox: Man/Auto **Boot:** 370/890 **Seats**: 5 **Doors:** 5

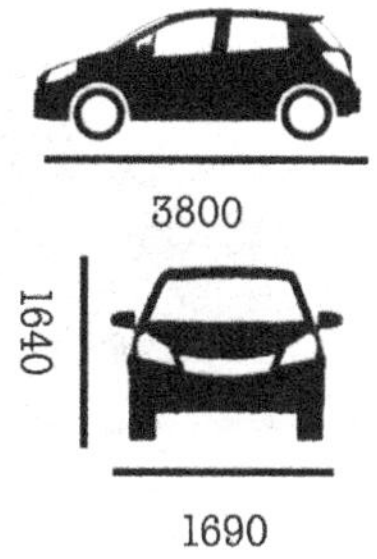

Daihatsu Materia 2007 - 2010

One of those Japanese boxy things. Looks interesting has plenty of kit and space. Wide doors and high seats make it OAP friendly, but it makes a decent stab at being a family car that's easy to live with. Well built, one perky, but reasonably economical engine. Parts are becoming a problem, ideally you won't need any as they tend to be extremely reliable.

Engines: 1.5 **CO2:** 169g/km **MPG**: 39.2 **Gearbox:** Man/Auto **Boot:** 170/845 **Seats**: 5 **Doors:** 5

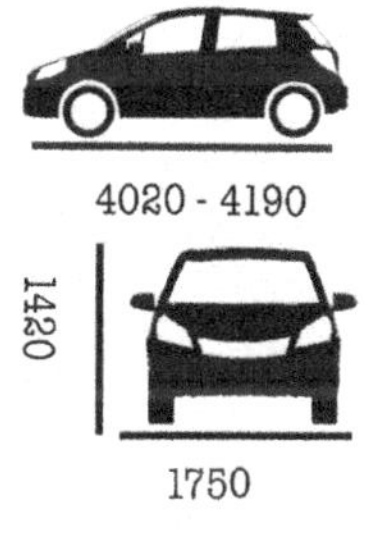

Fiat Brava/Bravo 1995 - 2002

Your choice, a three door or a five door with different names and some neat design twists (trick wheel jack). Still a Fiat though, doesn't make it bad, but means they have been neglected despite now being rarer than most contemporary Ferraris. Buy 1.6 petrol and TD100SX. Worn suspension, catalytic converter, blocked radiators, electric windows.

Engines: 1.2 • 1.6 • 1.9D **CO2:** 146-186g/km **MPG**: 36.2-51.2 **Gearbox:** Man/Auto **Boot:** 379/1164 **Seats**: 5 **Doors:** 3/5

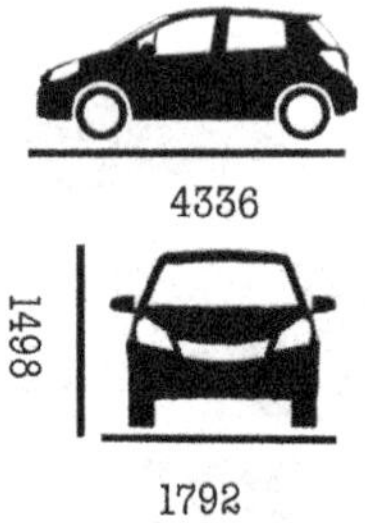

Fiat Bravo 2007 - 2014

Return of the name, not really a match for a Golf or Focus, but massively cheaper. Looks smart and there are decent diesels. Few around the classified adverts. Not a lot to look out for, just make sure that everything works. Worth a punt and better buy than a Punto. Reliable, just worn anti-roll bars, 1.6D EGR filter is clogged because many stop/starts.

Engines: 1.4 • 1.6D **CO2:** 115-167g/km **MPG**: 42.2-61.4 **Gearbox:** Man/Auto **Boot:** 400/1175 **Seats**: 5 **Doors:** 5

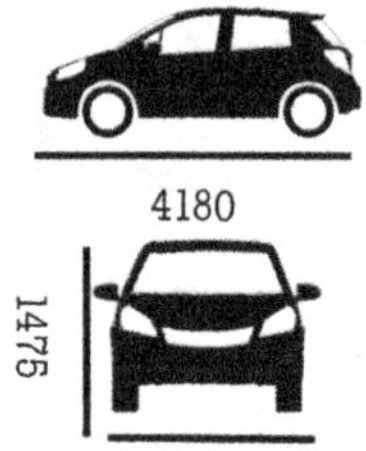

Fiat Stilo 2001 - 2007

Goodbye Brava and Bravo, but not many said hello to the Stilo. It was never a half bad hatch. Because it is a Fiat which isn't a retro darling like a 500, mostly worthless and that's a great reason to buy. The JTD80 is the diesel way to go, but the fairly tiny 1.2 petrol is something of a survivor. One of those Fiats with failing sensors and electrics to baffle you.

Engines: 1.4 • 1.6 • 1.8 • 2.4 • 1.9D **CO2:** 140g-233/km **MPG**: 35.3.0-52.3 **Gearbox:** Man/Auto **Boot:** 370/1120 **Doors:** 3/5

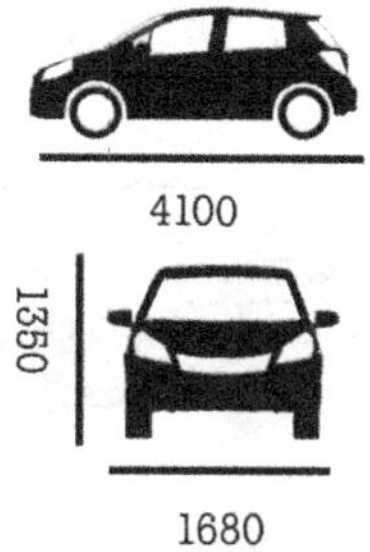

Ford Escort 1990 - 2001

Mark 7 lived on as a special edition in the Focus era. Flight and Finesse with 1.6 petrol and 1.8 Turbo Diesel. Like driving a museum exhibit, so oddly cool. Rust, neglect and abuse killed most of them. Looks and drives prehistoric, but a solid used buy. Blown head gaskets, bodywork rust serious, rusty fusebox, can be fatal.

Engines: 1.6 • 1.8D **CO2:** 167-175g/km **MPG**: 42.7-44.2 **Gearbox:** Man/Auto **Boot:** 380/1130 **Seats**: 5 **Doors:** 5

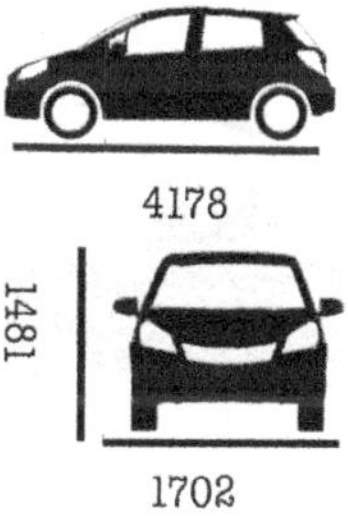

Ford Focus 1998 - 2004

Everyone loves a Focus and they are great to drive and easy to own. Hard to go wrong with any of them, groovy looking three door still a thing, oh and briefly a saloon. TDCI diesels are great Zetec spec and the petrols are all pretty reliable. Getting old and scrappy and starting to rust. Otherwise it is just wear and tear to the suspension.

Engines: 1.4 • 1.6 • 1.8 • 2.0 • 1.8D **CO2:** 140-188g/km **MPG**: 30.0-53.3 **Gearbox:** Man/Auto **Boot:** 350/1205 **Doors:** 3/5

Ford Focus 2005 - 2011

More of the same excellence in a less striking body. Revamped in 2008 and they went down the ECOnetic ultra green 70+mpg. Roomy inside rather than having anything really clever seatwise. Straightforward to fix and easy to live with. Buy one. No serious issues, just wear and tear, suspension and brakes, needs regular servicing.

Engines: 1.4 • 1.6 • 1.8 • 2.0 • 1.6D • 2.0D **CO2:** 119-179g/km **MPG**: 42.2-62.7 **Gearbox:** Man/Auto **Boot:** 364/1177

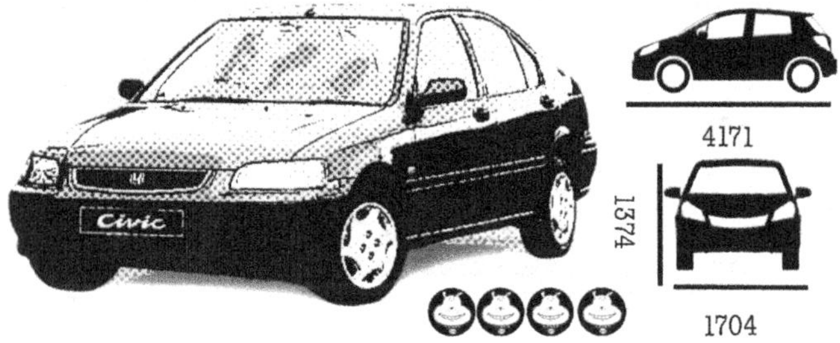

Honda Civic 1995 - 2001

Blueprint Banger. Really, owners will tell you just how brilliant these are. It is also a Rover 400, but better. Great petrol engines and comfortable ride. Everything else seems average from space to performance, but it is just relentlessly capable. That's a good thing. No real issues all, rust is probably more of an issue which can kill it.

Engines: 1.4 • 1.5 • 1.8 **CO2:** 164-206g/km **MPG**: 32.1-47.5 **Gearbox:** Man/Auto **Boot:** 225/600 **Seats**: 5 **Doors:** 3/5

Honda Civic 2000 - 2005

Bland, but a very good representation of blandness that won't let you down. Spacious enough for a small family and great petrol engines, plus a diesel. Hybrid 1.4 IMA based on USA saloon. Comfy seats and dead easy to drive. Mild revamp in 2004. Just keeps going. Nothing to worry about just squeaks and silly stuff provided it was looked after.

Engines: 1.4 • 1.6 • 1.7D **CO2:** 134-157g/km **MPG**: 44.8-56.5
Gearbox: Man/Auto **Boot:** 353/732 **Seats**: 5 **Doors:** 3/5

Honda Civic 2006 - 2011

Space pod styling and perhaps it doesn't have the engine range, room or versatility of its rivals, but at least it doesn't break down. Only the diesel can throw up some glow plug issues. Tuned to be a sporty ride which might be a mistake, unless you want the hot R one. Window and light seals let in water. That's as bad as it gets.

Engines: 1.4 • 1.8 • 2.2D **CO2:** 132-152g/km **MPG**: 44.1-55.4
Gearbox: Man/Auto **Boot:** 485/1352 **Seats**: 5 **Doors:** 3/5

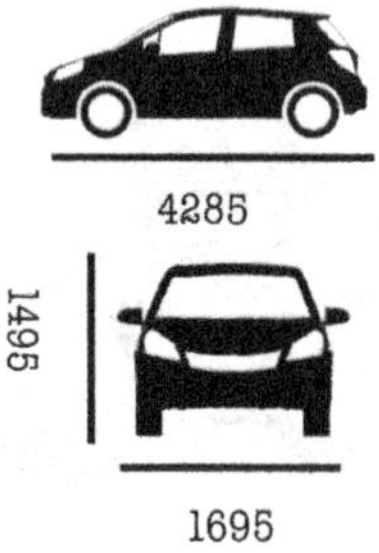

Honda Civic IMA 2003 - 2005

Early era Hybrid, was never going to have a Banger future and by now battery expired. American spec saloon witth Executive trim inc: leather, climate and electric windows. Complicated running gear and these early IMA batteries can fail, may not fully recharge. Best avoided at this stage as batteries well over a decade old. Also likes to rust.

Engines: 1.4H **CO2:** 116g/km **MPG**: 57.7 **Gearbox:** Auto **Boot:** 370/732 **Seats**: 5 **Doors:** 4

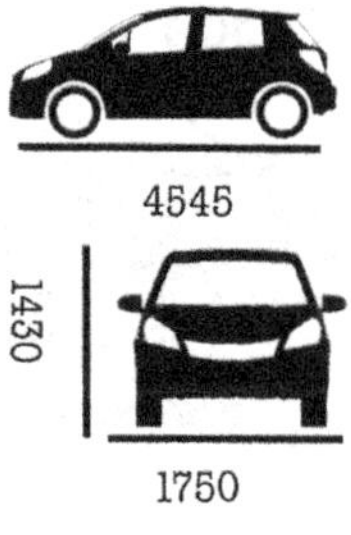

Honda Civic Hybrid 2006 - 2011

Honda's Prius which didn't catch on, partly because it was a saloon. Standard high specification which means alloys, climate, cruise and a CD player. Battery age is the issue, otherwise a high quality package, if a tad dull. Easily covers over 100K miles, but the battery? Truth now coming home to roost also the structure can rust.

Engines: 1.3H **CO2:** 109g/km **MPG**: 61.4 **Gearbox:** Auto **Boot:** 350 **Seats**: 5 **Doors:** 4

Honda Insight 2009 - 2014

Still not the future but constantly refined so that 96g/km emissions from 2011. Got UK focused suspension from 2010 as well. SE + ES spec levels with the EX at the top. Decade old batteries now. Still works with a flat AA, so get it checked. Price vs remaining battery life equation bothers the true Bangernomicist who should just pass.

Engines: 1.3H **CO2:** 96g/km **MPG**: 64.2 **Gearbox:** Auto **Boot:** 408/584 **Seats**: 5 **Doors:** 5

Hyundai Accent 2000 - 2005

When South Korean cars came good. Grim styling, but otherwise it does the carrying people and shopping thing to a sort of perfection. There was even a saloon, although no one has ever seen one. Just another utterly reliable Korean in need of a makeover. Nothing to worry about just the normal wear and tear to suspension and brakes.

Engines: 1.3 • 1.6 **CO2:** 144-160g/km **MPG**: 42.2-47.0 **Gearbox:** Man/Auto **Boot:** 481/934 **Seats**: 5 **Doors:** 3/5

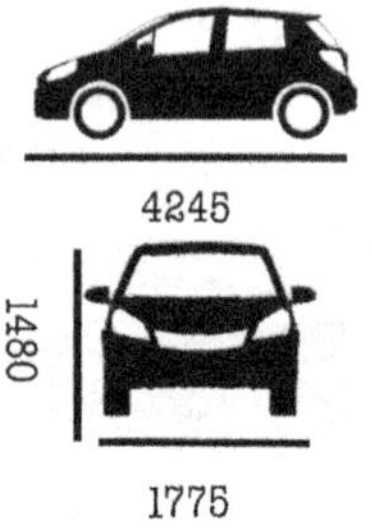

Hyundai i30 2007 - 2014

These are nice. They look OK, but mostly everything is standard from the 'Comfort' upwards, 'Premium' is just an absolute overload of stuff you don't need. The benefits of the extended warranty and servicing deals is private family owners should keep them in 'premium' condition. Check power steering pumps, low level rust. The odd ECU issue.

Engines: 1.3 • 1.6 • 1.5D **CO2:** 144-165g/km **MPG**: 40.9-51.4
Gearbox: Man/Auto **Boot:** 321 **Seats**: 5 **Doors:** 5

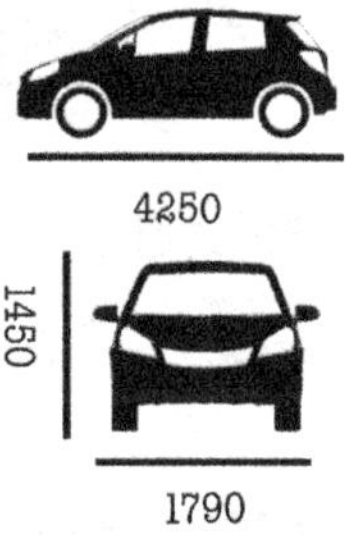

Kia Ce'ed 2007 - 2012

I refer you to the nearby i30 assessment. They are the same but different. Comparing specs is something you only do when buying new models. Overall better specs with the Kia, but you really need to be a spec level nerd. Essentially the Kia is a good little bus. Nothing to worry about just wear and tear in the usual brake and suspension places.

Engines: 1.4 • 1.6 • 1.6D **CO2:** 110-146g/km **MPG**: 46.3-65.7
Gearbox: Man/Auto **Boot:** 340/1300 **Seats**: 5 **Doors:** 3/5

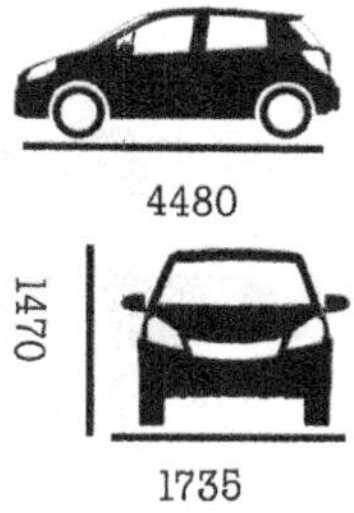

Kia Cerato 2004 - 2006

This is the pre-Ce'ed and it works well as a no frills and nonsense hatch, plus there was a long forgotten saloon. Quite a lot of safety kit, reasonably decent economy and it doesn't seem to like breaking down. Most sophisticated of the old school Kias. Nothing to worry about when it comes to reliability, just wear and indeed, tear.

Engines: 1.6 • 2.0 • 1.5D **CO2:** 129-180g/km **MPG**: 37.7-57.6
Gearbox: Man/Auto **Boot:** 538/1494 **Seats**: 5 **Doors:** 5

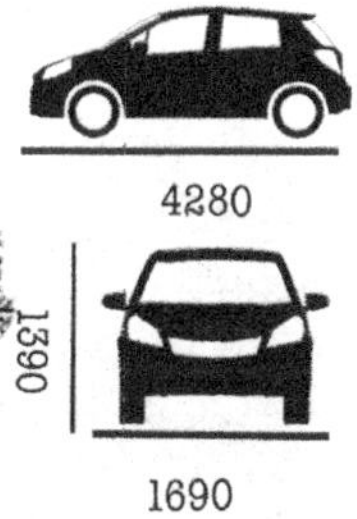

Kia Mentor 1994 - 2003

Just like the Escort, this just makes it into the Bangerpedia and is a similar sort of depressing '90s saloon. Even so it was based on the more interesting Mazda 323, which guarantees it won't break. Became a Mentor II then a new Mentor in 2001. That's enough Mentor. Nothing to see there, just the usual worn suspension and brakes.

Engines: 1.6 **CO2:** 190g/km **MPG**: 35.3 **Gearbox:** Man/Auto
Boot: 328 **Seats**: 5 **Doors:** 4

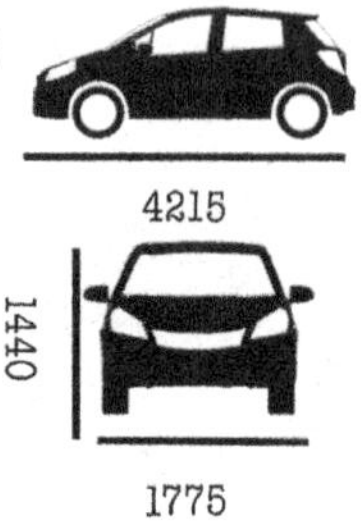

Kia Rio 2001 - 2005

Five door hatch that looks a bit estate-ish. Like all Kias of the era, not a looker. A dull old thing with engines. The trim levels were mucked about with for a bit, a minor revamp in 2002, but buyers ended up with a Hobson's choice of either an LE or LE+. It is a struggle to find problems which are yet again wear and tear related.

Engines: 1.3 **CO2:** 157g/km **MPG**: 42.8 **Gearbox:** Man **Boot:** 449 **Seats**: 5 **Doors:** 5

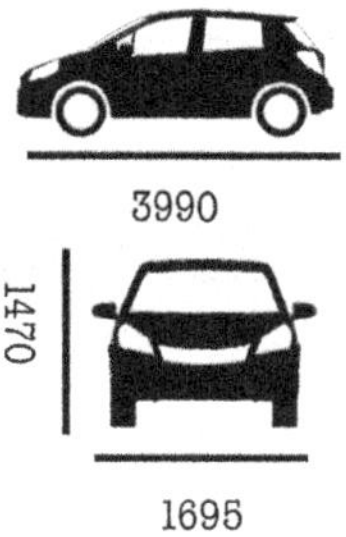

Kia Rio 2005 - 2011

Blimey, there was another one and at least it was all new. So bigger and more practical than before, but crucially just a tad shorter. The big news was a diesel engine, important at the time. Loads of kit of course and an on going reputation for staying in one piece. Nothing to really look at apart from the usuals, exhaust, brakes and suspension.

Engines: 1.4 • 1.6 • 1.5D **CO2:** 119-153g/km **MPG**: 47.9-62.8 **Gearbox:** Man/Auto **Boot:** 270/1145 **Seats**: 5 **Doors:** 5

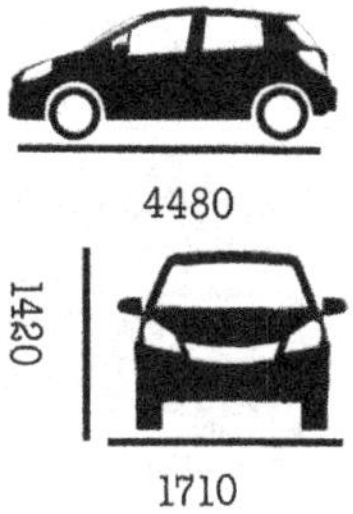

Kia Shuma 1999 - 2001

This is just a hatchback Mentor I'm afraid. There was a 1.5 and 1.8 and it had had some extras, but there was nothing to make it remotely desirable. When new it was cheap, when used it has to be really, really cheap as there is nowt else apart from reliability to recommend it. Might be rusty and need a brake overhaul and service.

Engines: 1.5 • 1.8 **CO2:** 185g/km **MPG**: 31.0-35.8 **Gearbox:** Man/Auto **Boot:** 328 **Seats**: 5 **Doors:** 5

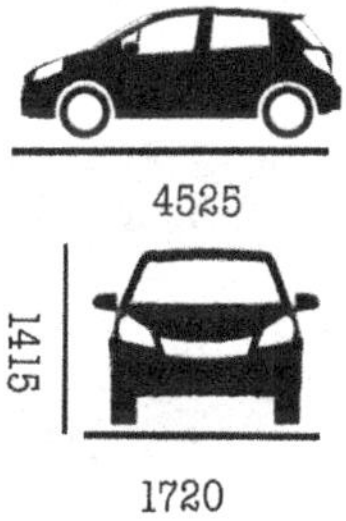

Kia Shuma II 2002 - 2004

Dear me, a follow-up which was around for five minutes but had a pretty major revamp, if anyone really cared. Loads of kit as usual with multi adjustable seats and for the time clever airbags and electronic brakeforce distribution. Does anyone care? You are only looking for rust and kicking the tyres really.

Engines: 1.6 **CO2:** 190g/km **MPG**: 35.3 **Gearbox:** Man/Auto **Boot:** 440/1207 **Seats**: 5 **Doors:** 5

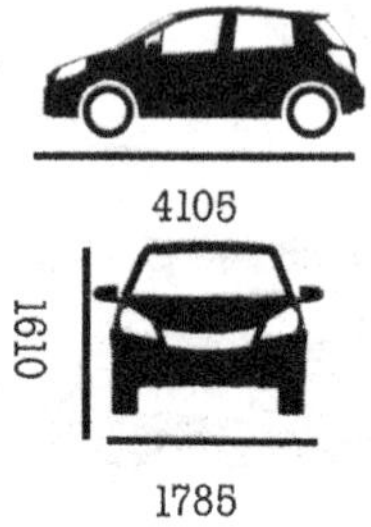

Kia Soul 2009 - 2013

Another one of those characterful Far Eastern boxes with a ton of useful space inside. There was an understandable model and engine line up: 1.6 petrol or diesel. Some might call it funky if they didn't know any better. Seem to be looked after shoppers, beware ex-minicabs, yes really. Hard to find anything to check, just general wear and tear.

Engines: 1.6 • 1.6D **CO2:** 137-153g/km **MPG**: 43.5-54.3 **Gearbox:** Man/Auto **Boot:** 222/700 **Seats**: 5 **Doors:** 5

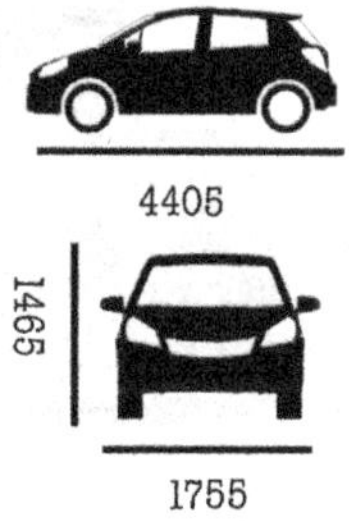

Mazda 3 2004 - 2008

Ford Focus without much personality . Split rear seats and things, plus if you really wanted a saloon, they did exist. The Mazda 3 sells because it is 100% reliable. 2006 revamp didn't change much but a 2.0d actually did. Smaller petrols are fine. Leaking shock absorbers, shot engine mounts, possible engine fault codes.

Engines: 1.4 • 1.6 • 2.0 • 1.6D • 2.0D **CO2:** 138-196g/km **MPG**: 34.5-56.5 **Gearbox:** Man/Auto **Boot:** 300/635

Mazda 3 2009 - 2013

More Focus fun in a bodywork that belongs to the Mazda. Still the exciting option of a saloon that will be even cheaper. Will easily soak up one hell of mileage and will inevitably be a diesel, 1.6 or 2.0. Have lots of time for the basic 1.6 petrol. Otherwise modern car problems such as blocked diesel particulate filters and checking the service history. **Engines:** 1.6 • 2.0 • 1.6D • 2.0D **CO2:** 119-159g/km **MPG**: 42.0-62.8 **Boot:** 340/1360 **Seats**: 5 **Doors:** 4/5

Mazda 323 1998 - 2003

These were great. Even if they didn't have the pop up headlamps anymore. Great to drive and tough as absolute nuts. There are the tiny petrols which will run all day. 1.8 and 2.0 are pokey and great of you intend on ambitious journeys. These go on forever, getting quite old and tired, will need shock absorbers, broken electrical switches.

Engines: 1.3 • 1.6 • 1.8 • 2.0 **CO2:** 179-203g/km **MPG**: 33.2-38.2 **Gearbox:** Man/Auto **Boot:** 356/697 **Doors:** 5

MG ZS 2001 - 2005

Here's a Rover 400 that's been driven grille first into a Demon Tweeks catalogue. The boot balances the spoiler and you can go for the 2.0 diesel, may be better than the head gasket troubles of the petrols. Looked different for the last year and doesn't deserve its own entry. Head gasket failure is the most common one, rust, difficult in getting some parts.

Engines: 1.8 • 2.5 • 2.0D **CO_2:** 150-227g/km **MPG**: 29.7-52.7
Gearbox: Man/Auto **Boot:** 370(hatch)/470(sal) **Doors:** 4/5

Nissan Almera 2000 - 2006

Less boring than the previous one, not that it helps much. Just well built, comfortable and businesslike. Perfect Banger material which is solid and does not attract attention. Lots of trim levels with a very strong cubbyhole game for all your bits and bobs. Gets overlooked, so worth finding before they rust away. Suspension springs and crankshaft sensors.

Engines: 1.5 • 1.8 • 2.2D **CO_2:** 160-180g/km **MPG**: 37.7-47.9
Gearbox: Man/Auto **Boot:** 328/490 **Seats**: 5 **Doors:** 5

Peugeot 306 1993 - 2001

A Pug faced Citroen ZX. Looked even prettier from '97 and got safer with ABS and airbags a year later. There were still saloons around in the late '90s, but these are all about the diesels with hatchback bodies. Became even harder to find as the range was pared down to just estates and cabs. Worn suspension, driveshafts, plus a lot are very tired and rusty.

Engines: 1.4 • 1.6 • 1.8 • 1.9D • 2.0D **CO2:** 141-198g/km **MPG:** 33.6-54.3 **Gearbox:** Man/Auto **Boot:** 338/637 **Doors:** 3/5

Peugeot 307 2001 - 2007

One of those forgettable Car of the Year winners. Quite high rise, so lots of room inside and as with all modern vehicles there are a bazillion options. There are six engine choices. 2.0 HDi is the one you want. The 1.6 petrol will do otherwise. Check the electricals to be on the safe side. Diesel filters clogged, suspension, electrical issues including engine ECU.

Engines: 1.4 • 1.6 • 2.0 • 1.6D • 2.0D **CO2:** 129-219g/km **MPG:** 35.3-57.6 **Gearbox:** Man/Auto **Boot:** 341/1328 **Doors:** 3/5

Peugeot 308 2007 - 2014

A 307+1. Lots of options again, go for the SE if you want the full house version. As ever the 1.6 petrol and 2.0 diesel are the best picks irrespective of the trim. In 2011 it got...we don't care about the new lights and other things because it is outside of the Bangerpedia brief. Water leaks into cabin, ECU failures, hairy electrics, blocked diesel filters.

Engines: 1.4 • 1.6 • 1.6D • 2.0D **CO2:** 120-159g/km **MPG**: 42.2.-62.7 **Gearbox:** Man/Auto **Boot:** 430/1398 **Doors:** 3/5

Proton Gen 2 Hatch 2004 - 2012

This was supposed to be the budget car future, but no one is really sure why. A Gen 2 was just another five door hatch that was nice to drive with 1.3 or 1.6 petrol engines and acres of interior grey plastic. At one point in '08 there was a 1.6 saloon. Essentially a zero budget used car. Ultra reliable, but getting some parts is becoming difficult.

Engines: 1.3 • 1.6 **CO2:** 164-172g/km **MPG**: 39.2-40.9 **Gearbox:** Man/Auto **Boot:** 370/700 **Seats**: 5 **Doors:** 5

Renault Grand Modus 2008 - 2012

Sort of a midget Espace, optional rear flip up central seat allowing the others to slide inwards. 1.2 petrol will do along with a 1.5 diesel. Did well in crash tests. Plenty of issues with used ones when it comes to electricals, rusty suspension towers, ignition coils and automatic gearbox. Electric roof broken. Caution needed.

Engines: 1.2 • 1.6 • 1.5D **CO2:** 120-140g/km **MPG**: 47.8-62.8 **Gearbox:** Man/Auto **Boot:** 305/1454 **Seats**: 5 **Doors:** 5

Renault Megane 1999 - 2002

Complicated line up with the less pretty second generation model. Loads of spin offs but we are sticking to the hatchback here. 1.4 or 1.6 petrol and the 1.9 diesels are the power packs behind what is one of the safest old bangers out there. Otherwise it is gearbox and electrics. Clutch repair difficult, coil packs and flywheel sensors.

Engines: 1.4 • 1.6 • 1.8 • 2.0 • 1.9D **CO2:** 139-180g/km **MPG**: 36.7-54.3 **Gearbox:** Man/Auto **Boot:** 348/1210 **Doors:** 3/4/5

Renault Megane 2002 - 2008

These looked interesting almost like science fiction people transporters as 3 and 5 hatches plus a saloon from '03-'06. Nissan related, so feels better built, with loads of engine options. A simple 1.6 petrol or the 1.5 dCi are the power picks. The range was revamped in 2006. Watch electrics. EGR valve can get blocked, water leaks into engine bay.

Engines: 1.4 • 1.6 • 2.0 • 1.5D • 2.0D **CO2:** 119-209g/km **MPG**: 43.0-62.8 **Gearbox:** Man/Auto **Boot:** 330/1190 **Doors:** 3/4/5

Renault Megane 2008 - 2016

More of the same with a new generation featuring a great big boot and tip top safety features. The 1.5 dCi remains the pick of the bunch and the simple small petrols. Check the recall history, electricals and like all moderns the EGR filter is going to block up if you just bimble about.

Engines: 1.4 • 1.6 • 2.0 • 1.5D • 1.9D • 2.0D **CO2:** 134-190g/km **MPG**: 38.7-55.4 **Gearbox:** Man/Auto **Boot:** 372/1129

Rover 45 1999 - 2005

In a previous life a Rover 400 with that fancy grille. Still old tech rather than cutting edge. Saloon and hatches and poncy model grades, Classic, Club and Connoisseur, but all very endearing especially as a V6 and diesel. Four cylinder blown head gaskets, not so endearing. Getting rusty and parts are more of a challenge to source.

Engines: 1.4 • 1.6 • 2.0 • 2.0D **CO2:** 150-168g/km **MPG**: 30.1-52.5 **Boot:** 370(hatch)/470(saloon)

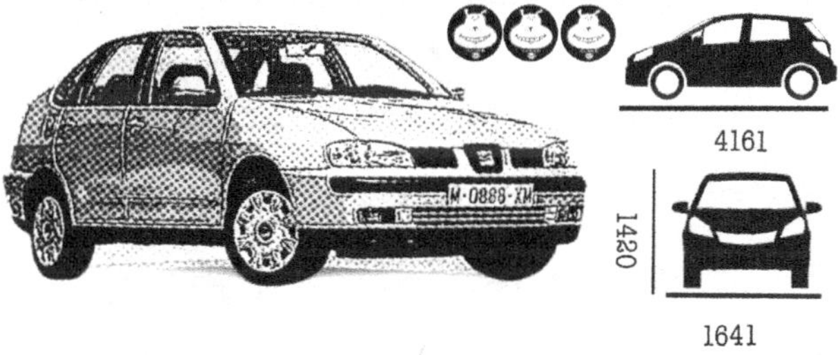

SEAT Cordoba 1994 - 2001

Polo related Spaniard with a big old boot. Air con from '97, miracle if it is still working, the saloon and hatch are very practical load luggers. 1.9 diesel what an oil burner should be, otherwise 1.4 petrol is fine. Revamp in 1999 so models looked more alike. Oil leaks, suspension wear, mass air flow sensor failure. Rust making itself known.

Engines: 1.6 • 1.9D **CO2:** 135-189g/km **MPG**: 36.2-57.5 **Gearbox:** Man/Auto **Boot:** 455/762 **Seats**: 5 **Doors:** 5

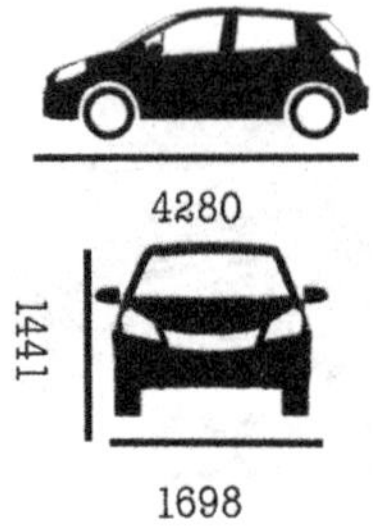

SEAT Cordoba 2003 - 2006

Back as an Ibiza with a boot with a 1.4 petrol and 1.9 diesel. Not the most popular SEAT ever made. SEs and Stylance were the top of the trim tree buys. The VW parts mean that these are pretty reliable if very rare motors mostly privately owned. Mostly dull. Timing chain tensioners, worn suspension and minor electricals like central locking

Engines: 1.4 • 1.9D **CO2:** 135-156g/km **MPG**: 43.5-54.3 **Gearbox:** Man/Auto **Boot:** 485 **Seats**: 5 **Doors:** 4

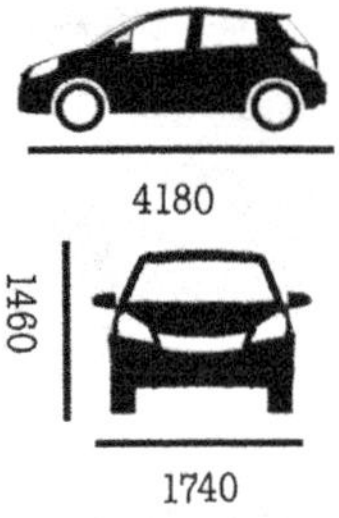

SEAT Leon 2000 - 2005

This one of set of models, so a.k.a. Golf, Bora, A3, Octavia, the Leon is quite perky and a Toledo without a boot. Cupras may offer value performance, but stick to the 1.4 and 1.8 petrols or a 1.9 TDi as the easy to look after day to day hacks. Water leaks into the cabin and boot. Faulty mass air sensors and ECUs cause rough running.

Engines: 1.4 • 1.6 • 1.8 • 1.9D **CO2:** 140-204g/km **MPG**: 41.5-54.3 **Gearbox:** Man/Auto **Boot:** 340/1050 **Doors:** 5

SEAT Leon 2005 - 2012

Complicated modern styling along with 1.6 and 2.0 FSI petrol engines and 1.9 to 2.0 diesels and then eco engines with a 1.4 petrol and Ecomotive 1.9 diesel. That was boring. You can get well into 60mpg, which is slightly more interesting. Leon used to be great, now just another hatch. Ok if cheap. ABS pump, sensor issues and stability control.

Engines: 1.4 • 1.6 • 2.0 • 1.9D • 2.0D **CO2:** 147-169g/km **MPG**: 37.2-50.4 **Gearbox:** Man/Auto **Boot:** 1166 **Seats**: 5 **Doors:** 5

Skoda Roomster 2006 - 2015

You can have a pointless, 'is it an MPV?' debate if you have nothing better to do, but it's just a small alt-van with windows which is roomy. Clever reconfigurable interior with sliding, reclining and removable rear seats. The 1.4TDI is great and if you don't do miles the 1.2 petrol. Leaky sunroofs, power steering failure, manual gearbox problems.

Engines: 1.2 • 1.4 • 1.6 • 1.9D **CO2:** 149-170g/km **MPG**: 39.8-51.4 **Gearbox:** Man/Auto **Boot:** 178/537 **Doors:** 5

Subaru Impreza 2000 - 2005

Yes there is the mental Scooby we all know and love, but there was simply a cooking 1.6 and 2.0 hatchback and saloon all with four wheel drive. So not that frugal and when they break the bills tend to be high. So could be a one use banger. Revamped in 2005 with just a 2.0. Blowing turbos, worn suspension and brakes.

Engines: 2.0 **CO2:** 212g/km **MPG**: 31.7 **Gearbox:** Man/Auto **Boot:** 353 **Seats**: 5 **Doors:** 4

Subaru Impreza 2007 - 2010

Not the purposeful looker anymore simply an ugly old hatch with some extra off road talents. Basic 4WD available with 1.5 and 2.0 flavours. Remains an oddity in a conventional world. Only occasional caravanners and unmade road dwellers need apply. Suspension, brakes and every thing that needs doing is always expensive.

Engines: 1.5•2.0 **CO2:** 176-199g/km **MPG**: 33.6-37.7 **Gearbox:** Man/Auto **Boot:** 301/1257 **Seats**: 5 **Doors:** 5

Suzuki Baleno 1995 - 2002

Utterly forgettable 1.6 saloon and three door hatch that despite revamps never became handsome. What it has going for it is Suzuki attention to detail and reliability. 1.8 engine also in the mix and not really much to recommend the Baloney except that it is a competent small family car. Might be a bit rusty now, just the usual wear and tear issues.

Engines: 1.6 • 1.8 **CO2:** 171-184g/km **MPG**: 36.7-39.8 **Gearbox:** Man/Auto **Boot:** 346 **Seats**: 5 **Doors:** 4

Suzuki Liana 2001 - 2007

Accidental TV star available as a good old fashioned notchy saloon (deleted in 2006) and a rather estate looking hatch. Life In A New Age wasn't exactly that, just popular with the Old Age private buyer. 2004 upgraded interior and exterior. Bigger than a contemporary Ford Focus and much more reliable. Trying to find reliability downsides is very difficult.

Engines: 1.6 **CO2:** 171g/km **MPG**: 39.8-40.9 **Gearbox:** Man/ Auto **Boot:** 348/595 **Seats**: 5 **Doors:** 4/5

Suzuki SX4 2006 - 2014

One of those crossovery things, sort of Rover Streetwise done proper and optional 4×4. Fiat diesels and petrols as well as a Suzuki 1.6. 4Grip is the 4×4 version which chips in when you really need it, otherwise you are supposed to just enjoy the high rise driving position. Also saloon '09 -'11. Nothing major goes wrong, there were recalls, but minor.

Engines: 1.6 **CO_2:** 143g/km **MPG**: 45.6 **Gearbox:** Man/Auto **Boot:** 270/1045 **Seats**: 5 **Doors:** 4/5

Toyota Auris 2007 - 2013

If you thought the Corolla was a dull consumer durable the Auris rightly continued in the same vacuous vein. Unbreakable 1.4 and 1.6 petrols and diesels in three or five door formats. Excellent basic air con and remote locking kit, you won't go wrong with one of these. However automatic gearbox can be problematic, clutch failure .

Engines: 1.33 • 1.6 • 1.4D **CO_2:** 133-153g/km **MPG**: 43.5-48.7 **Gearbox:** Man/Auto **Boot:** 354/761 **Seats**: 5 **Doors:** 5

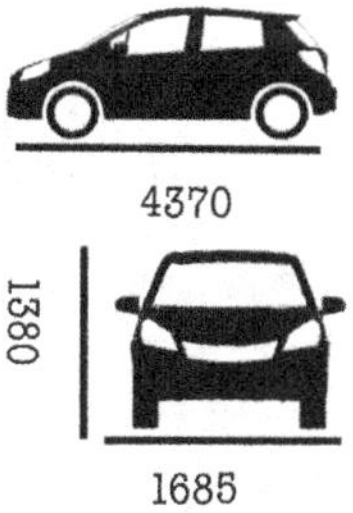

Toyota Corolla 1997 - 2002

Legend is an overused word, but the Corolla is so much more. A worldwide used car phenomenon. Liftbacks, saloons and all that with unbreakable petrols plus a diesel. Avensislike styling from 2000, saloon deleted and common rail diesel added. Made in Blighty. Rather rusty now, lambda sensors fail eventually, overall just worn out bits.

Engines: 1.4 • 1.6 • 2.0D **CO2:** 156-168g/km **MPG**: 40.4-48.7

Gearbox: Man/Auto **Boot:** 289/960 **Seats**: 5 **Doors:** 3/5

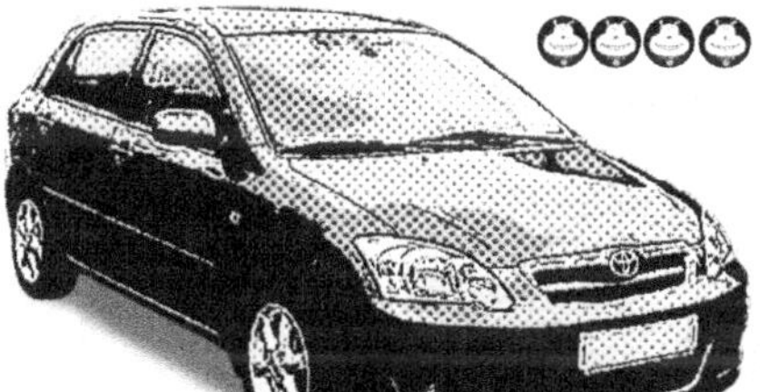

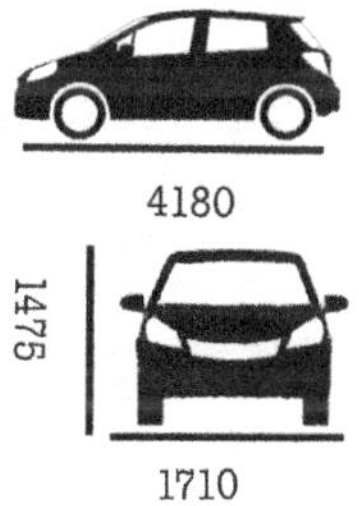

Toyota Corolla 2002 - 2007

Last Banger gasp of the legend. Mark 10 no less. Four petrols and two diesels. 2004 revamp to make it look even duller and there was a 1.4 diesel to replace the 2.0 one. Sadly you couldn't rely on the name badge. Passed away in the UK. See Auris. Recalls related to airbags they did not make. Just the usual worn out parts and possibly neglect.

Engines: 1.4 • 1.6 • 1.4D • 2.0D **CO2:** 151-168g/km

MPG: 40.4-49.6 **Gearbox:** Man/Auto **Boot:** 289 **Doors:** 5

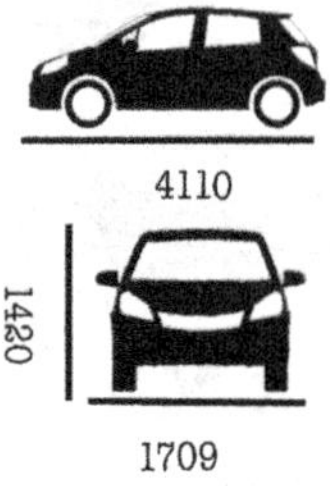

Vauxhall Astra 1998 - 2005

Mark 4, quite possibly the best Astra in all of human history. Massive range, seven engines and extremely roomy inside. Easy to drive and live with, but open to company car abuse. Quite tough in the real used car world, underrated which keeps prices marginal. Corroded ECUs, cockpit leaks, failing Air conditioning and water pumps.

Engines: 1.4 • 1.6 • 1.8 • 2.0 • 1.7D • 2.0D **CO2:** 154-172g/km
MPG: 38.2-49.6 **Gearbox:** Man/Auto **Boot:** 370/1210

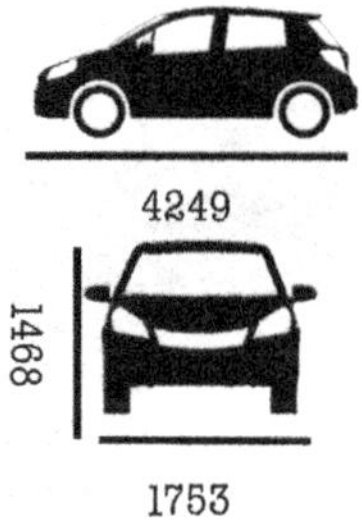

Vauxhall Astra 2004 - 2010

Next generation Astra is still pretty good and the pick of the bunch is the 1.6 petrol as an all rounder and the 1.7 ecoFLEX if you do big miles. There have been engine issues, but sorted out under recall and natural selection. Oil leaks, crankshaft oil seal. Worn suspension bushes and rusty turrets. Niggly electrical, lights, wipers, locks.

Engines: 1.4 • 1.6 • 1.8 • 2.0 • 1.7D • 1.9D **CO2:** 135-187g/km
MPG: 36.2-56.5 **Gearbox:** Man/Auto **Boot:** 302/1270

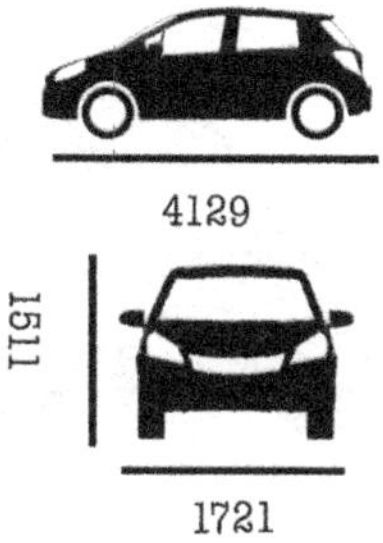

Volkswagen Beetle 1999 - 2010

Here's a Golf in a frumpy dress. Retropointless as there is less room and mpg than the original. Bought as a last resort, or as a personal transportation pod, either slow 1.6, or amusing V5. Stick to the 1.9 diesel for all round usefulness and minimal pointlessness. Lots of minor electrical failures, windows, locks etc. Engine ECU. Now bubbling rust.

Engines: 1.4 • 1.6 • 1.8 • 2.0 • 2.3 • 1.9D **CO2:** 143-185g/km

MPG: 36.7-53.3 **Gearbox:** Man/Auto **Boot:** 209/769

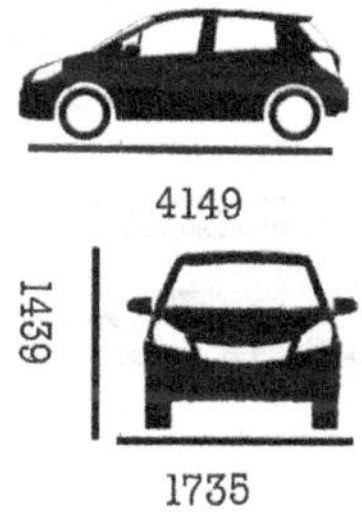

Volkswagen Golf 1997 - 2004

VW had something of a wobble with the last model, this one was spot on when it came to fit and finish and although quite old, have held up. Buy a five door boring 1.9 diesel or an interesting V5 three door. Hard to go badly wrong, but the GTIs weren't all that. Rust on sills and arches. Engine ECU cause rough running, brake calipers and water pumps.

Engines: 1.4 • 1.6 • 1.8 • 2.0 • 2.3 • 1.9D **CO2:** 140-211g/km

MPG: 32.1-57.6 **Gearbox:** Man/Auto **Boot:** 380/1270

Volkswagen Golf 2004 - 2008

Here's a blunt fronted model with a whole bunch of engines and specs, much like any other car, except this is a Golf. Experts will say avoid the SDi non-turbo, but this is proper basic unburstable banger. Otherwise this sophisticated hatch works well with a 1.4 TSI. Door seal failure, diesel turbo oil seal failure, poor air con.

Engines: 1.4 • 1.6 • 1.6D • 2.0D **CO2:** 143-178g/km **MPG**: 38.2-53.3 **Gearbox:** Man/Auto **Boot:** 350/1305

Volkswagen Golf 2008 - 2013

Golf 6 just sneaks into the Bangerpedia and is rather modern. Much more of the same winning formula. Again the 1.4TSI engine is one of the most sensible and economical units in a world that is turning against diesels. It is much more of a computer on wheels that needs reboots and updates. Coil pack failures on petrols, suspension bushes.

Engines: 1.4 • 1.6 • 2.0 • 1.9D • 2.0D **CO2:** 119-144g/km **MPG**: 45.6-62.8 **Gearbox:** Man/Auto **Boot:** 350/1305

Volkswagen Golf Plus 2005 - 2009

Fat Golf. Not MPV enough to be classed as one of those, just accommodates more big people. The rear seats though slide and fold flat, so it is a sort of Golf commercial. There is a double height boot floor, you can use the folding rear seat as a table and all that. It is an uglier Golf, but if it is better value, so that's why you buy. Lookouts as for the slim Golf.

Engines: 1.4 • 1.9D • 2.0D **CO2:** 151-174g/km **MPG**: 38.2-60.1
Gearbox: Man/Auto **Boot:** 395/1450 **Seats**: 5 **Doors:** 5

Volkswagen Golf Plus 2009 - 2013

Fat Golf 2. We were torn about this one, but on the basis that these are not that popular it should be a bargain Banger buy. Again, this is a no seats out MPV, but the rears recline. The double height boot and fold flat seats if a buyer needs a justification, just more Golf than normal. That's all. Normal Golf checkpoints. Plug in to a computer for an ECU check.

Engines: 1.4 • 1.6D **CO2:** 115-151-g/km **MPG**: 43.5-50.4
Gearbox: Man/Auto **Boot:** 395/1450 **Seats**: 5 **Doors:** 5

Volkswagen Bora 1999 - 2005

The Golf with a boot never went away. So it is less practical and more boring looking, that's the trade off. Mark 4 Golf is a rather good basis for something so conventional. V5 and V6 not very conventional the 1.9TDi restores the buying balance. Rust on sills and arches. Engine ECU cause rough running, brake calipers and water pumps.

Engines: 1.6 • 1.8 • 1.9D **CO2:** 143-192g/km **MPG**: 39.8-53.3 **Gearbox:** Man/Auto **Boot:** 455/785 **Seats**: 5 **Doors:** 4

Volkswagen Jetta 2006 - 2010

Another Golf with a boot and dusting off another old name. Pointless when there is the Passat, but where there's a niche. More upmarket than a Golf inside and a boot that's even bigger than a Bora. Find a cared for private example and enjoy, otherwise it could be company car hammered. Diesel pumps leaks, steering rack failures, worn clutches.

Engines: 1.4 • 2.0 • 1.9D • 2.0D **CO2:** 157-178g/km **MPG**: 34.0-48.7 **Gearbox:** Man/Auto **Boot:** 527 **Seats**: 5 **Doors:** 4

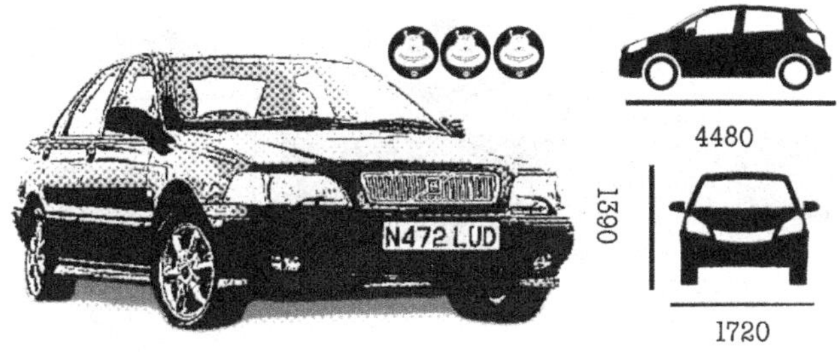

Volvo S40 1996 - 2004

Most people bought the estate. Those who wanted a posh saloon went 3-Series or C-Class. Mitsubishi Lancer related, so tough as they come if not the high quality saloon you might expect, just a dull worker. Suits the 1.9D engine, bodywork starts to rot. Corroded loom causes poor electrics, air con/power steering can fail, brake calipers break.

Engines: 1.6 • 1.9D **CO2:** 142-185g/km **MPG**: 35.8-52.3 **Gearbox:** Man/Auto **Boot:** 415 **Seats**: 5 **Doors:** 4

Volvo S40 2004 -2012

S60 but smaller, probably the best pre-executive buy in the whole of the Bangeruniverse. As ever the 2.0 diesel is all most people need, but if the miles are marginal then a 1.6 remains the most cost effective way of running an old school Volvo. Ignition broken or failing, particulate filter blocked on diesels. ABS control units.

Engines: 1.6 • 1.8 • 2.0 • 2.5 • 1.6D • 2.0D • 2.4D **CO2:** 154-217g/km **MPG**: 35.8-52.3 **Boot:** 357/883 **Doors:** 4

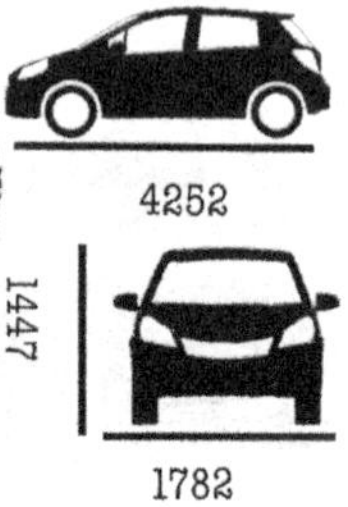

Volvo C30 2007 - 2012

Hatchback baby Swede that is obviously safe, stylish and very well built. Makes a change from the Golf orthodoxy, but it is limited because there are only three doors and hardly a huge boot. Yes it is different. 1.6 diesel is hyperfrugal, otherwise a 1.6 petrol is the dullish Banger answer. Air conditioning condensers.

Engines: 1.6 • 1.8 • 2.0 • 2.5 • 1.6D • 2.0D • 2.4D **CO2:** 129-174g/km **MPG**: 38.2-49.6 **Boot:** 364/1010 **Doors:** 3

Ex Rep Bangers for the Family

I may have suggested that those Happy Shoppers, then the Family Bangers were the hardest working motors known to humanity. Actually it may these Ex Reps which actually used to have real jobs.

The Ford Mondeo was built for this and the Mondeo 2 was arguably the best version of them all which had the equipment, comfort and engines to earn a serious living. Ditto the Vauxhall Vectra that was just about to become the Insignia. The Nissan Primera, after a good run was now starting to fade away and a clue to the future was the irresistible rise of the Qashqai. SUVs were coming, meanwhile the old fashioned saloon was still something of thing in the fleet market.

That explains the Chevrolet Epica, Chrysler Sebring and Daewoo Leganza to name just a few notchbacks. Actually there were loads of them, Yanks like the Dodge Avenger, but also Euro weirdos in the shape of the Fiat Marea. Otherwise the boot remained an option for what was their last significant decade as an option.

Back in the old days, company representatives really did load these up with samples and photocopier parts. They were selling and fixing stuff that kept the country moving. When the vehicles reached the end of their useful motoring lives, usually at the end of the company contract hire period, they were recycled at auctions. Bought by the car trade these Ex-Reppers ended up at a car lot or open air car supermarket.

Another route to the car supermarket was via hire companies who in some cases just ran them for a few months or thousand miles. As these motors accounted for the bulk product of car manufacturers, they might be registered and sold with zero to not many miles simply to boost sales figures. Incredible isn't it? That's when real motorists bought them often from their local dealers.

Pretty much these are the quintessential Bangers. Not fashionable, clever, or desirable, just built to do a job and there are loads to choose from.

Chevrolet Epica 2007 - 2010

Reconstituted Daewoo saloon and obviously not remotely epic. Kept it simple with one petrol and a diesel. There was a shed load of equipment for the lucky buyers who went for either the LS or LT. Air conditioning, suspension and sundries can fail due to miles and neglect.

Engines: 2.0 • 2.0D **CO2:** 169-205g/km **MPG**: 34.5-46.3
Gearbox: Man/Auto **Boot:** 480 **Seats**: 5 **Doors:** 4

Chrysler Sebring 2007 - 2009

A cross-pollinated saloon with a Mitsubishi undercarriage and even a VW diesel and related to the Dodge Avenger, but less scary. Just another forgettable family car. It still ought to be a lot of car for not much money. Has loads of safety kit and comfy interior. Engines are tough but need a service every 9K miles. Check all electrical gadgets.

Engines: 2.0 • 2.4 • 2.0D **CO2:** 170-211g/km **MPG**: 54.0.-58.0
Gearbox: Man/Auto **Boot:** 385 **Seats**: 5 **Doors:** 4

Citroen C5 2001 - 2004

The super clever Citroen returns, which isn't always a good thing and Hydractive 3 suspension is great to drive, but not straightforward to sort out a decade later. Most would buy one of the diesels, even though the potential for disappointment is huge. Shock absorbers, ECU controlling suspension, ABS, weak clutch, cracked alloy wheels.

Engines: 1.8 • 2.0 • 3.0 • 2.0D • 2.2D **CO2:** 147-226g/km **MPG**: 29.4-50.4 **Gearbox:** Man/Auto **Boot:** 456/882

Citroen C5 2004 - 2008

Revamped with added complication (lane departure system, xenon headlamps, maximum speed limiter) and size. Bags of room like the last one and a new 1.6 diesel. Seven standard airbags so it is as safe as houses. Great when it all works, nightmare when it doesn't. Automatic gearbox failure. Sealed suspension failing. Weak Clutch.

Engines: 1.8 • 2.0 • 3.0 • 1.6D • 2.0D • 2.2D **CO2:** 142-187g/km **MPG**: 36.7-52.3 **Gearbox:** Man/Auto **Boot:** 456/882

Citroen C5 2008 - 2016

Remade for a new era, but buyers are even more difficult to convince in the face of more dependable Germans. Looks a bit different from before and has some big modern diesels that are wonderful. Like before though, complicated and will destroy you financially. All the same checks and potential problems, might be less wrong because a bit newer.

Engines: 1.8 • 2.0 • 1.6D • 2.0D • 2.2D • 2.7D

CO2: 149-198g/km **MPG**: 33.6-50.4 **Boot:** 439 **Doors:** 4

Daewoo Leganza 1997 - 2002

These are almost extinct now, just a 2.0 petrol saloon with bucket loads of standard kit. It looks like a very low rent Lexus GS, but isn't that refined to drive. Lots of room and on the whole reliable. Never held its value, so pay a £1 and drive it into the ground. Nothing to look for, just wear and tear items like brakes and suspension.

Engines: 2.0 **CO2:** 215g/km **MPG**: 30.7 **Gearbox:** Man/Auto

Boot: 400 **Seats**: 5 **Doors:** 4

Dodge Avenger 2007 - 2009

Undercover cop car looks, but this is another version of the Chrysler Sebring which comes with VW diesel engine to keep Europeans happy, at the time. Lots of room inside and a split fold rear seat, plus a fold flat front which is handy. Few around obviously. Not much seems to go wrong, just the usual wear and tear items like tyres, brakes and suspension.

Engines: 2.0•2.4•2.0D **CO2:** 170-211g/km **MPG**: 31.7-45.6

Gearbox: Man/Auto **Boot:** 382 **Seats**: 5 **Doors:** 4

Fiat Marea 1997-2002

Stylish thing which is nice to drive, but the fact it is a Fiat meant no one took it seriously new or used. Petrol and diesel engines complemented by not very many standard bits of kit. If any are around, should cost buttons. Chocolate ones. Suspension drop links, sunroof leaks, instrument panel failures, needs red master key otherwise worthless.

Engines: 2.0•1.9D **CO2:** 149-234g/km **MPG**: 28.8-50.4

Gearbox: Man/Auto **Boot:** 438 **Seats**: 5 **Doors:** 4

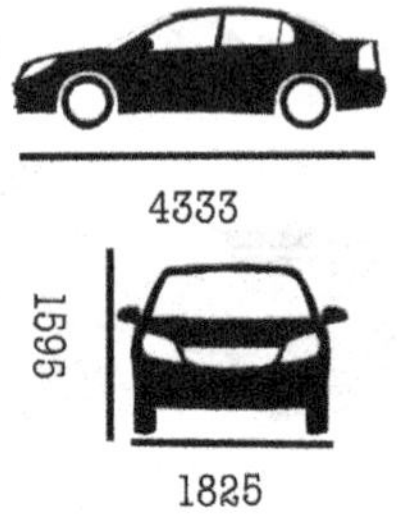

Ford C-Max 2003 - 2010

Officially this is a compact MPV, but really it is a pumped up, less aerodynamic Focus and could take a big fridge/freezer whole whereas an estate might struggle. It seats five which is all anyone really needs. Good to drive and loads of engines. As a Ford it is mostly cheap to run and fix. Slipping clutch, air con failure, Xenon lights expensive to replace.

Engines: 1.6 • 1.8 • 2.0 • 1.6D • 1.8D • 2.0D **CO2:** 154-169g/km

MPG: 39.8-49.5 **Gearbox:** Man/Auto **Boot:** 473/1633

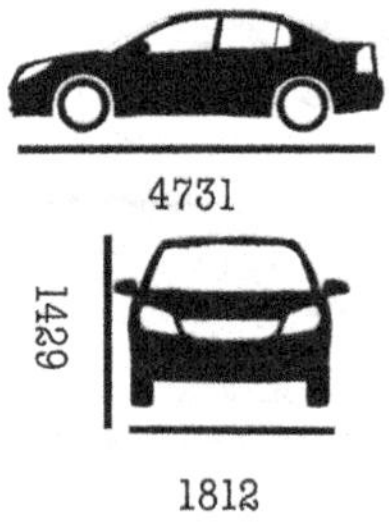

Ford Mondeo 2000 - 2007

Second generation Mondeo is probably the best incarnation. Big boot, interior and great to drive. The TDCi engines made all the difference. Like most Fords the Quickclear screen is brilliant on a frosty morning. Constantly updated, buy later editions and Zetecs or Ghias for the win. Suspension bushes, diesel engines ECU issues and oil leaks. Brake pipes rot.

Engines: 1.8 • 2.0 • 2.5 • 3.0 • 2.0D • 2.2D **CO2:** 154-187g/km

MPG: 36.2-48.7 **Gearbox:** Man/Auto **Boot:** 500/1370

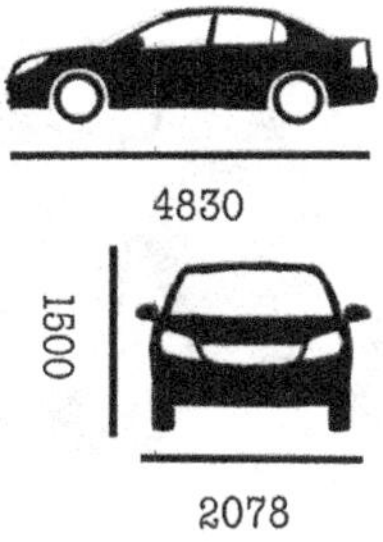

Ford Mondeo 2007- 2014

Mark 4 is a great example of how to make a working class company car. Last of the breed which is absolutely huge on the inside. Still available as a hatchback and saloon. 2.0 petrol if the miles are low, 2.0 diesel otherwise. All models have air con, big spec Titaniums around in numbers. Air con failure, remote locking, minor ECU, electrical issues.

Engines: 1.8 • 2.0 • 2.3 • 2.5 • 1.8D • 2.0D • 2.2D **CO2:** 154-177g/km **MPG**: 38.2-48.7 **Boot:** 528/1448 **Doors:** 4/5

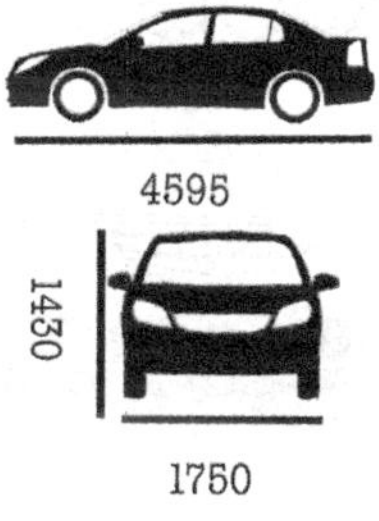

Honda Accord 1999 - 2003

Fabulous family transport which is big enough, nice to drive and built to such a high standard. This is more Banger royalty. No diesel, just superb petrol engines that will go on and on. Rust is the only enemy and serious owner neglect. It really is the wear and tear items like brakes and suspension. Some parts though may cost a bit or be difficult to find.

Engines: 1.8 • 2.0 • 2.3 **CO2:** 203-209g/km **MPG**: 32.1-32.8 **Gearbox:** Man/Auto **Boot:** 430 **Seats**: 5 **Doors:** 4

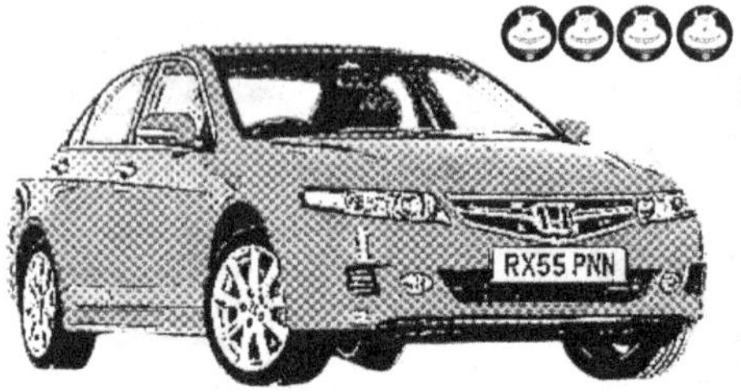

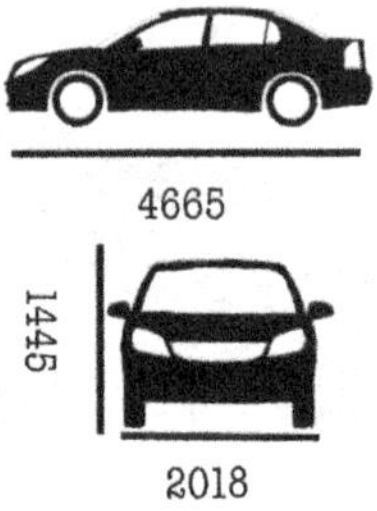

Honda Accord 2003 - 2008

More greatness and a proper quality package. Safe, well made and always going to work provided it is looked after. Buyers got a diesel delivered in 2004, otherwise the petrols are fine if you are only pottering about. A luxury car in a modest package. Perfect. ABS brake sensors, wheel bearings and tailgate release, mostly on the Tourer.

Engines: 2.0•2.4•2.2D **CO2:** 143-214g/km **MPG**: 31.4-52.3

Gearbox: Man/Auto **Boot: 459 Seats**: 5 **Doors:** 4

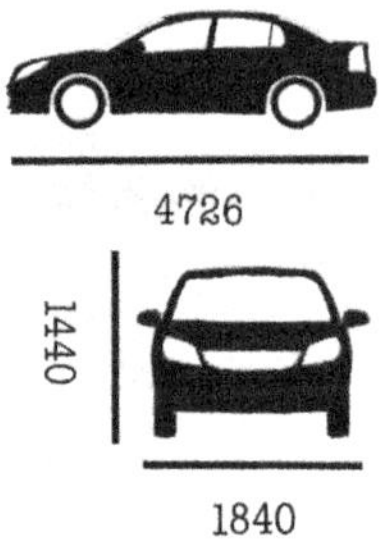

Honda Accord 2008 - 2011

More effective executive salooning courtesy of the eighth Accord incarnation. Three engines a 2.2 diesel and a couple of petrols. Parts are not too expensive and outside of regular servicing it should need to many as it continues the near perfect reliability record. But getting complicated. Clutch wear, dual mass flywheel problems, blocked diesel filters.

Engines: 2.0•2.4•2.2D **CO2:** 148-209g/km **MPG**: 32.1-50.4

Gearbox: Man/Auto **Boot:** 460 **Seats**: 5 **Doors:** 4

Hyundai Elantra 2001 - 2006

Budget effort available as both a hatch and a saloon up to '04. Large boot and split rear seats. The petrols 1.6 and 2.0 will do, but the 2.0 diesel isn't over complicated. A mostly unremarkable model which is becoming something of a rarity. Utterly reliable. Just kick the tyres, look at the brakes and suspension, that's it.

Engines: 1.6 • 2.0 • 2.0D **CO2:** 160-196g/km **MPG:** 34.4-46.3
Gearbox: Man **Boot:** 368/801 **Seats**: 5 **Doors:** 5

Mazda 626 1999 - 2002

Scaling up the 323 worked, here was a dull, but very worthy saloon and hatchback. The restyle in 1999 was meant to help but it was still dreary. It's an old school minicab because it is roomy enough and needs just basic servicing to keep on working, probably forever. These are remarkably tough, just neglect and worn out parts such as suspension and brakes are the killers.

Engines: 1.8 • 2.0 • 2.0D **CO2:** 152-195g/km **MPG**: 35.8-54.3
Gearbox: Man/Auto **Boot:** 502 **Seats**: 5 **Doors:** 4/5

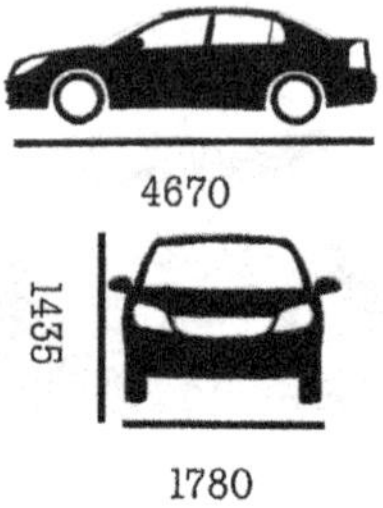

Mazda 6 2002 - 2007

Revised styling and a bunch of brand new engines. Great to drive and overall Mazdability, that's reliability, makes it a best buy from the 2000s. Excellent levels of kit from the S model upwards. Neat one touch folding rear seats make living with one easy as you get a nice flat floor. Properly enjoyed as a hatch. Clutch failure and neglected air con.

Engines: 1.8 • 2.0 • 2.3 • 2.0D **CO2:** 172-212g/km **MPG**: 34.9-43.5 **Gearbox:** Man/Auto **Boot:** 492/1662 **Doors:** 4/5

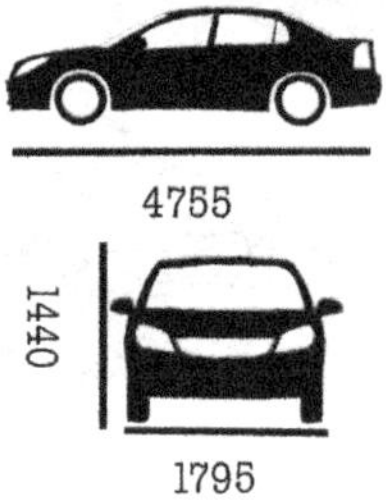

Mazda 6 2007 - 2012

Handsome thing. Saloon cancelled in '09. 1.8 petrol and 2.0 Diesel and aim for TS2 spec and that should be everything a Mazda driver needs. 2010 facelift is just inside our Bangerpedia remit. Overall these later cars seem to be ultra reliable, which is good. Otherwise check for leaking air con and the short recall history especially on petrol.

Engines: 1.8 • 2.0 • 2.5 • 2.0D **CO2:** 147-168g/km **MPG**: 39.8-51.4 **Gearbox:** Man/Auto **Boot:** 510 **Seats**: 5 **Doors:** 4/5

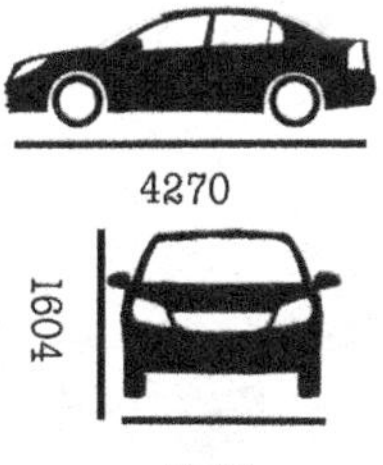

Mercedes B Class 2005 - 2011

Scaled up A-Class that failed to really hit the prestigious family spot. Never mind because used buyers get a ton of space, safety and some style. Small range of the usual engines then from 2009 the ultra eco Blue Efficiency petrols that make a lot more sense now. Rusty brake discs, smokey diesels, blocked filters, central locking and air con failure.

Engines: 1.5 • 1.7 • 2.0 • 2.0D **CO2:** 122-146g/km **MPG**:37.6-60.1 **Boot:** 554/2245 **Doors:** 3/5

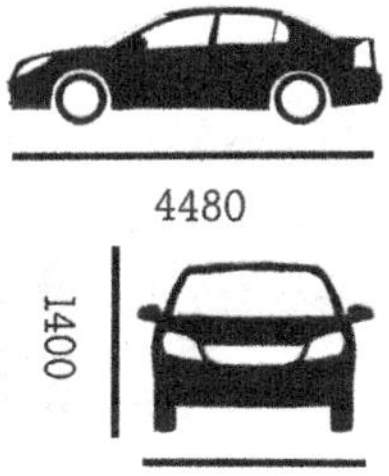

Mitsubishi Carisma 1999 - 2004

The perennial gag is that it doesn't have any. Another of those anonymous '90s Japanese hatches and saloons, which just got on with working. Lots of standard equipment with a Classic, Equippe, Elegance or Sport. Low running costs make up for lack of personality. Not much to look for, just wear and tear, incredibly fault free.

Engines: 1.6 • 1.8 • 1.9D **CO2:** 168-176g/km **MPG**: 38.2-46.3 **Gearbox:** Man/Auto **Boot:** 430 **Seats**: 5 **Doors:** 4/5

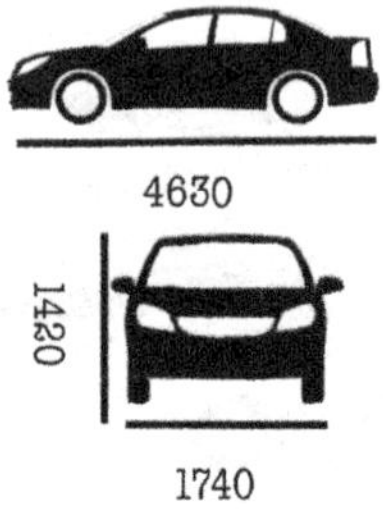

Mitsubishi Galant 1997 - 2003

Japan's car of the year '97 apparently. It featured a GDI engine, a gas direct injection unit which was jolly efficient even if Euro fuel wasn't always pure enough for it. Not a big hit, few around, classy and solid, proper alternative to one of those German saloons. Parts are very expensive when required. Check gearbox condition, brakes and suspension.

Engines: 2.0 • 2.4 • 2.4 **CO2:** 203-207g/km **MPG**: 29.7-33.2

Gearbox: Man/Auto **Boot:** 470 **Seats**: 5 **Doors:** 4

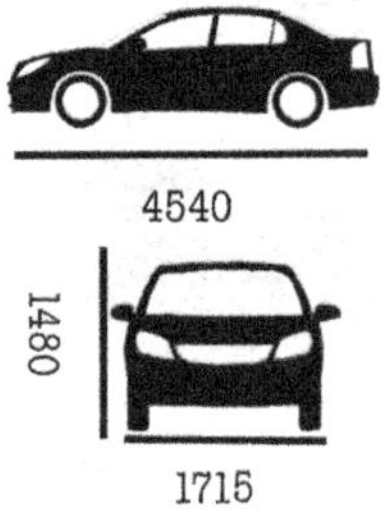

Mitsubishi Lancer 2005 - 2008

Apart from the hardcore rally ones, there are the 1.6 and 2.0 petrol ones which could not be more straightforward. Reasonable amount of room and split rear seat, there are better alternatives out there, but they may not be as good value or reliable. Apart from checking the electricals, it is the wear and tear parts. That's it.

Engines: 1.6 • 2.0 **CO2:** 163-207g/km **MPG**:32.5-41.5

Gearbox: Man/Auto **Boot:** 430 **Seats**: 5 **Doors:** 4/5

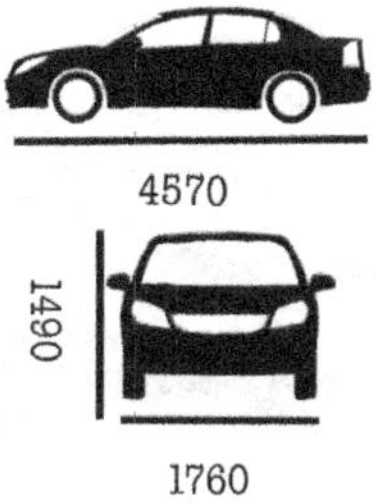

Mitsubishi Lancer 2008 - 2011

Not any less underwhelming than it was before, just the choice of Sportback hatches in various GS trims and a couple of top spec saloons. Like all Mitsubishis parts are always more expensive than a cooking Ford or Vauxhall. Just in case you wondered, the saloon left us in 2010. Not much to look for just electricals plus wear and tear components.

Engines: 1.5 • 1.8 • 2.0D **CO2:** 162183g/km **MPG**: 36.7-44.8 Gearbox: Man/Auto Boot: 344/1394 Seats: 5 Doors: 4/5

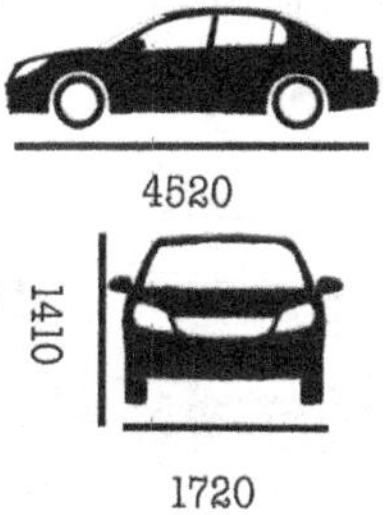

Nissan Primera 1999 - 2002

Here was the UK's favourite minicab in the '90s based on the fact that it would just keep on working, probably forever. Lots of standard kit of course. The engines are great and it is only going to be consumables like tyres and exhausts that may need replacing. Older examples are starting to rot, suspension wear, air con busted, old alloys corrode.

Engines: 1.6 • 1.8 • 2.0 • 2.0D **CO2:** 179-195g/km **MPG**: 34.8-43.5 **Gearbox:** Man/Auto **Boot:** 490/1000 **Doors:** 4/5

Nissan Primera 2002 - 2006

Suddenly it looked all modern. The Primera also became a lot more complicated especially inside where most had LCD screens. Almera based and lots of standard fixtures like climate air con and plenty of air bags. Reliable, but fancy electronics go on the blink, windows and central locking. Turbo diesel failure and crankshaft sensor issues.

Engines: 1.8 • 2.0 • 2.2D **CO2:** 164-208g/km **MPG**: 33.2-47.9
Gearbox: Man/Auto **Boot:** 460/780 **Seats**: 5 **Doors:** 5

Nissan Qashqai 2007 - 2013

Welcome to the future. SUV looks, but otherwise it's just a great big hatchback with a huge boot. There was a 4×4 option, which is rare and a +2 from '08 added seats, but they are for children.Diesel Particulate filters blocked, early model shock absorber failure. Clutches take a lot of stick. Water leaks into the cabin.

Engines: 1.6 • 2.0 • 1.5D **CO2:** 145-199g/km **MPG**: 34.4-52.3
Gearbox: Man/Auto **Boot:** 410-660 **Seats**: 5 **Doors:** 5

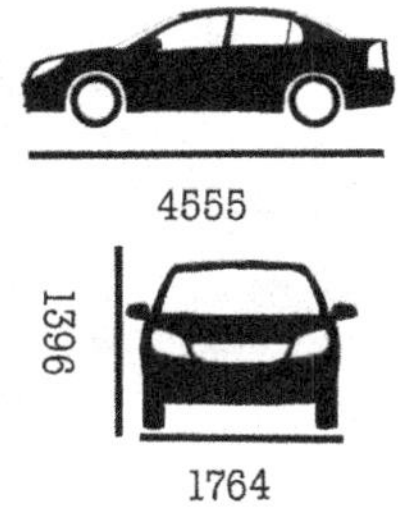

Peugeot 406 1996 - 2004

Old school Pug saloon, but all the better for that. Good to drive and simple enough to own the 1.9 Turbo is prehistoric but arguably better than most flaky modern oil burners. Otherwise the 2.0 HDi is perfect for the motorway. Revised in 2001, but our sort of Banger relic. Corroded radiators, rusty brake discs, engine ECU issues, suspension bushes.

Engines: 1.8 • 2.0 • 2.0D **CO2:** 147-197g/km **MPG**: 34.0-51.3 **Gearbox:** Man/Auto **Boot:** 526 **Seats**: 5 **Doors:** 4

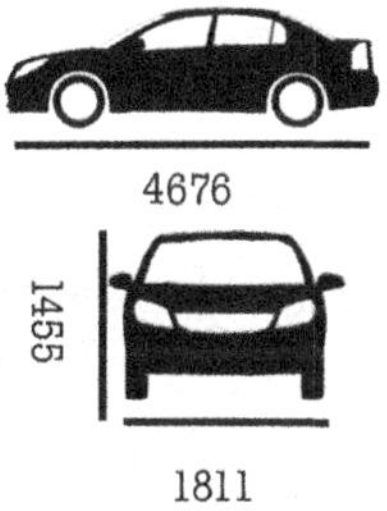

Peugeot 407 2004 - 2011

A great big saloon that was comfy and nice to drive. Lots of spec levels and plenty of complicated electrics. The upside is countless airbags and general safety kit. Overall an awful lot to go wrong, so not the prime Banger to buy unless cheap and with the fullest service history. Dashboard failure, clutch failure, ABS sensors, diesel throttle issues.

Engines: 2.0 • 1.6D • 2.0D • 2.2D • 2.7D **CO2:** 140 - 206g/km **MPG**: 30.6-53.2 **Gearbox:** Man/Auto **Boot:** 407 **Doors:** 4

Proton Impian 2001 - 2007

It has just the one thing going for it and that's a Mitsubishi engine. Otherwise just a dullard saloon from planet Proton which has proved cheap to run with reasonable 40mpg economy. The Lotus connection remains an utter irrelevance. Try and find one. Not a lot to look for, just brakes, tyres and suspension and whether serviced ever.

Engines: 1.6 **CO2:** 161g/km **MPG**: 42.2 **Gearbox:** Man/Auto **Boot:** 440 **Seats**: 5 **Doors:** 4

Proton Wira 2000 - 2004

No one loves these on the grounds that they are dull and dated. Mitsubishi engines help, there are a couple of petrols and diesels, but this is an '80s throwback without much charm. Will get you from A to B though. As ever, nothing to trouble a buyer except abject neglect. So tyres, brakes and regular records of an oil change are all reassuring.

Engines: 1.3•1.5•1.6 **CO2:** 198-211g/km **MPG**:32.7-35 **Gearbox:** Man/Auto **Boot:** 427.6 **Seats**: 5 **Doors:** 4

Renault Laguna 1994 - 2001

Constantly updated and not a half bad hatch. Most of them have been driven into the tarmac and largely neglected. The basic diesels and petrols are the ones which have proved to be pretty dependable. Struggle to see these for sale now. Rust now a big issue, corroded heating system, failed air con, electricals always a worry, gearboxes weak, locks break.

Engines: 1.6 • 2.0 • 3.0 • 1.9 • 2.2D **CO2:** 149-185g/km **MPG**: 24.1-51.4 **Gearbox:** Man/Auto **Boot:** 453 **Seats**: 5 **Doors:** 5

Renault Laguna 2001-2007

The Laguna II was needlessly complicated which just made used buying much more of a hassle. 2005 revisions, did something or other, important to remember that survivors with an MOT might well be fine. Buy cheap. Weak clutches, EGR valves blocked, electrics terrible, corroded and cracked alloy wheels, air con on strike.

Engines: 2.0 • 3.0 • 1.9D • 2.2D **CO2:** 154-200g/km **MPG**: 33.2-47.9 **Gearbox:** Man/Auto **Boot:** 430 **Seats**: 5 **Doors:** 5

Renault Laguna 2007 - 2014

Laguna III isn't pretty and yet again is a world of painful complication. Less buying options as the badly selling range was constantly pared down to the essentials. Buyers may risk their sanity/bank balance. Better than the old ones possibly less broken, but still look closely at the electrics, EGR valves and gearbox operation.

Engines: 2.0 • 1.5D • 2.0D **CO2:** 136-158g/km **MPG**: 35.8-55.4 **Gearbox:** Man/Auto **Boot:** 450/1377 **Seats**: 5 **Doors:** 5

SEAT Exeo 2009 - 2013

The Volkswagen Group indulging in a bit of recycling by taking an old (2004) Audi A4 and sticking a SEAT badge on it. So long as the asking price is low enough the family tree doesn't matter as the engines and build quality are great. Nothing to really worry about, just modern day issues like blocked particulate filters, otherwise wear and tear.

Engines: 2.0 • 2.0D **CO2:** 141-179g/km **MPG**: 36.7-51.4 **Gearbox:** Man/Auto **Boot:** 445 **Seats**: 5 **Doors:** 4

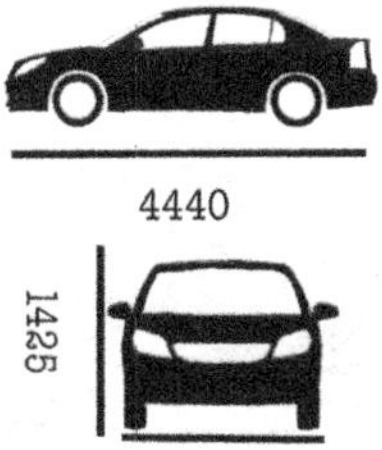

SEAT Toledo 1999 - 2005

Great big booted saloon related to so many VWs we haven't got space to list them. Looked good, obviously practical and excellent engines. Cheap to fix and service which are the real plus points. Used car prices are a great leveler, but a tidy one of these, but prices should be marginal. Diesels can have ECU issues, worn clutches, minor electricals inside.

Engines: 1.8 • 2.3 • 1.9 **CO2:** 135-211g/km **MPG**: 32.1-56.5 **Gearbox:** Man/Auto **Boot:** 460 **Seats**: 5 **Doors:** 4

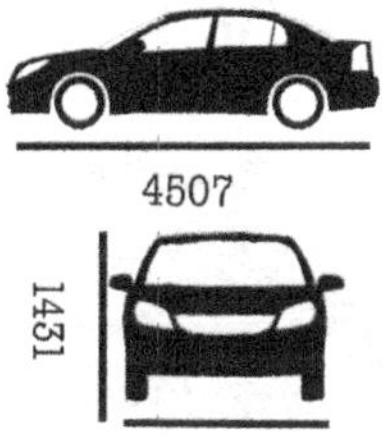

Skoda Octavia 1998 - 2005

A saloon style hatchback which remains a rather excellent way to herd your family around. Smart looking and the usually excellent line up of VW engines. Who really needs a Golf when you can have one of these? 1.9 diesels and 1.6 petrols are easy to deal with. Diesel ECU failure, water pumps, air mass sensors and ignition coils across the engine range. Electrical failures in the cockpit.

Engines: 1.4 • 1.6 • 2.0 • 1.9D • 2.0D **CO2:** 143-168g/km **MPG**: 40.4-53.3 **Gearbox:** Man/Auto **Boot:** 560/1350 **Doors:** 5

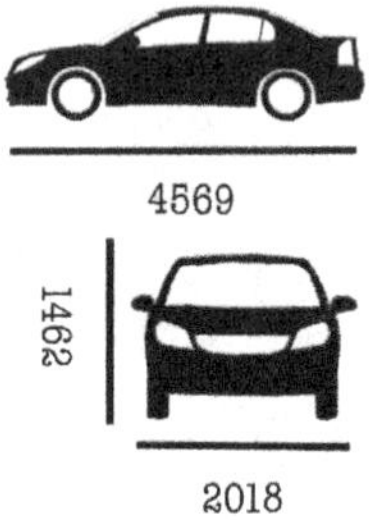

Skoda Octavia 2004 - 2012

Spacious and useful, that's all you really need to know. There was a very broad range with high performance vRS stuff and the super eco Greenlines, Cheap and simple to service. Styling revamp in 2009. Popular choice as diesel taxis, like many Skodas. Electricals and water leaks, climate control, ABS and ESP sensors and ECUs.

Engines: 1.4 • 1.8 • 2.0 • 1.9D • 2.0D **CO2:** 132-154g/km

MPG:42.8-57.6 **Gearbox:** Man/Auto **Boot:** 560/1350

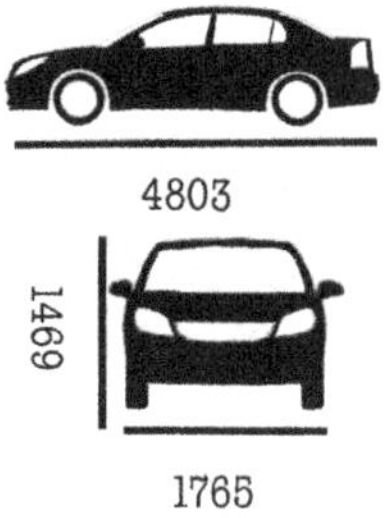

Skoda Superb 2002 - 2008

Like all Skodas based on something else, this was a Passat, but what a wonderfully huge and useful saloon it turned out to be. Hard working 1.9 diesel later replaced by more current 2.0 TDI, restyled in 2006. Look out for electric issues and general wear and tear, so shot suspension, brakes and possible ECU failures. Another regular taxi.

Engines: 1.8 • 2.0 • 2.8 • 1.9D • 2.0D • 2.5D **CO2:** 151-213g/km

MPG: 32.8-50.4 **Gearbox:** Man/Auto **Boot:** 462 **Doors:** 4

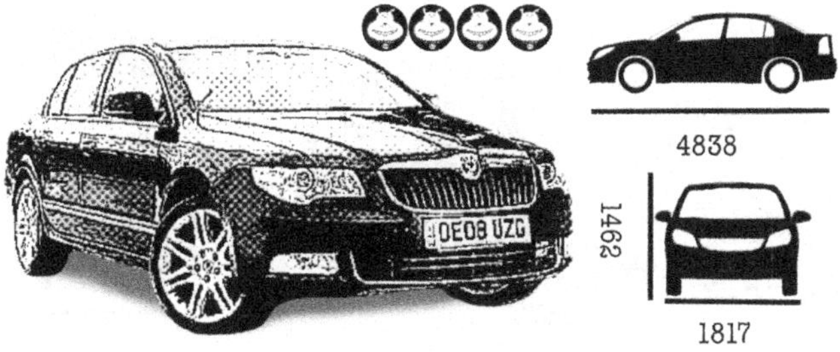

Skoda Superb 2008 - 2015

Well this was clever, it came with the so-called Twindoor, so it's got a boot and a hatch, when you press the right button. There's an umbrella in the door which was a novelty, not a used deal breaker though. Most on sale are diesels and yet again many lived a life in the taxi business. ABS sensor failure, DSG gearboxes malfunction, keyless entry issues.,

Engines: 1.4 • 1.8 • 3.6 • 1.9D • 2.0D **CO2:** 155-180g/km **MPG**: 39.2-47.9 **Gearbox:** Man/Auto **Boot:** 565/1670

Subaru Legacy 1999 - 2003

Legendary four wheel drive saloon, but you don't have to get the mad one, the 2.0 litre engine GL is fine, but the 2.5 uses a little more fuel. Uglification revamp in 2001. Fixing them is expensive. Problems with the four wheel drive system only if neglected. Everything costs a fortune if broken. It will be wear and tear on the running gear.

Engines: 2.0 • 2.5 **CO2:** 215-228g/km **MPG**: 30.4-31.4 **Gearbox:** Man/Auto **Boot:** 433 **Seats**: 5 **Doors:** 4

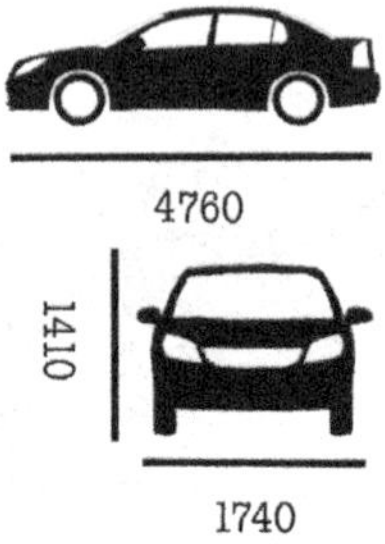

Subaru Legacy 2003 - 2010

Still an odd buy for most people, what could be used as justification was the arrival of a diesel engine although that didn't happen until '08. Had a decent boot though. Meanwhile it was restyled a bit in '07 but the saloon was running out of road. Costs even more to fix now, brakes and suspension are the least of your worries.

Engines: 2.0 • 3.0 • 2.0 **CO2:** 148-233g/km **MPG**: 23.3-50.4 **Gearbox:** Man/Auto **Boot:** 433 **Seats**: 5 **Doors:** 4

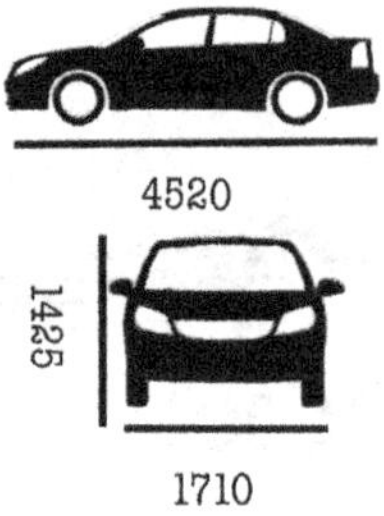

Toyota Avensis 1997 - 2003

The Brit built Carina E was fabulous and here was a saloon and Liftback based on it, but with an awful lot more safety equipment. Within a couple of years there was a diesel, the D4-D. Proper company cars that led hard lives but still managed to not break. Can still do a job. Clutch and gearbox failures over time, brake fluid leaks, just wear and tear.

Engines: 1.6 • 1.8 • 2.0 • 2.0D **CO2:** 158-181g/km **MPG**: 36.7-47.9 **Gearbox:** Man/Auto **Boot:** 510 **Seats**: 5 **Doors:** 4/5

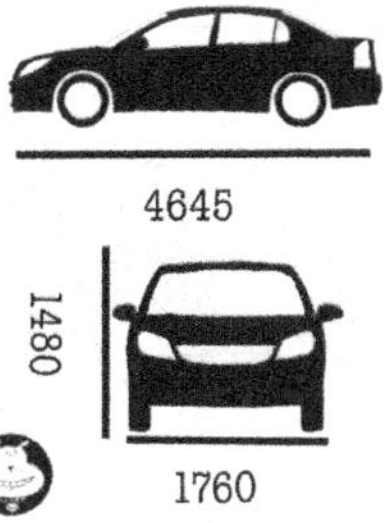

Toyota Avensis 2003 - 2009

Effectively a Japanese Passat, solidly built and a real feeling of quality, with guaranteed reliability, but pretty blunt styling. Revamped in 2006 and a bit sharper to drive and some more thumping diesels. Clutch failure, worn rear suspension, also double check interior electricals.

Engines: 1.8 • 2.0 • 2.0D • 2.2D **CO2:** 155-191g/km **MPG**: 34.9-48.7 **Gearbox:** Man/Auto **Boot:** 510/1320 **Doors:** 4/5

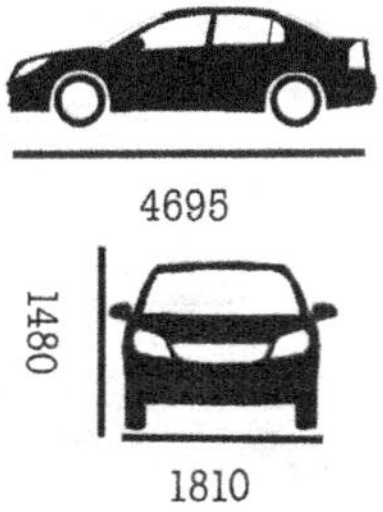

Toyota Avensis 2009 - 2011

Anonymous but effective mover of people and whatever they want to pack inside. Solid and sensible family saloon which has light controls and is easy to drive and own, with a ton of airbags and the usual stability control. Best of the late model saloons (few left) which may be electrically complicated, but thankfully it is a Toyota. There have been a few recalls, relating to suspension, otherwise it's wear/tear.

Engines: 1.6 • 1.8 • 2.0D • 2.2D **CO2:** 135-154g/km **MPG**: 33.2-55.4 **Gearbox:** Man/Auto **Boot:** 509/1320 **Seats**: 5

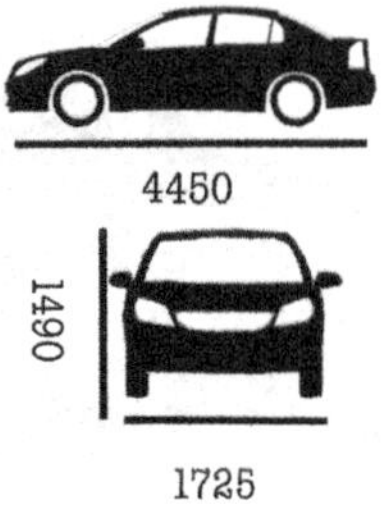

Toyota Prius 2000 - 2004

Here's how to do electrification properly, partially. Complicated tech doesn't make for good Bangers, but these are the exceptions. Good MPG and electric running makes it the town centre taxi of choice. Battery condition is the biggest worry for buyers. Can you get a year or two out of it? Minor electrical issues in the cockpit.

Engines: 1.5 **CO2:** 114g/km **MPG**: 57.6 **Gearbox:** Auto **Boot:** 457/ **Seats**: 5 **Doors:** 5

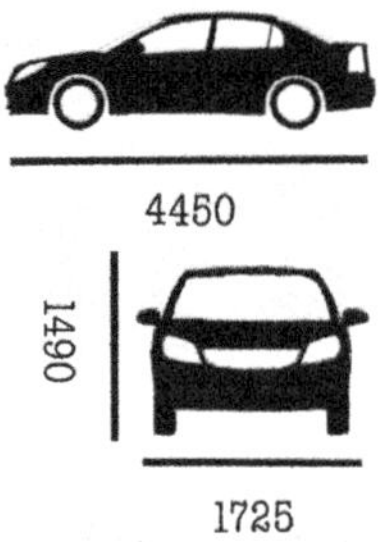

Toyota Prius 2004 - 2009

More practical hatch with a bigger boot, so an even better minicab. Better, with more complicated propulsion, but more responsive. Overall it returns around 50mpg. A solid enough battery, engine thing and it remains all about the battery when used. How long will it last? Seem to be better than predictions., caution still required, especially if an ex-taxi.

Engines: 1.5 **CO2:** 106g/km **MPG**: 65.7 **Gearbox:** Auto **Boot:** 408 **Seats**: 5 **Doors:** 5

Toyota Prius 2009 - 2015

One of the most successful hybrids of its type and an overly familiar sight in cities, the third generation continues where the others left off. It is truly awesome as a no worries used buy, but so many have been trashed, it is essential to check the service and insurance history. If the battery is playing up the petrol engine would run all the time and if you could do a run it would struggle to 40mpg.

Engines: 1.8 **CO2:** 89g/km **MPG**: 72.4 **Gearbox:** Auto **Boot: 445 Seats**: 5 **Doors:** 5

Vauxhall Vectra 1995 - 2002

Designed to thrash up and down motorways. Pretty basic stuff overall with a lot of engine and trim options. The problem is that most have been run pretty ragged and neglected, so '90s ones are properly finished. Rust obviously, leaking coolant, variously broken suspension parts, failing water pumps. Clutch remains a complicated job.

Engines: 1.6 • 1.8 • 2.2 • 2.6 • 2.0D • 2.2D **CO2:** 161-236g/km **MPG**: 28.8-46.3 **Gearbox:** Man/Auto **Boot:** 460 **Doors:** 4/5

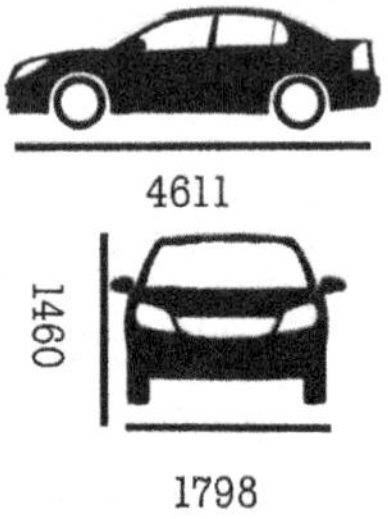

Vauxhall Vectra 2003 - 2009

New model certainly better to drive and better use made of the interior. Most will go for the 50mpg diesels and there was even a sporty V6 one. LPGs exist if you can find them. Company cars can be worn out with noisy, oily engines and suspension EGR valve, fuel injector failure, snapped timing chains.

Engines: 1.8 • 2.2 • 2.8 • 1.8D • 1.9D • 3.0D **CO2:** 157-180g/km **MPG**: 37.3-48.7 **Gearbox:** Man/Auto **Boot:** 500/1050

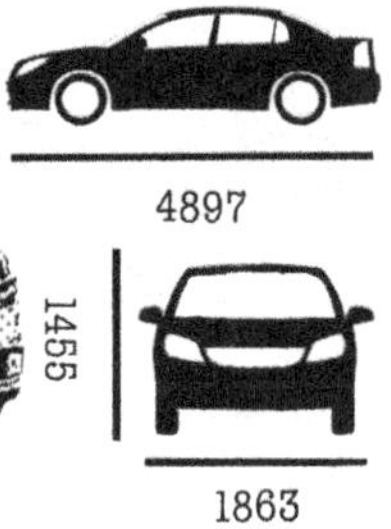

Vauxhall Insignia 2009 - 2017

We couldn't leave this out. Quite meaty and purposeful. This model is refined, comfortable and very roomy. Standard kit is very comprehensive throughout the eight trim levels. Most company car drivers bought Tech Lines with all the fancy Bluetooth and DAB. Dual Mass Flywheel failure, dashboard electrics, power steering leaks.

Engines: 1.6 • 1.8 • 2.0 • 2.8 • 2.0D **CO2:** 154-184g/km **MPG**: 36.2-48.7 **Gearbox:** Man/Auto **Boot:** 520/1465 **Doors:** 4/5

Volkswagen Passat 2000 - 2005

One of those affordable, but still premium quality saloons. All the engines, especially the 1.9 TDI diesel are excellent and there is always something for everyone. Mainly a working car so important not to buy a tired one with a patchy history. Worn turbocharger, needs proof of timing belt changes, leaking air con into cabin, check for rust.

Engines: 1.8 • 2.0 • 2.3 • 2.8 • 4.0 • 1.9D • 2.5D **CO2:** 151-317g/km **MPG**: 21.4-50.4 **Gearbox:** Man/Auto **Boot:** 480 **Doors:** 4

Volkswagen Passat 2005 - 2010

Upgraded and bigger than before with a bewildering number of engine options. Easy to understand why most family buyers are going to opt for the estate. Perfect company car, which is why some caution always required when it is time to buy. Oil pump failure, ECU failures especially after water leaks into cabin, electrical faults and flat batteries.

Engines: 1.4 • 1.8 • 2.0 • 3.2 • 2.0D **CO2:** 153-155g/km **MPG**: 36-48.7 **Gearbox:** Man/Auto **Boot:** 565/683 **Doors:**4

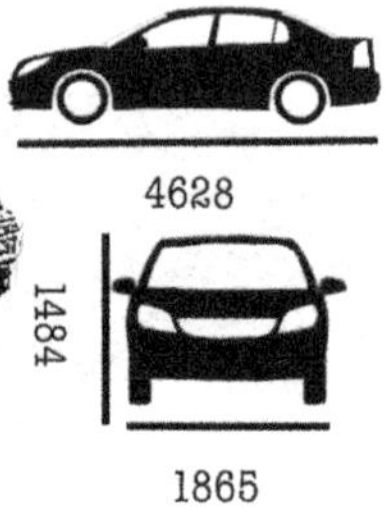

Volvo S60 2000 - 2009

This Scandinavian beauty makes sense as the best Volvo that isn't an estate. Lots of engine options from slow to sensible, slightly mad and just frugal. In 2006 there was some detail changes to make it just a bit sharper. Safe as houses of course, some are quite shabby. Faulty diesel injectors, failed ECUs and poor clutches.

Engines: 2.0 • 2.4 • 2.5 • 2.4D **CO2:** 174-212g/km **MPG:** 31.7-42.8 **Gearbox:** Man/Auto **Boot:** 394/808 **Doors:** 4

Shooting Brake Bangers * - Matters of Estate

Apparently these are out of fashion because everyone wants a great big Stupid Utility Vehicle. Good news as that means the rest of us can buy one for nowt. Well, maybe not that much.

For a while though the estate, or shooting brake in old money, was all the average family, or working bloke needed for awkward loads, canoes and camping gear. That can mean there is a good chance of buying one which has led a very hard life. It is up to the buyer to sort out the well kept estates from the real states. Buyers will never go wrong with a V-Class Volvo or any Passat which can lug away almost forever.

Banger buyers don't care, or rather should not care too much about aesthetics but at least the addition of a conservatory at the back can quite often improve a stylishly compromised motor. Case in point is the Jaguar X-Type. You will still break down of course but you feel better about your automotive choices as you wait for the recovery vehicle.

Otherwise, estates are rather handsome, which is reason enough to buy. The style over true cargo substance explained the success of the BMW 3 Series Touring and the Audi A4 Avants. An estates with a big engine is of course a laugh and it was perfectly suited to diesel power, until it was demonised. Estates are more economical and in many cases easier to live with than a People Carrier, or SUV. If you need space and not seats then those adapted vans are often slow and not mpg friendly, or the answer. Estates can be.

Banger Estates are great and here is the absolute proof over the following pages. A Fiat Stilo Multiwagen may be buttons, the Ford Focus, just about perfect.

*Anoraktechnicality I do realise that these are not shooting brakes which were two door, sporty cars usually with a tailgate or split rear doors. Think old coach built stuff from the '30s to '50s and the wonderfully plastic Reliant Scimitar, or the Italian Lancia HPE. Oddly the Alfa 159 Sportwagen, or a Chrysler 300C are spiritually shooting brake, but I could be wrong.

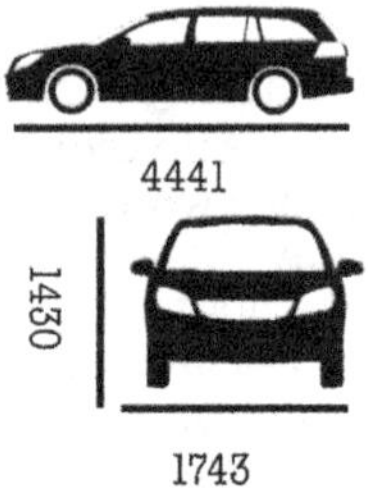

Alfa Romeo 156 Sportwagon 2000 - 2006

Great range of engines, from V6, to 2.4 JTD diesel, but boot is smaller than the four door, so we are in 'lifestyle' rather than 'workhorse' territory. Has a lipped tailgate, 60:40 split rear seat but a flat loading bay with reversible flooring, so surprisingly practical. Failing oil pump and air flow sensors. Air con problems, Selespeed gearbox malfunctions.

Engines: 1.6 • 1.8 • 2.0 • 3.2 • 1.9D • 2.4D **CO2:** 161-212g/km **MPG**: 32.1-46.3 **Gearbox:** Man/Auto **Boot:** 360/1180

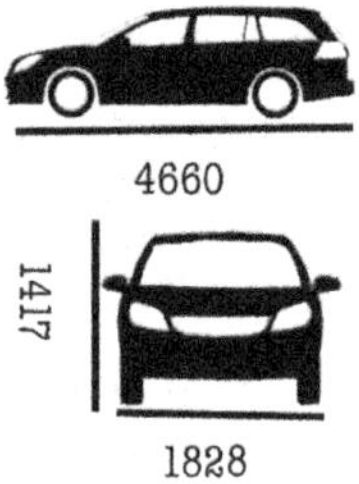

Alfa Romeo 159 Sportswagon 2005 - 2011

Better built and bigger than the last Sportwagon. Now larger inside than the saloon, however nothing that clever going on, the seats just split, fold and flip forward. It does leave a fairly flat and wide area and there are 12V power sockets in the back. Lots more spec levels from 2009. Water pump failures, dual mass flywheels, steering rack breakages.

Engines: 1.8 • 2.2 • 3.2 • 1.9D • 2.4D **CO2:** 138-260g/km **MPG**: 25.7-53.3 **Gearbox:** Man/Auto **Boot:** 445/1235 **Seats**: 5

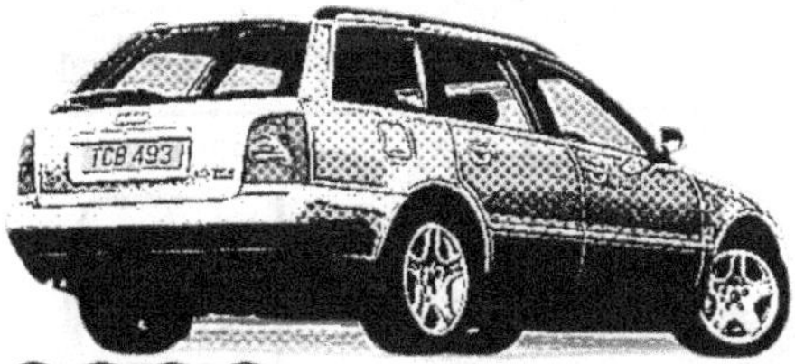

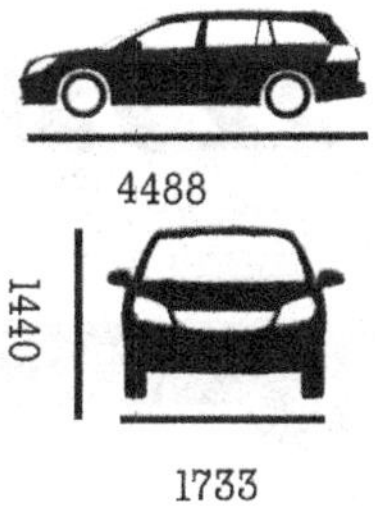

Audi A4 Avant 1996 - 2001

Default compact prestige estate from the '90s onward. '99 a revamp tweaked the suspension and steering for the better. Good size load area, that is wide and flat, nothing innovative, just carpeted quality. Hardly huge, it has oddments space but won't take a wardrobe. Even a Delcor one. Wear and tear issues of brakes and suspension, check service history.

Engines: 1.6 • 1.8 • 2.4 • 2.6 • 1.9D • 2.5D **CO2:** 142-288g/km **MPG**: 23-53 **Gearbox:** Man/Auto **Boot:** 390/1430

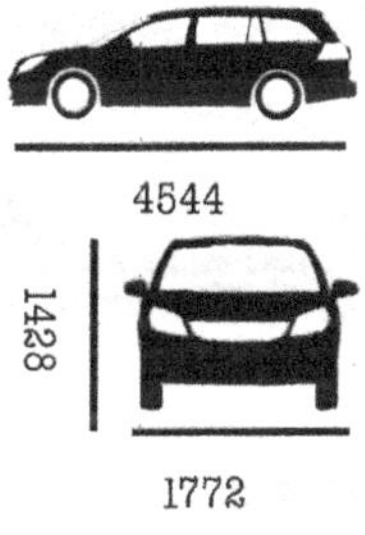

Audi A4 Avant 2001 - 2004

Estate game not moved on very much, just more of the high quality sameness. A more socially acceptable level of standard equipment and arrival of the 2.0 FSI was economical and perky. Diesels are great if still legal. Early model ignition coils, water pumps, automatic gearbox can be unreliable which is obviously an expensive fix.

Engines: 1.6 • 1.8 • 2.0 • 2.6 • 3.0 • 1.9D • 2.5D **CO2:** 149-252g/km **MPG**: 26.9-51.4 **Gearbox:** Man/Auto **Boot:** 442/1184

Audi A4 Avant 2005 - 2008

Mid period estates are dull but well finished with the usual mind blowing and wallet saving range of engines. Obviously the four cylinder TDi diesels are universally excellent and the fresh FSi petrol units are very economical too. If mpg is not paramount then the V6 multivalve 2.6, or 2.8 are great. Clutch failing and central locking.

Engines: 1.6 • 1.8 • 2.0 • 3.2 • 1.9D • 2.0D • 2.5D • 3.0D **CO2:** 153-252g/km **MPG**: 34.0-49.7 **Boot:** 460/720

Audi A4 Avant 2008 - 2015

These late period estates represents prime diesel technology although they are starting to become very complicated. The 2.0-litre TDI engines are great otherwise it is the TFSI petrols as the obvious go anywhere for less choices. Bigger boot than before with or without the rear seats folded. Blocked Particulate Filters, worn brakes and suspension.

Engines: 1.8 • 2.0 • 3.2 • 2.0D • 2.7D • 3.0D **CO2:** 120-252g/km **MPG**: 34.0-61.4 **Gearbox:** Man/Auto **Boot:** 480/962

Audi A6 Avant 1998 - 2004

Executive transportation as even Audi described it as a 5 door coupe with the usual ton of engine and drivetrain options. Avoid the Tiptronic clutchless manual on the grounds of complication. Fuel pumps and air flow meters can fail, wheel bearings, plus with old age electric windows and central locking.

Engines: 1.8 • 2.0 • 2.04 • 2.7 • 3.0 • 4.2 • 1.9 • 2.5D **CO2:** 154-312g/km **MPG**: 21.7-49.6 **Boot:** 455/1591 **Seats**: 5

Audi A6 Avant 2005 - 2011

Ramped up practicality. There loads of useful hooks, eyes, stowage nets, ski bag, optional reversible flooring and rear bumper protector. In 2008 got a major refresh to include LED tail lights and a bunch of new engines. Engine ECU failure means rough running. Air conditioning on the blink, brakes and suspension.

Engines: 2.0 • 2.8 • 3.0 • 2.7D • 3.0D **CO2:** 149-219g/km **MPG**: 30.1-49.6 **Boot:** 565/1660 **Seats**: 5 **Doors:** 5

BMW 3 series Touring 1999 - 2005

Averagely spacious as it isn't a real estate. Has floor hooks and straps and storage under the floor and the clever split tailgate glass for convenience. Standard SE specification is good making it the perfect small executive. Power loss and smoke, diesel issues, broken air mass meters and cracked inlet manifolds. Worn balljoints.

Engines: 1.8 • 2.0 • 2.2 • 2.5 • 3.0 • 1.8D • 2.0D • 3.0

CO2: 151-218g/km **MPG**: 31.0-50.4 **Boot:** 435/1345

BMW 3 series Touring 2005 - 2011

Generation 5. Build quality does not seem to be as excellent as the old days. You do need to choose your engine carefully as the 2.0 diesel from '07 caused a huge amount of grief. Split tailgate and underfloor compartments as before plus a cargo net. Efficient Dynamics, stop/start, brake regeneration things to go wrong. Avoid the 2.0d then.

Engines: 1.8 • 2.0 • 2.2 • 2.5 • 3.0 • 1.8D • 2.0D • 3.0D **CO2:** 151-218g/km **MPG**: 31.0-57.6.**Boot:** 460/1385 **Seats**: 5 **Doors:** 5

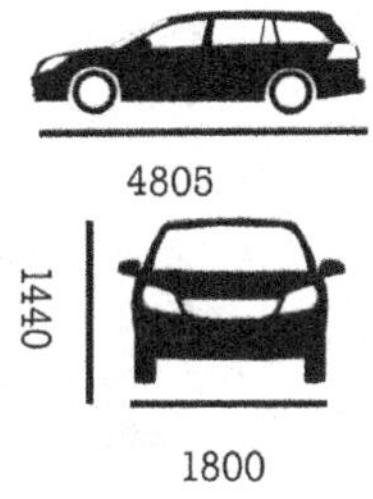

BMW 5 Series Touring 1997 - 2004

The last of the fixable BMWs, unarguably one of the greatest BMWs. More practical than before, can tow 2000kg, has self-leveling suspension, a load net and the split fold rear seats lie flat. The diesels are great especially the bigger sixes and the petrol sixes will last forever, if not longer. Electronics on old ones, suspension bushes. Diesels may need new injectors.

Engines: 2.0 • 2.2 • 2.5 • 3.0 • 3.5 • 4.0 • 2.5D • 3.0D **CO_2:** 179-295g/km **MPG**: 23.0-42.2 **Boot:** 408/1521 **Seats**: 5 **Doors:** 5

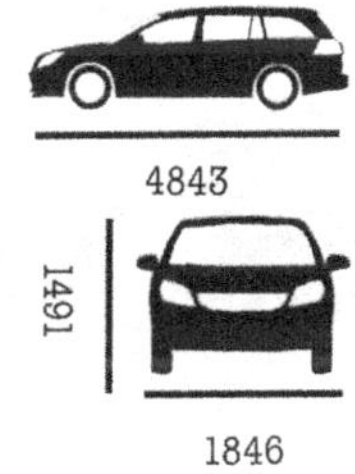

BMW 5 Series Touring 2003 - 2010

Becoming slightly unfixable, needs some effort. There's no load lip, the seats fold, but not flat and overall, the other posh rivals are bigger. That's why it is a Touring. Big facelift in 2007 saw the economical upgrades including regenerative braking and the TV may even work as it is digital. Self leveling suspension costs a bomb.

Engines: 2.5 • 3.0 • 2.0D • 3.0D **CO_2:** 165-176g/km

MPG: 38.2-45.6 **Boot:** 500/1650 **Seats**: 5 **Doors:** 5

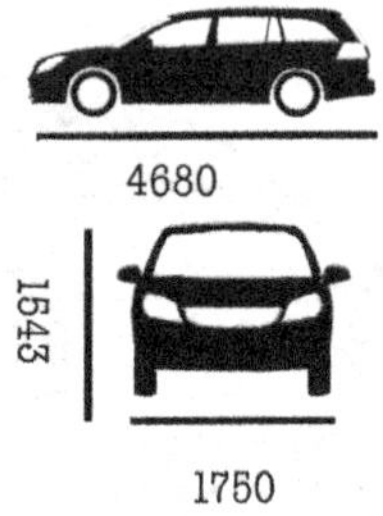

Cadillac BLS 2006 - 2010

Caddy's first estate made from Saab 9-3 underparts (built in Sweden)and a 1.9 diesel also found in a Vauxhall Vectra, actually marginally better. Designed to be comfy and cruiserish. Decent standard equipment, (Elegance has everything) offers value. Few sold and the downsides as a Saab 9-3 and has electrical issues, warning lights.

Engines: 2.0 • 2.8 • 1.9D **CO2:** 151-259g/km **MPG**: 26-49

Gearbox: Man/Auto **Boot:** 419/1285 **Seats**: 5 **Doors:** 5

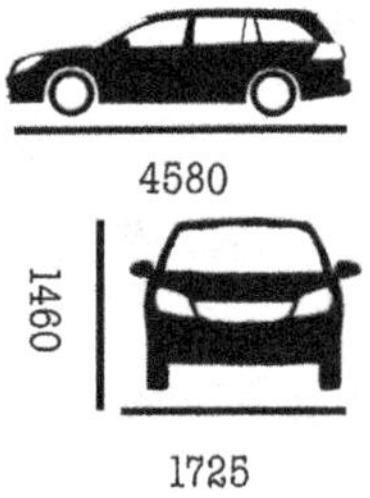

Chevrolet Lacetti SW 2005 - 2011

Here's a basic estate with 1.6 and 1.8 petrol options. Not a bad load bay especially with the 60/40 rear seats folded down. There's access to a 12V power socket and an utterly unbelievable 25 storage cubby holes. Pretty decent amount of space. Usual wear and tear brakes and suspension. Electricals worth checking.

Engines: 1.4 • 1.6 **CO2:** 171-178g/km **MPG**: 39.2-39.7

Gearbox: Man/Auto **Boot:** 400/1410 **Seats**: 5 **Doors:** 5

Chrysler 300C Touring 2005 - 2011

Mercedes E Class sourced diesel is the sensible choice and V6 is petrol only. Big tailgate lifts part of the roof for great access. Extra storage under the high floor. Revamped in 2008 with a spruced up interior. Finished to a high standard. Diesel electronic throttle sensor failure, oil leaks, starter motors break and engine oil leaks.

Engines: 3.5 • 5.7 • 6.1 • 3.0D **CO2:** 215-288g/km **MPG**: 23.3-34.9 **Gearbox:** Man/Auto **Boot:** 630/1602 **Doors:** 5

Citroen Xsara 2000 - 2004

Proper old school basic Frenchie fairly reliable and pretty frugal on the old fuel whether you go for a 1.4 or 2.0 HDi. Estate is roomy and comfortable. Does not seem to have many reliability downsides provided it has not been run into the ground. ECU issues, some late '90s diesel engine issues can kill the engine. Diesel clutches can be weak.

Engines: 1.4 • 1.6 • 2.0 • 1.9D • 2.0D **CO2:** 138-184g/km **MPG**: 40.9-54.3 **Gearbox:** Man/Auto **Boot:** 408/1198

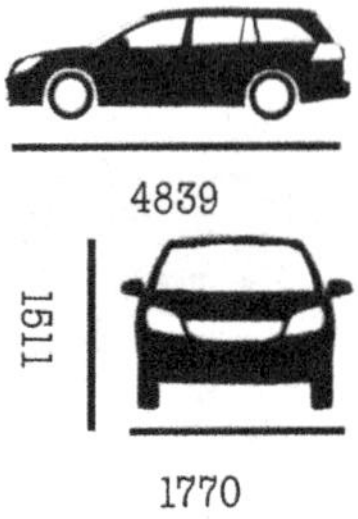

Citroen C5 2001 - 2008

Split tailgate good, no struts makes loading easier. Reversible carpet on a flat wide floor with concealed storage bins. Hydractive 3 suspension means a flat floor but complication. Lots more in 2004 with lane departure system, Xenon headlamps, maximum speed limiter. Automatic gearbox and clutch weak. Sealed suspension fails.

Engines: 1.8 • 2.0 • 3.0 • 1.6D • 2.0D • 2.2D **CO2:** 142-187g/km

MPG: 36.7-52.3 **Gearbox:** Man/Auto **Boot:** 563/1658

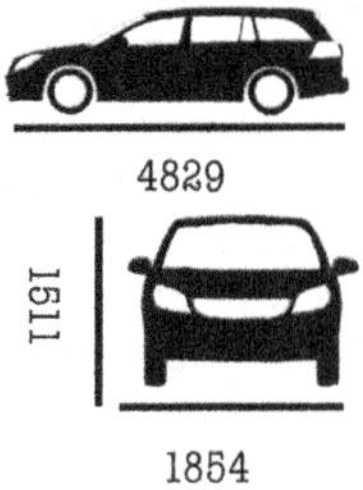

Citroen C5 Tourer 2008 - 2016

Not as big as most rivals like a Mondeo. Door bins not as big as you would think and just the one up front cup holder. Large tailgate is great still hampered by a small loading lip. As always, complicated and will destroy you financially. Just not that clever anymore. All the same checks and potential problems, might be less wrong because a bit newer.

Engines: 1.8 • 2.0 • 1.6D • 2.0D • 2.2D • 2.7D

CO2: 149 -198g/km **MPG**: 33.6-50.4 **Boot:** 505/1462

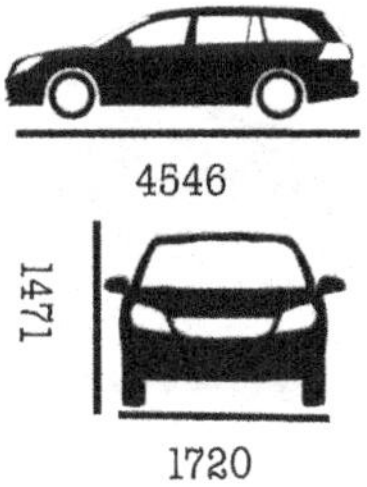

Daewoo Nubira 1997 - 2002

Estate car with a long flat loading bay is all any buyer wants. It doesn't deliver much else except an unrefined 1.6 engine, but it is reliable. SX includes air con and alloys. Quite likely to be neglected and worn out, but still working. You could be paid to take it away. Nothing to look for, just wear and tear items like brakes and suspension.

Engines: 1.6 • 2.0 **CO2:** 198-214g/km **MPG**: 32.8-36.2
Gearbox: Man/Auto **Boot:** 549/1838 **Seats**: 5 **Doors:** 5

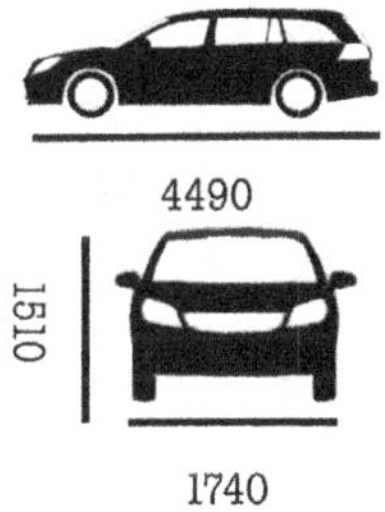

Fiat Marea Weekend 1997 - 2003

Stylish thing with a huge boot that got huger with the rear seats folded. It is nice to drive, but the fact it is a Fiat meant no one took it seriously except those who actually ran them and most are very happy. If any are around, should cost buttons. Chocolate ones. Suspension drop links, sunroof leaks, instrument panel failures, needs red master key.

Engines: 2.0 • 1.9D **CO2:** 149-234g/km **MPG**: 28.8-50.4
Gearbox: Man/Auto **Boot:** 501/1549 **Seats**: 5 **Doors:** 5

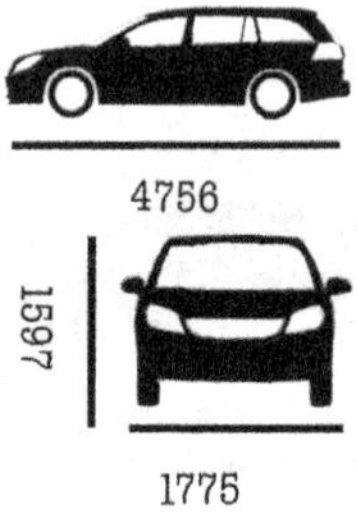

Fiat Croma 2005 - 2007

Reinvention of the Croma as a great big estate only, surprising Giuigiaro didn't make it look less lumpy. Based on Saab 9-3 and the dull Opel Vectra. Fiat probably sold three. Lots of equipment, tons of space and suited to the turbo diesels though there were petrols. Electrics go on the blink, windows and locks. Smokey diesels.

Engines: 2.2 • 1.9D • 2.4D **CO2:** 160-212g/km **MPG**: 32.8-46.3 **Gearbox:** Man/Auto **Boot:** 500/1610 **Doors:** 5

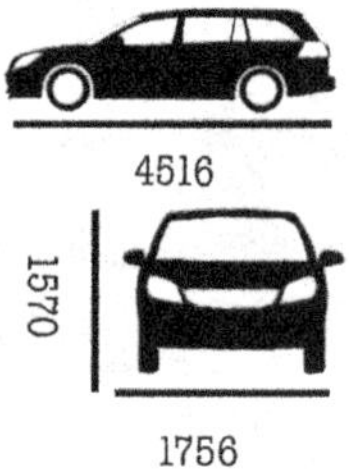

Fiat Stilo Multiwagen 2003 - 2007

The rear seats slide and tilt, but when the 60/40s split then fold they don't end up flat. Useful low loading height though and flat floor. Because it is Fiat mostly worthless, that's a great reason to buy. The JTD80 is the diesel way to go, but the tiny 1.2 petrol is something of a survivor. One of those old Fiats with failing sensors and electrics to baffle you.

Engines: 1.4 • 1.6 • 1.8 • 2.4 • 1.9D **CO2:** 140g-233/km **MPG**: 35.3.0-52.3 **Gearbox:** Man/Auto **Boot:** 510/1480

Ford Focus 1998 - 2004

Long wheelbase means more luggage space, are great to drive and easy to own. Low loading lip and well packaged suspension TDCI diesels are great Zetec spec and the petrols are all pretty reliable. Getting old a patchy now though and starting to rust. Getting old and patchy now, also quite rusty. Otherwise it is just wear and tear to the suspension.

Engines: 1.4 • 1.6 • 1.8 • 2.0 • 1.8D **CO2:** 140-188g/km **MPG**: 30.0-53.3 **Gearbox:** Man/Auto **Boot:** 521/1580 **Seats**: 5

Ford Focus 2005 - 2011

Same excellence in a less striking body, plus a bit more room in the back. Revamped in 2008 with ECOnetic ultra green 70+mpg. Roomy inside rather than having anything really clever seat wise. Straightforward to fix and easy to live with. Buy one. No serious issues, just wear and tear, suspension and brakes and usually OK provided proper servicing.

Engines: 1.4 • 1.6 • 1.8 • 2.0 • 1.6D • 2.0D **CO2:** 119-179g/km **MPG**: 42.2-62.7 **Gearbox:** Man/Auto **Boot:** 455/1439

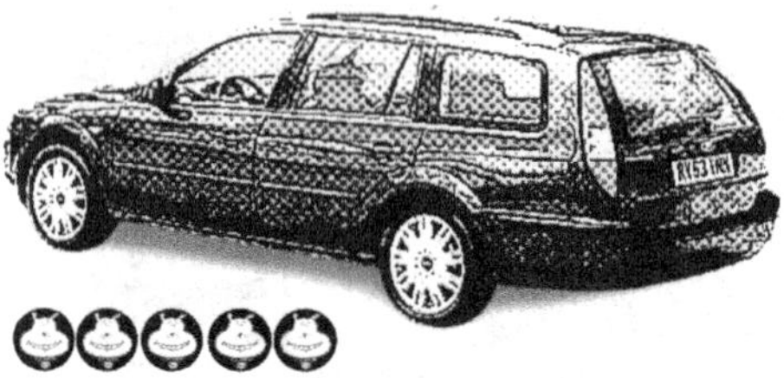

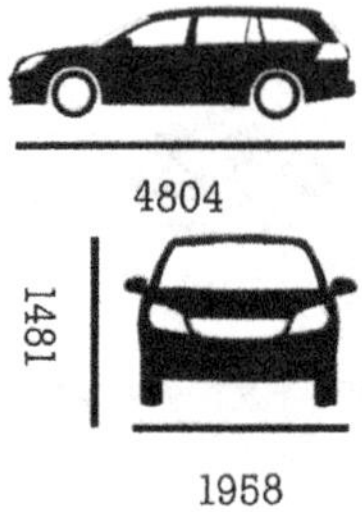

Ford Mondeo 2000 - 2007

Second generation Mondeo is probably the best incarnation. Really big boot and interior, but it does nothing clever, just has folding seats and lots of oddment space and cubbyholes. The TDCi engines made all the difference. Constantly updated, buy later editions and Zetecs or Ghias for the win. Air con failure, remote locking, minor ECU, electrical issues.

Engines: 1.8 • 2.0 • 2.5 • 3.0 • 2.0D • 2.2D **CO2:** 154-187g/km **MPG**: 36.2-48.7 **Gearbox:** Man/Auto **Boot:** 540/1700

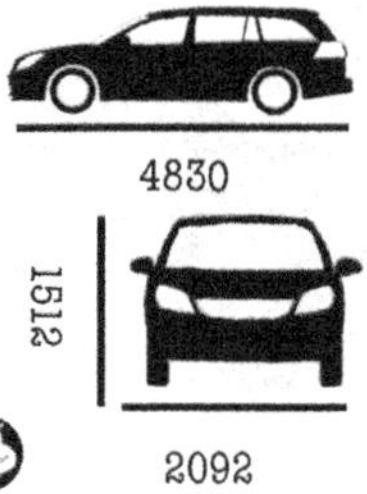

Ford Mondeo 2007 - 2014

Very large luggage area which is flat wide and has no lip. This is boosted by folding the seats has optional cargo net and a sliding floor. 2.0 petrol if the miles are low, 2.0 diesel otherwise. All models have air con, big spec Titaniums around in numbers. Restyled a bit in 2010. Air con failure, remote locking, minor ECU, electrical issues.

Engines: 1.8 • 2.0 • 2.3 • 2.5 • 1.8D • 2.0D • 2.2D **CO2:** 154-177g/km **MPG**: 38.2-48.7 **Boot:** 554/1685 **Doors:** 5

Honda Accord Tourer 2003 - 2008

High quality but hardly the biggest load area, it is at least user friendly. One touch collapsable fold flat seats at the touch of a button and pull of a lever. The power tailgate is very useful as it opens at the press of the key fob. Buyers got a diesel delivered in 2004 otherwise the petrols are great. ABS brake sensors, wheel bearings and tailgate release.

Engines: 2.0 • 2.4 • 2.2D **CO2:** 143-214g/km **MPG**: 31.4-52.3

Gearbox: Man/Auto **Boot:** 626/1707 **Seats**: 5 **Doors:** 5

Honda Accord Tourer 2008 - 2015

Bigger vehicle, but the load space is now awkward and not that big at all. Three engines a 2.2 diesel and a couple of petrols and overall has a near perfect reliability record. But getting increasingly complicated. Not that big inside hatch, almost a reverse Tardis. Clutch wear, dual mass flywheel problems, blocked diesel filters.

Engines: 2.0 • 2.4 • 2.2D **CO2:** 148-209g/km **MPG**: 32.1-50.4

Gearbox: Man/Auto **Boot:** 396/672 **Seats**: 5 **Doors:** 5

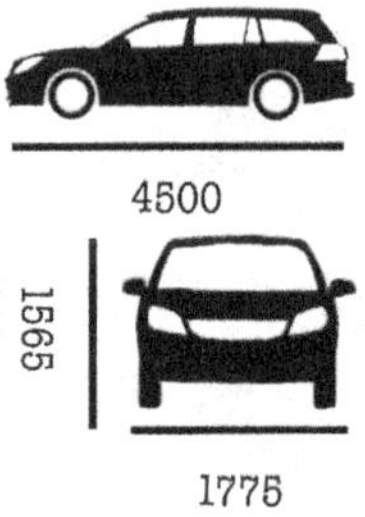

Hyundai i30 2008 - 2012

Also see the Kia Ceed SW, which is actually a bit bigger. Big square load area is good, although compared to the hatchback there is more rear legroom. Mostly everything is standard, there is air conditioning in the Comfort and the 'Premium' has climate and partial leather. Power steering pumps, very low level rust. The odd ECU issue.

Engines: 1.3 • 1.6 • 1.5D **CO2:** 144-165g/km **MPG**: 40.9-51.4 **Gearbox:** Man/Auto **Boot:** 415/1395 **Seats**: 5 **Doors:** 5

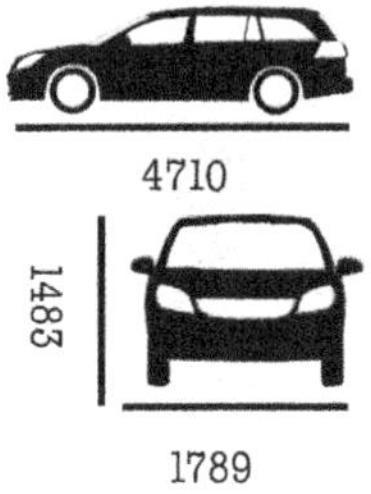

Jaguar X-Type 2004 - 2010

Ford related, but a patchy reliability record with dual mass flywheels on the diesel and wheel hubs to highlight just a couple of faults. Slightly more boot space than the saloon, but the folding seats do not lie flat. Facelift in 2008. Automatic gearboxes can fail, four wheel drive transfer box expensive. Otherwise clutch failures. Oil and coolant leaks.

Engines: 2.5 • 3.0 • 2.0D • 2.2D **CO2:** 149-234g/km **MPG**: 29.5-49.1 **Gearbox:** Man/Auto **Boot:** 455/1415 **Seats**: 5

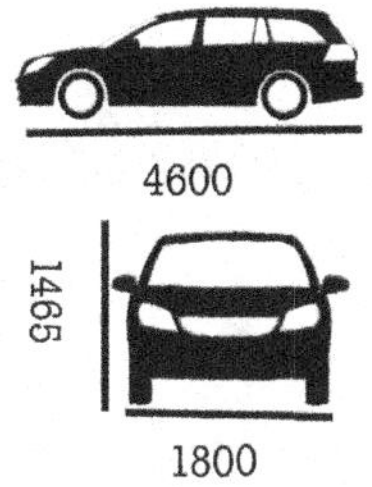

Kia Ceed SW 2007 - 2012

See also i30, the same but different, Ceed a bit smarter, slightly larger in luggage area. That big, square tailgate has hinges in the roof so it is possible to get most things in. Fold flat seats and more storage under the floor. The diesels are good and it was facelifted in 2009. Nothing to worry about just wear and tear in the usual brake and suspension areas.

Engines: 1.4 • 1.6 • 1.6D **CO2:** 110-146g/km **MPG**: 46.3-65.7

Gearbox: Man/Auto **Boot:** 534/1664 **Seats**: 5 **Doors:** 5

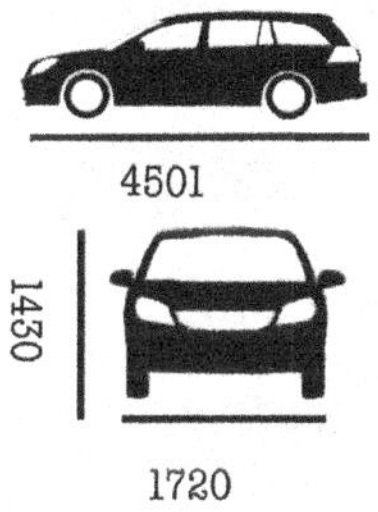

Lexus IS Sportcross 2001 - 2005

More space in the boot than the saloon of course and initially available with the 3.0 litre engine, then the marginally more economical 2.0. Everything is standard and the reliability record is exceptional, not least because there is no diesel. No issues, just wear and tear as usual to fix the brakes, suspension and any missing parts.

Engines: 2.0 • 3.0 **CO2:** 233-256g/km **MPG**: 26.2-29.1

Gearbox: Man/Auto **Boot:** 340/1046 **Seats**: 5 **Doors:** 5

Mazda 6 Sport Tourer 2002 - 2007

Smart looking thing with a decent boot and once you fold the seats, tons more space. 1.8 petrol and 2.0 Diesel make sense and buyers ought to aim for TS2 spec and that should be everything a Mazda driver needs. Overall these cars seem to be ultra reliable, which is good. One of the best potential Bangers out there. Check clutch, air con.

Engines: 1.8 • 2.0 • 2.3 • 2.0D **CO2:** 174-192g/km **MPG**: 34.9-43.5 **Gearbox:** Man/Auto **Boot:** 505/1712 **Doors:** 5

Mazda 6 Sport Tourer 2008 - 2012

Still a handsome thing. Like the last one, based on the Ford Mondeo, but more stylish with loads of room for luggage. One clever design detail is the load cover going up with the tailgate. 2010 facelift meant it was a bit sharper to drive with more economical engines. Leaking air con. Check short recall history especially on petrol.

Engines: 1.8 • 2.0 • 2.5 • 2.2D **CO2:** 147-168g/km **MPG**: 39.8-51.4 **Gearbox:** Man/Auto **Boot:** 519/1751 **Doors:** 5

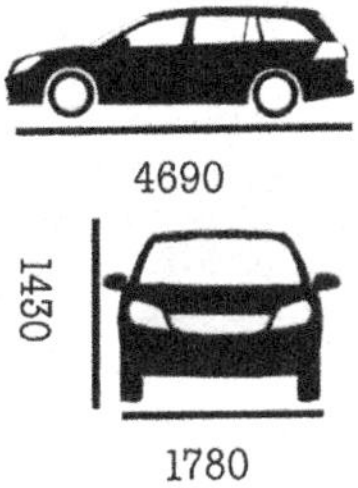

Mazda 626 2000 - 2002

Dull, but very worthy estate. The restyle in 1999 was meant to help but still dreary. More than enough room in the boot and with the rear seats folded. Decent level of standard equipment. Nothing special at all. Basic servicing to keep it working, probably forever. Remarkably tough, just neglect and worn out parts such as suspension and brakes.

Engines: 1.8 • 2.0 • 2.0D **CO2:** 152-195g/km **MPG**: 35.8-54.3

Gearbox: Man/Auto **Boot:** 484/844 **Seats**: 5 **Doors:** 5

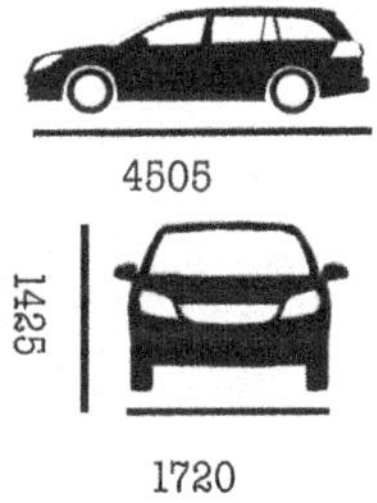

Mercedes C-Class 1996 - 2001

Rear seats split/fold and roof bars are standard, but that was all you needed back then. The 1997 revamp was considerable with new engines and silly subtle styling things like dark tail lights, if that was your bag. The diesels are great of course. Still rusted. Indeed rusty suspension parts, wipers cost a bomb to fix, engine oil leaks.

Engines: 1.8 • 2.0 • 2.3 • 2.4 • 2.8 • 2.2D • 2.5D **CO2:** 163-225g/km **MPG**: 26.6-45.6 **Gearbox:** Man/Auto **Boot:** 465/1510

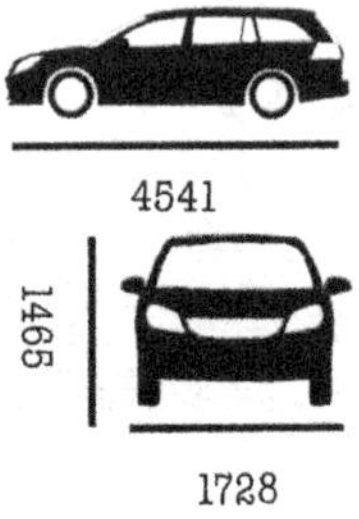

Mercedes C-Class 2000 - 2007

Tailgate is electric and under the floor is a pull out box, there's a safety net and the headrests flip up so that the seats are completely flat to the floor. Huge range still, but 'Twinpulse' engines apparently would save the planet and your wallet. Still some rust. Electrics patchy check windows +central locking, turbo oil seals + condition/age of glow plugs.

Engines: 1.8 • 2.5 • 3.0 • 3.5 • 2.1D • 3.0D **CO2:** 169-183g/km
MPG: 35.3-44.1 **Gearbox:** Man/Auto **Boot:** 470/1385

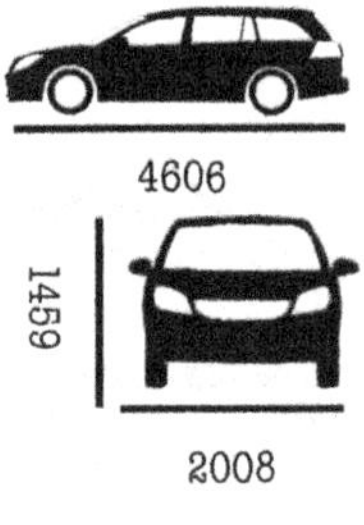

Mercedes C-Class 2008 - 2014

Solidly serious and bigger than before with a longer rear deck. This model is also electronically more complicated. C220 Cdi is all that anyone needs, but the petrols are properly economical and future proof. Changes outside of the Bangerpedia remit include 2011 and 2013 revamps. Instruments can go blank. Diesel wear, suspension + brakes.

Engines: 1.6 • 1.8 • 2.5 • 3.0 • 3.5 • 2.1D • 3.0D **CO2:**127-187g/km **MPG**: 40.9-58.9 **Gearbox:** Man/Auto **Boot:** 485/1500

Mercedes E-Class 1996 - 2003

Es were good, but these were disappointingly rusty. Usual diesels and petrols and substantially revamped in 1999, so better to buy one of those if you must. The common rail diesels are good, eventually illegal and the petrols reliable. Oil leaks, rust to brake lines, suspension bushes, engine mountings. More Mercedes branded rust.

Engines: 2.0 • 2.4 • 2.8 • 3.2 • 4.3 • 2.1D • 2.7D • 3.2D **CO2:** 164-271g/km **MPG**: 25.0-45.6 **Boot:** 538/1974

Mercedes E-Class 2002 - 2008

Bigger floor is wider, deeper rear seats fold flat optional sliding floor. Weight saving aluminium panels, bonnet and front wings, worth remembering when it comes to repairs. Mid life quality upgrade 2006. V6 petrols better if you don't do the miles. Fuel pump failures, tired fuel injectors, catalytic converter issues, broken suspension springs

Engines: 1.8 • 3.0 • 3.5 • 5.0 • 2.2D • 3.0D **CO2:** 169-202g/km **MPG**: 33.6-42.8 **Gearbox:** Man/Auto **Boot:** 690/1950

Mercedes E-Class 2009 - 2016

Ultra modern estate that was annoyingly complicated.. It remained the diesel exec of choice and that's the vast majority of used options. Colossal inside, complicated self-leveling rear suspension, electric tailgate and load cover. Four cylinder turbo petrol probably the easiest and dullest to own. Worn diesel injectors that smoke, also oil leaks.

Engines: 5.5 • 2.2D • 3.0D **CO2:** 139-256g/km **MPG**: 25.9-53.3 **Gearbox:** Auto **Boot:** 540 **Seats**: 5 **Doors:** 5

MG ZT-T 2001 - 2004

Warning triangle built into the tailgate and illuminated spare, so you are all prepared for a breakdown. Initially shoved along by the V6 2.5 litre engine in two states of tune, a 160 and 180bhp. BMW diesel excellent. Handful of V8s. 2005 last gasp facelift. Rotten sills and wheelarches, blown head gasket, diesel pump failure. Corroded suspension.

Engines: 1.8 • 2.5 • 2.0D **CO2:** 163-225g/km **MPG**: 30.0-48.8 **Gearbox:** Man/Auto **Boot:** 400/1222 **Seats**: 5 **Doors:** 5

MINI Clubman 2007 - 2014

More practical five door, two as the tailgate, two at the front and reverse hinged one on the off side. Longer and higher, so there is a tad more legroom and almost proper boot space. Three seats in the back, which is a squeeze. A real oddity that not everyone 'got', consequently good value. A pretty minimarginal estate. Clutch and Gearbox need checking.

Engines: 1.4 • 1.6 • 1.6D **CO2:** 118-139g/km **MPG**: 48.7-64.2
Gearbox: Man/Auto **Boot:** 260/930 **Seats**: 5 **Doors:** 5

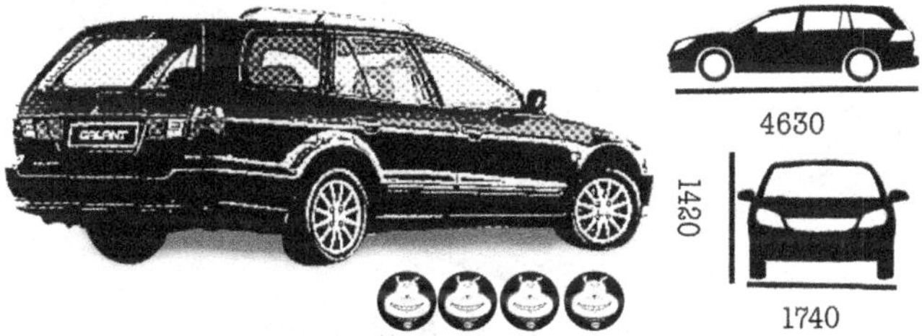

Mitsubishi Galant 1997 - 2003

When a buyer needs plenty of space and has miles to cover this is perfect. Contained a GDI engine, a gas direct injection unit which was jolly efficient even if Euro fuel wasn't always pure enough for it. Not a big hit, few around, classy and solid, proper alternative to one of those German estate cars. Check gearbox condition, brakes and suspension.

Engines: 2.0 • 2.4 • 2.4 **CO2:** 203-207g/km **MPG**: 29.7-33.2
Gearbox: Man/Auto **Boot:** 419/1399 **Seats**: 5 **Doors:** 5

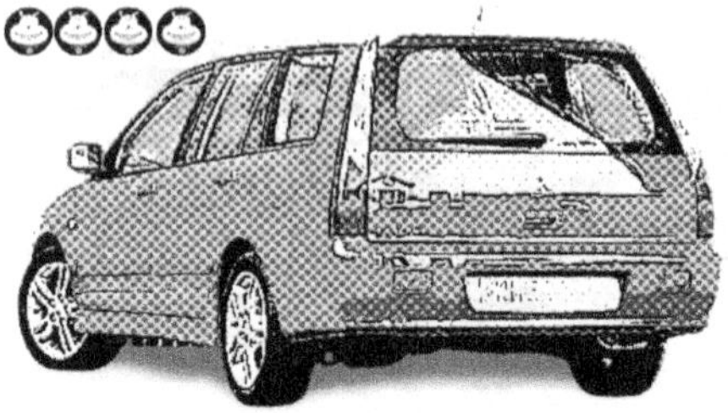

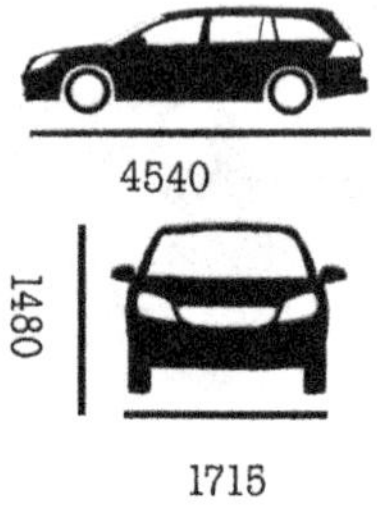

Mitsubishi Lancer 2005 - 2008

Standard smallish estate with split rear seats which do not fold flat. Underneath the floor is a waterproof storage area and that's it. There are 1.6 and 2.0 petrol engines which are tough and reliable. Dreary inside, but the level of standard kit is good which includes air conditioning. Not much to look for, just electricals plus wear and tear components.

Engines: 1.6 • 2.0 **CO2:** 163-207g/km **MPG**:32.5-41.5
Gearbox: Man/Auto **Boot:** 344/1079 **Seats**: 5 **Doors:** 5

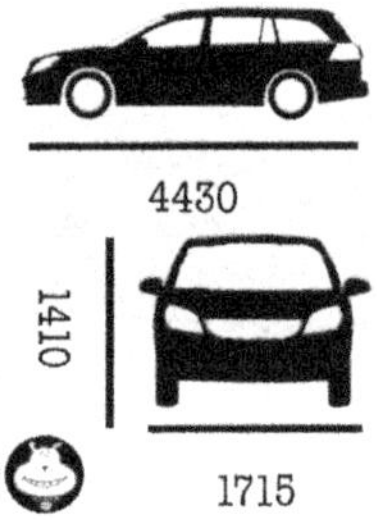

Nissan Primera 1998 - 2002

Blandness in estate format with lots of standard kit including airbags, ABS brakes and climate control, except entry model. The engines are great, go for the 1.8 or 2.0 petrol or the turbo diesel. It will only be consumables like tyres and exhausts that may need replacing. Bodies rot, suspension wear, air con busted, old alloys corrode.

Engines: 1.6 • 1.8 • 2.0 • 2.0D **CO2:** 179-195g/km **MPG**: 34.8-43.5 **Gearbox:** Man/Auto **Boot:** 430/1651 **Seats**: 5

Nissan Primera 2002 - 2006

Becoming a lot more complicated especially inside where most had LCD screens. Almera based and lots of standard fixtures like climate air con and plenty of air bags. 2004 uprate included sat nav, which can go on the blink. Reliable, but electronics such as windows and central locking go on the blink. Turbo diesel failure and crankshaft sensor issues.

Engines: 1.8 • 2.0 • 2.2D **CO2:** 164-208g/km **MPG**: 33.2-47.9 **Gearbox:** Man/Auto **Boot:** 465/1670 **Seats**: 5 **Doors:** 5

Peugeot 206W 2002 - 2006

The tailgate contains a rear window that can be independently opened and closed. Has sturdy load cover, luggage nets and bag hooks. The range was reduced to one trim level (Verve) in mid 2005 with only the 1.4 and 1.6 petrol and HDi diesel. Worn lower ball joints, broken engine mounts, rattly sunroofs, ECU problems with auto gearbox.

Engines: 1.1 • 1.4 • 1.6 • 1.4D • 1.6D • 2.0D **CO2:** 113-166g/km **MPG**: 45.6-65.7 **Gearbox:** Man/Auto **Boot:** 313/1136

Peugeot 207 SW 2007 - 2012

If you need a supermini with a tad more more room for your shopping, this'll do. The rear screen opens independently so you can chuck stuff in. Lots of safety kit as well as space and more than OK to drive. Basic S spec will do as it has air and the Outdoor has a pointless, 'I'm not a real SUV' body kit.

Engines: 1.4 • 1.6 • 1.4D • 1.6D **CO2:** 126-150g/km

MPG: 44.8-60.1 **Gearbox:** Man/Auto **Boot:** 430/1365

Peugeot 306 SW 1997 - 2002

Partly Citroen ZX with less dreary bodywork. Connected to the basic 1.9 and the later Hdi diesels it was brilliant. These days a 1.6 petrol is all any buyers needs to keep on motoring, they won't be cool though as the air con pipes on the higher spec ones rot away. Electrics are the let down. Worn suspension, driveshafts, rust, mostly worn out.

Engines: 1.4 • 1.6 • 1.8 • 1.9D • 2.0D **CO2:** 141-198g/km **MPG**: 33.6-54.3 **Gearbox:** Man/Auto **Boot:** 442/1512

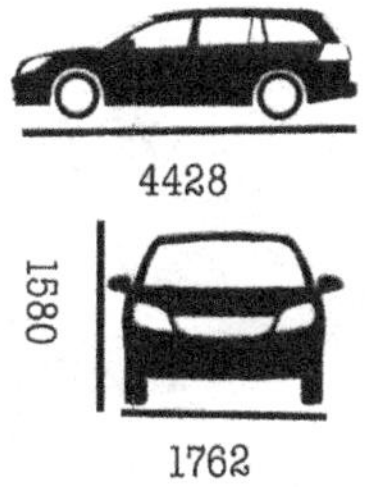

Peugeot 307 SW 2002 - 2008

Can take 5 or 7 depending on the model or whether the extra seats have been hoofed out. Move the rear seats around in the 5 and there is a truly massive load space. Got a revamp in '07. Mostly it is a modestly sized estate with flaws. Clutch failure. ECU for diesels. Starting issues. Water leaking into the cabin. Electrical issues and total failures.

Engines: 1.6 • 1.6D **CO2:** 129-178g/km **MPG**: 36.7.-56.5
Gearbox: Man/Auto **Boot:** 503/1805 **Seats**: 5/7 **Doors:** 5

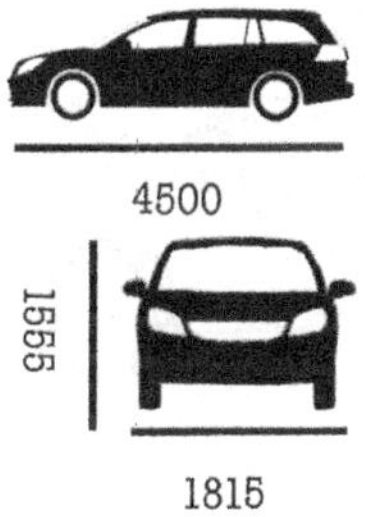

Peugeot 308 SW 2008 - 2011

Comes with modern complications. Use it as a five seater rather than a seven seat SE, if you want to carry their stuff. As ever the 1.6 petrol and 2.0 diesel are the best picks irrespective of the trim. Always going to be a bit flakey and mechanically and electrically. Water leaks into cabin, ECU failures, hairy electrics, blocked diesel filters.

Engines: 1.4 • 1.6 • 1.6D • 2.0D **CO2:** 120-159g/km
MPG: 42.2.-62.7 **Gearbox:** Man/Auto **Boot:** 674/2149

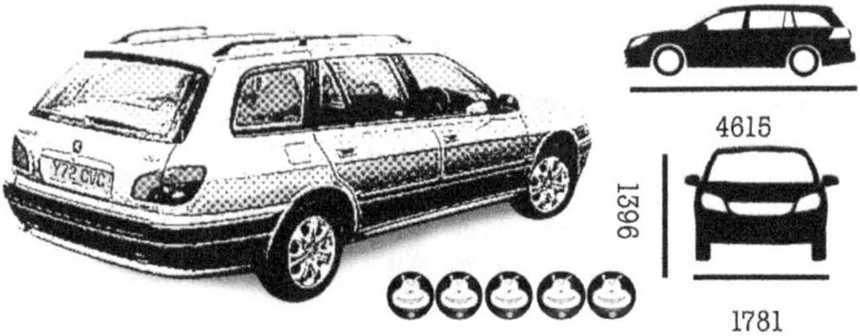

Peugeot 406 1996 - 2004

It is usefully long with a low sill with flipping and folding the seats leaves a large vanlike amount of space. 1.9 Turbo is prehistoric but better than most flaky modern oil burners. Otherwise the 2.0 HDi is perfect for the motorway. Revised in 2001, but our sort of banger relic. Corroded radiators, rusty brake discs, engine ECU issues, suspension bushes.

Engines: 1.8 • 2.0 • 2.0D **CO2:** 147-197g/km **MPG**: 34.0-51.3
Gearbox: Man/Auto **Boot:** 526/1741 **Seats**: 5 **Doors:** 5

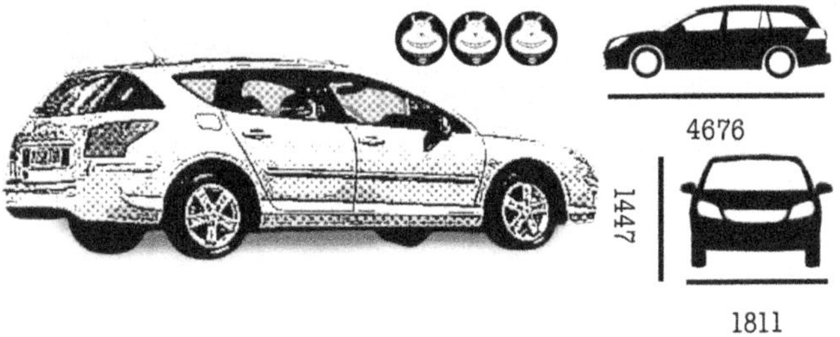

Peugeot 407 SW 2004 - 2011

Land Yacht estate with lots of clever touches. Low load height, wide bay and the passenger seat folds forward so that ladders can fit inside, plus split rear seats that fold flat. Under floor storage, side pods, it has it all. One upside is countless airbags and general safety kit. Dashboard failure, clutch failure, ABS sensors, diesel throttle issues.

Engines: 2.0 • 1.6D • 2.0D • 2.2D • 2.7D **CO2:** 140 - 206g/km
MPG: 30.6-53.2 **Boot:** 430/1365 **Seats**: 5 **Doors:** 4

Renault Clio Sports Tourer 2008 - 2012

Quite well put together and has the usual sky high levels of safety. Almost too many engines to choose from. Underfloor storage, tug the handles in the boot for a flat floor. Buy what you can afford and what makes sense to run. Diesel EGR valves and turbos, air con faulty, electricals such as central locking and ignition issues.

Engine: 1.2 • 1.4 • 1.5D • 1.9D • 2.0D **CO2:** 123-179g/km

MPG: 43.5-62.8 **Gearbox:** Man/Auto **Boot:** 524/1600

Renault Laguna Sport Tourer 2001 - 2007

Extra rear seats silly and best left to children. Laguna II was needlessly complicated with lots of electronic tricks which just made used buying much more of a hassle. important to remember that survivors with an MOT might well be fine. Weak clutches, EGR valves blocked, electrics terrible, corroded and cracked alloy wheels, air con on strike.

Engines: 2.0 • 3.0 • 1.9D • 2.2D **CO2:** 154-200g/km **MPG**: 33.2-47.9 **Gearbox:** Man/Auto **Boot:** 430 **Seats**: 5 **Doors:** 5

Renault Laguna Sport Tourer 2008 - 2012

Laguna III is a world of painful complication. One button push folds down rear seats totally flat and lots of hooks and eyes to secure loads in place. Otherwise only makes some sense if cheap and there is a comprehensive service history. Buyers may risk their sanity/bank balance. Look closely at the electrics, EGR valves and gearbox operation.

Engines: 2.0 • 1.5D • 2.0D **CO2:** 136-158g/km **MPG**: 35.8-55.4 **Gearbox:** Man/Auto **Boot:** 508/1593 **Doors:** 5

Renault Megane Sport Tourer 2003 - 2006

House brick shaped Megane II has split rear seats that fold flat but not exactly at the press of one button. For its size though it actually has more space than the Laguna equivalent. Lots of cubby holes and you can get an air conditioned glovebox. Incredibly safe. Watch electrics. EGR valve can get blocked, water leaks in engine bay.

Engines: 1.4 • 1.6 • 2.0 • 1.5D • 2.0D **CO2:** 119-209g/km

MPG: 43.0-62.8 **Gearbox:** Man/Auto **Boot:** 524/1600

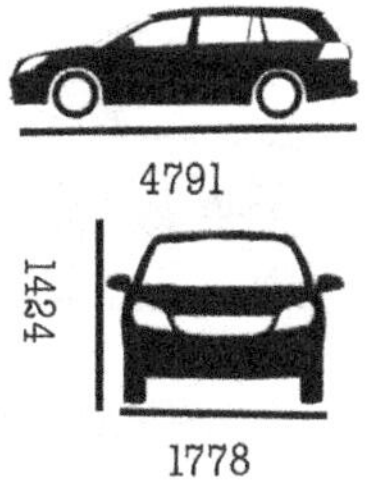

Rover 75 Tourer 2001 - 2004

Trad styling, but not spacious split rear seats drop one handed and fold. Prepared for breakdowns with the warning triangle on the tailgate and an illuminated spare wheel compartment. 3 petrols, a diesel and carefully crafted trim levels. Rotten sills and wheelarches, blown head gasket, diesel pump failure. Corroded suspension.

Engines: 1.8 • 2.5 • 2.0D **CO2:** 163-225g/km **MPG**: 30.0-48.8 **Gearbox:** Man/Auto **Boot:** 400/1222 **Seats**: 5 **Doors:** 5

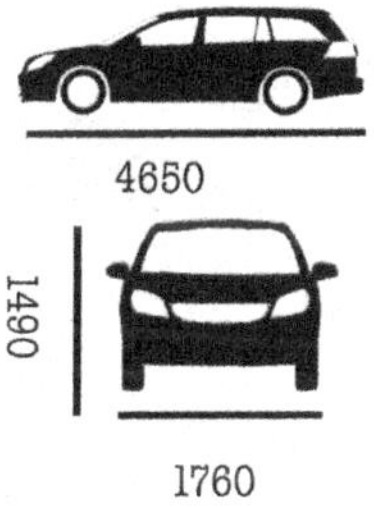

Saab 9-3 Sportswagon 2005 - 2011

Looks supercool, not superspacious. Rear seats split and fold, not totally flat plus there is a ski hatch with underfloor storage and 12V power outlets. Lots of engines and specification combinations. Go for the 2.0 litre petrol or 2.2 diesel. Safe and sound buy. Electrics, windows and locking not great, engine ECU rough running. Check aircon.

Engines: 1.8 • 1.8B • 2.0 • 2.0B • 2.8 • 1.9D **CO2:** 157-216g/km **MPG**: 35.8-48.7 **Gearbox:** Man/Auto **Boot:**419/1273

Saab 9-5 Estate 1998 - 2010

Huge inside and very comfortable. Split/fold rear seats obviously, with a flat floor which slides out beyond the bumper to help shoving stuff in. 2.0 and 2.3 and top of the range V6. 2001 revisions outside and in with diesel engines. 2005 facelift: headlamps with goth level mascara!. Coil pack failure, air con issues, automatic 'box + suspension bushes.

Engines: 2.0 • 2.3 • 1.9D • 2.0B • 2.3B **CO2:** 184-212g/km **MPG**: 31.7-41.5. **Gearbox:** Man/Auto **Boot:** 416/1490

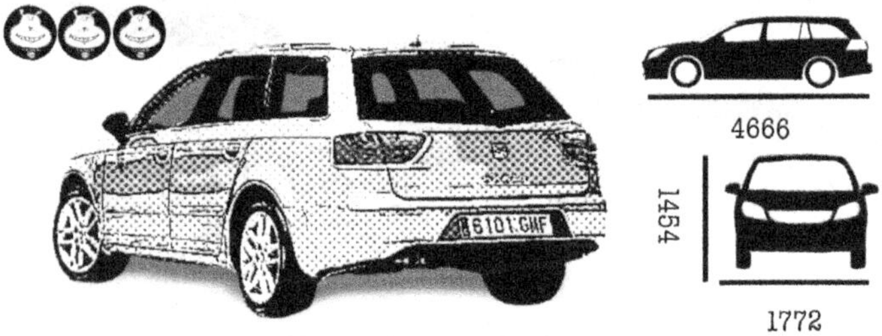

SEAT Exeo ST 2009 - 2013

VW recycling by taking old (2004) Audi A4 & sticking a SEAT badge on it. Then they added all the specification that was missing from the German like cruise control and Bluetooth. Rear seats split and fold flat at the pull of a lever, plus there are hooks and eyes to keep everything in place. Might be blocked particulate filters, otherwise wear & tear.

Engines: 2.0 • 2.0D **CO2:** 141-179g/km **MPG**: 36.7-51.4 **Gearbox:** Man/Auto **Boot:** 442/1354 **Seats**: 5 **Doors:** 5

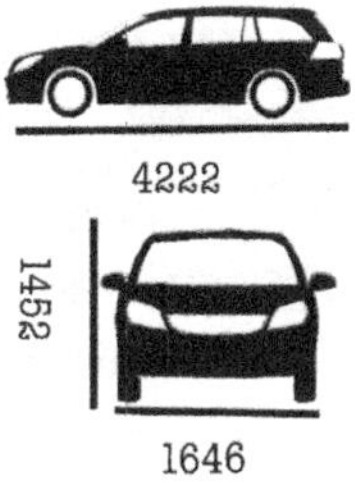

Skoda Fabia 2001 - 2007

With rear seats folded as big as posh lifestyle estates. You can enjoy it even more with a 1.9 SDI which won't ever let you down. Otherwise the little petrols 1.2 and 1.4 are fine. Revamped in 2004 with a refreshed up interior. Comfort and Ambiente are the middle order specs to buy. Coolant and oil leaks, timing belt tensioner, head gasket failure.

Engines: 1.2 • 1.4 • 2.0 • 1.4D • 1.9D **CO2:** 124-187g/km **MPG**: 36.2-61.4 **Gearbox:** 5spd **Boot:** 426/1225 **Seats**: 5 **Doors:** 5

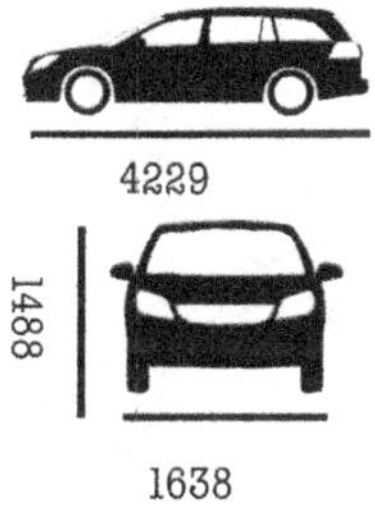

Skoda Fabia 2008 - 2014

Big tailgate, low loading lip, 60/40 split seats that do not fold completely flat. Has in built restraining straps. 1.4TDI GreenLine was the super eco one. Big revamp in 2010 with even more efficient engines and most irrespective of fuel deliver nearly or well over 50mpg. Reliable so long as they are properly serviced. No real problems, just fit and finish.

Engines: 1.2 • 1.4 • 1.6 • 1.4D • 1.9D **CO2:** 109-180g/km **MPG**: 36.2-61.4 **Gearbox:** 5spd/Auto **Boot:** 480/1460

Skoda Octavia 1998 - 2005

Spacious plus it has metal guide rails in the floor so you can slide stuff in and it is a metre wide. There was a very broad range with high performance vRS stuff, best to stick with the basic 1.9 diesels and 1.6 to 2.0 petrols which are easy to deal with. Diesel ECU failure, water pumps, air mass sensors and ignition coils. Electrical failures in the cockpit.

Engines: 1.4 • 1.6 • 2.0 • 1.9D • 2.0D **CO2:** 143-168g/km

MPG: 40.4-53.3 **Gearbox:** Man/Auto **Boot:** 560/1620

Skoda Octavia 2005 - 2013

Great big luggage compartment making it considerably huger than many traditional offerings. A bit high, but flat and there is underfloor storage as well, plus loads of oddment storage. Styling revamp in 2009. Popular choice as diesel taxis, like many Skodas. Electricals and water leaks, climate control, ABS and ESP sensors and ECUs.

Engines: 1.4 • 1.8 • 2.0 • 1.9D • 2.0D **CO2:** 132-154g/km

MPG:42.8-57.6 **Gearbox:** Man/Auto **Boot:** 560/1620

Skoda Superb 2009 - 2015

Absolutely huge boot and overall load area. Low loading lip, plus there is tailgate illumination which is certainly helpful. Plus a rechargeable torch and the novelty of an umbrella in the door. Most on sale are diesels and yet again many lived a tough old life in the taxi business. ABS sensor failure, DSG gearboxes malfunction, keyless entry issues.

Engines: 1.4 • 1.8 • 3.6 • 1.9D • 2.0D **CO2:** 155-180g/km

MPG: 39.2-47.9 **Gearbox:** Man/Auto **Boot:** 633/1865

Subaru Impreza Sports Wagon 2000 - 2005

Yes there is the mental Scooby we all know and love, but there was simply a cooking 1.6 and 2.0 estate all with four wheel drive and roof bars. Not that frugal and when they break the bills tend to be high. Could be a one to use as a Banger then discard. Revamped in 2005 with just a 2.0. Blowing turbos, worn suspension and brakes.

Engines: 1.6 • 2.0 **CO2:** 210-242g/km **MPG**: 27.7-31.7

Gearbox: Man/Auto **Boot:** 348 **Seats**: 5 **Doors:** 5

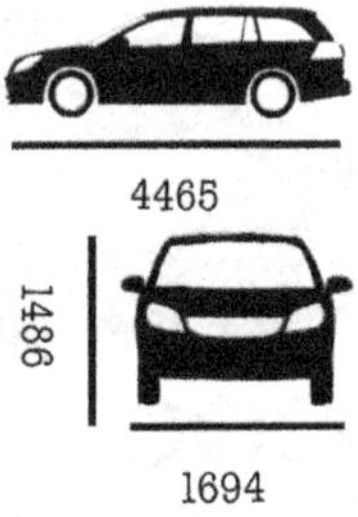

Subaru Impreza Sports Wagon 2005 - 2008

Basic 4WD available with 1.5 and 2.0 flavours. Remains an oddity in a conventional world and the same rules apply, so do you really need something this complicated? Only occasional carvanners and unmade road dwellers need apply. Nothing to get excited about. Suspension, brakes and everything that needs doing is always expensive.

Engines: 1.5 • 2.0 **CO2:** 176-212g/km **MPG**: 31.7-37.7

Gearbox: Man/Auto **Boot:** 301/1266 **Seats**: 5 **Doors:** 5

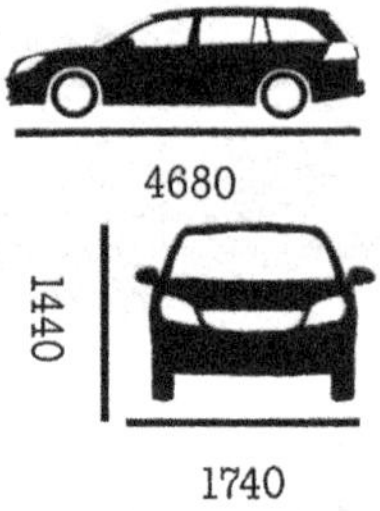

Subaru Legacy 1993 - 2003

Suspension re-engineered for a flatter floor, so a nice load area that gets bigger with the rear seats folded. Thirsty petrol engines, 2.0, 2.5 and 3.0. There were some really old American built 2.0 models coming to the UK in 1999. 2001 restyle didn't make it any prettier. Everything costs a fortune if broken. Wear and tear on the running gear.

Engines: 2.0 • 2.5 **CO2:** 215-228g/km **MPG**: 30.4-31.4

Gearbox: Man/Auto **Boot:** 433/1575 **Seats**: 5 **Doors:** 5

Subaru Legacy 2003 - 2009

Always well built, these ones are better finished now. Still an odd buy for most people, what could be used as justification was the arrival of a diesel engine although that didn't happen until '08. Restyled a bit in '07 and the 3.0 engine upgraded. Costs a lot to run. Costs even more to fix now, brakes and suspension are the least of your worries.

Engines: 2.0 • 3.0 **CO_2:** 212-233g/km **MPG**: 28.8-31.4

Gearbox: Man/Auto **Boot:** 459/1628 **Seats**: 5 **Doors:** 5

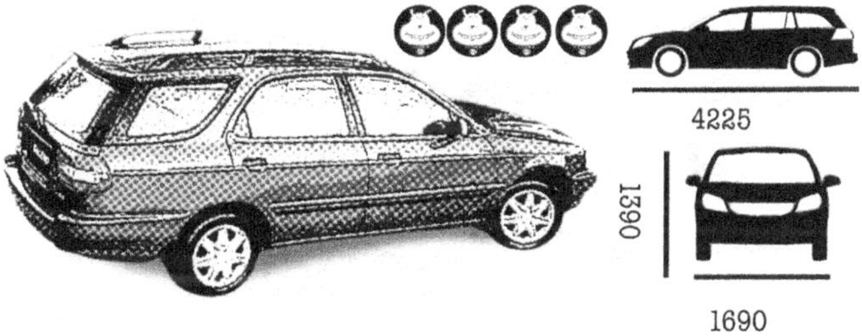

Suzuki Baleno 1999- 2005

Only feature going for it is the low loading height and tailgate as part of the rear bumper. Otherwise it's Suzuki attention to detail & reliability. 1.6 petrol, but later a 1.8 engine also in the mix and not much to recommend the Baloney except that it is a competent small family car. Might be a bit rusty now, just the usual wear and tear issues.

Engines: 1.6 • 1.8 **CO_2:** 171-184g/km **MPG**: 36.7-39.8

Gearbox: Man/Auto **Boot:** 374/1376 **Seats**: 5 **Doors:** 5

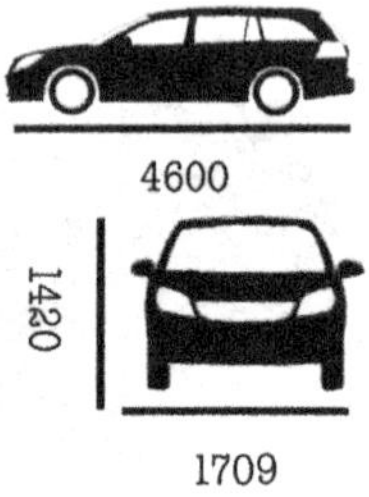

Toyota Avensis Tourer 1998 - 2003

Brit built Avensis is fabulously reliable, if dull and here was an estate, but with an awful lot more safety equipment. These are proper company cars that led hard lives but still managed to not break. There is a decent amount of room, especially with the rear seats folded. Clutch and gearbox failures over time, brake fluid leaks, plus wear and tear.

Engines: 1.6 • 1.8 • 2.0 • 2.0D **CO2:** 158-181g/km **MPG**: 36.7-47.9 **Gearbox:** Man/Auto **Boot:** 530/1000 **Doors:** 5

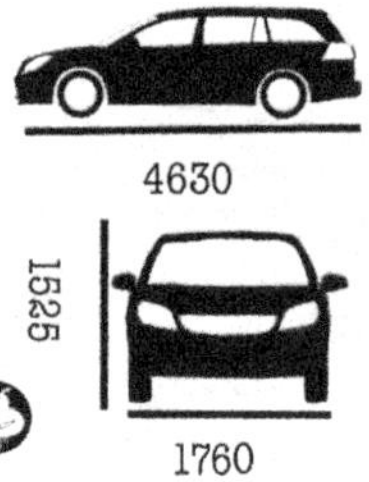

Toyota Avensis Tourer 2003 - 2008

Effectively a Japanese Passat Estate, solidly built and a feeling of quality, with guaranteed reliability, but pretty blunt styling. Revamped in 2006, a bit sharper to drive and some more thumping diesels to make life better. All it needs to do is work to make sense as a Banger buy. Clutch failure, worn rear suspension, also double check electricals inside.

Engines: 1.8 • 2.0 • 2.0D • 2.2D **CO2:** 158-193g/km

MPG: 34.9-48.7 **Gearbox:** Man/Auto **Boot:** 520/1500

Toyota Avensis Tourer 2009 - 2015

Anonymous but effective mover of people and whatever they want to pack inside. This is a spacious bus which has a big, square sided load bay with load rails above the basic T2. Best of the late estates which may be electrically complicated, but thankfully is a Toyota. There have been a few recalls, relating to suspension, otherwise it is wear/tear.

Engines: 1.6 • 1.8 • 2.0D • 2.2D **CO2:** 150-193g/km

MPG: 34.9-50.4 **Gearbox:** Man/Auto **Boot:** 509/1609

Toyota Corolla 1997 - 2002

Nothing to really set what is a compact model apart from rivals. Being a Corolla though means that it will be exceptionally reliable. Powered by unbreakable small petrols and a diesel. Avensislike styling from 2000 and a fancy common rail diesel. Rather rusty some of them, lambda sensors possibly. Mostly just worn out parts issues.

Engines: 1.4 • 1.6 • 2.0D **CO2:** 156-168g/km **MPG**: 40.4-48.7 **Gearbox:** Man **Boot:** 308/663 **Seats**: 5 **Doors:** 5

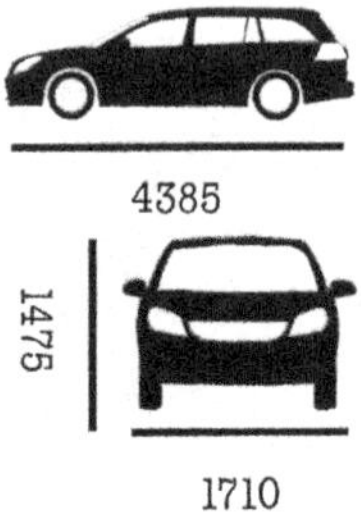

Toyota Corolla 2002 - 2007

Last Banger gasp of the legend, making it to Mark 10 no less. Four petrols and two diesels provide the motivation. 2004 revamp made it look even duller and a 1.4 diesel to replace the 2.0 one. Sadly you couldn't rely on the name badge. Passed away in the UK. Recalls related to airbags they did not make. Just the usual wear and tear problems.

Engines: 1.6 • 1.4D • 2.0D **CO2:** 152-172g/km **MPG**: 39.2-49.6 **Gearbox:** Man/Auto **Boot:** 402/975 **Doors:** 5

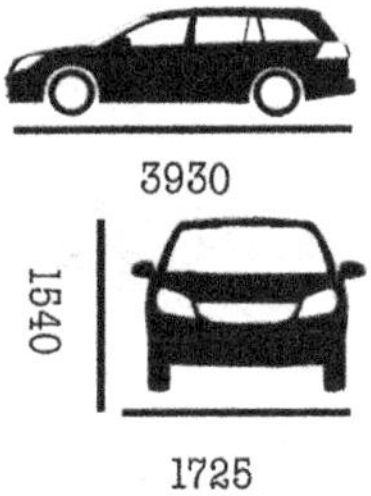

Toyota Urban Cruiser 2009 - 2012

Compact but roomy this is a boxy MPV with 60:40 split rear seats which also fold and recline, but not in the four wheel drive model. It isn't remotely even a pretend off roader. Has climate, alloys and keyless entry amongst other gadgets. Just a pretentious hatchback. Reliable though. Wear and tear items, suspension and brakes.

Engines: 1.33 • 1.4D **CO2:** 129-130g/km **MPG**: 51.4-57.6 **Gearbox:** Man/Auto **Boot:** 388/749 **Seats**: 5 **Doors:** 5

Vauxhall Astra 1998 - 2004

Practical load bay that is low wide and flat with plastic floor rails to make sliding baggage on board easier and a ski hatch to keep the rear seats upright. Massive range, seven engines. Easy to drive and live with, but company car abuse and can be worn out if used as a van. Corroded ECUs, cockpit leaks, failing Air conditioning and water pumps.

Engines: 1.4 • 1.6 • 1.8 • 2.0 • 1.7D • 2.0D **CO2:** 154-172g/km **MPG**: 38.2-49.6 **Gearbox:** Man/Auto **Boot:** 370/1210

Vauxhall Astra 2004 - 2010

Big, low loading bay, protected by stainless steel moulding, floor now has aluminium loading rails. Rear seats 60:40 split with optional: 40:20:40 and there was even a divider system called FlexOrganiser. The 1.6 petrol as an all rounder and the 1.7 ecoFLEX if you do big miles. Worn suspension bushes and rusty turrets. Niggly electrical, lights, wipers, locks.

Engines: 1.4 • 1.6 • 2.0 • 1.7D • 2.0D **CO2:** 135-187g/km **MPG**: 36.2-56.5 **Gearbox:** Man/Auto **Boot:** 470/1550 **Seats**: 5

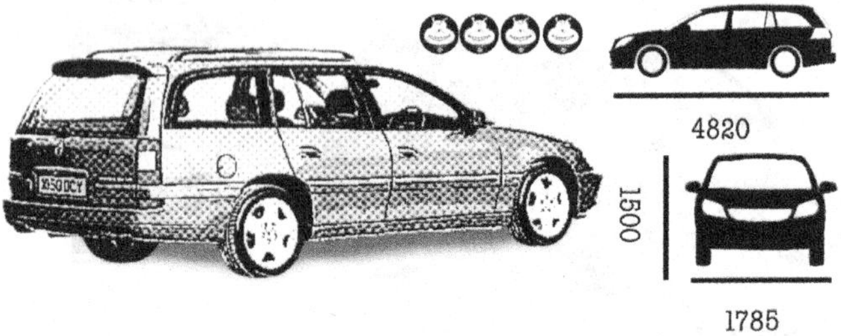

Vauxhall Omega 1994 - 2003
Made in Germany with a truly massive boot, so they were well screwed together and had a Griffin badge rather than an Opel one. Better looking from 1999 and the more powerful DTi diesels the ones to have. Motorway missile with super spec Elite and CDX. Worn suspension bushes, oil leaks, fuel injector seals, air conditioning and heating system.

Engines: 2.2 • 2.6 • 3.2 • 2.2D **CO2:** 192-284g/km **MPG**: 23.9-39.8 **Gearbox:** Man/Auto **Boot:** 541/1801 **Doors:** 5

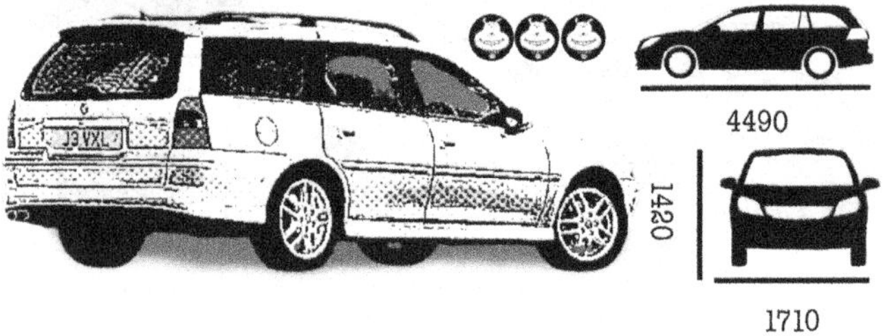

Vauxhall Vectra 1996 - 2002
Big tailgate and opening with folding rear seats. Loads of engine and spec options from DTI diesels to V6 petrol. Design model has all the basics which includes cargo net, ski hatch, plus an under floor compartment. Rust obviously, leaking coolant, variously broken suspension parts, failing water pumps. Clutch remains a complicated job.

Engines: 1.6 • 1.8 • 2.2 • 2.6 • 2.0D • 2.2D **CO2:** 161-236g/km **MPG**: 28.8-46.3 **Gearbox:** Man/Auto **Boot:** 460/1489

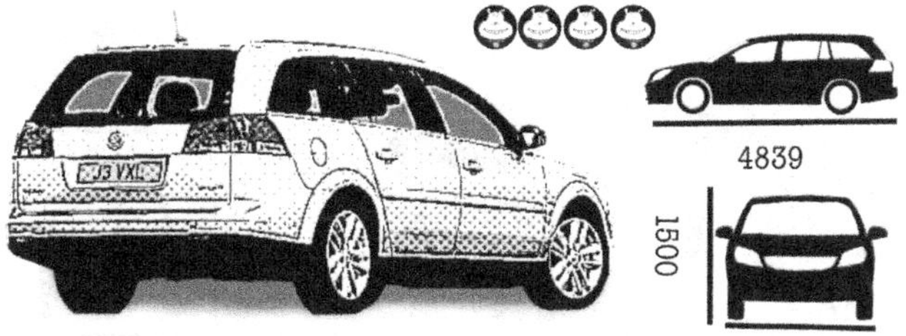

Vauxhall Vectra 2003 - 2008

Beyond the 60/40 split rears the front seat even folds flat to cope with longer loads. FlexOrganiser system can box areas off very easily. There is even an electronic tailgate on the top specifications and it can be locked in position. Company cars can be worn out with noisy, oily engines and suspension EGR valve, fuel injector failure, snapped timing chains.

Engines: 1.8 • 2.2 • 2.8 • 1.8D • 1.9D • 3.0D **CO2:** 157-180g/km **MPG**: 37.3-48.7 **Gearbox:** Man/Auto **Boot:** 540/1850

Vauxhall Insignia Sports Tourer 2009 - 2017

Refined, comfortable, roomy, with the rear seats folded. Has a wide tailgate, electrical on the top models. Revamped in 2013. Standard kit is comprehensive throughout the eight trim levels. Company car drivers bought Tech Lines with all the fancy Bluetooth and DAB. Dual Mass Flywheel failure, dashboard electrics, power steering leaks.

Engines: 1.6 • 1.8 • 2.0 • 2.8 • 2.0D **CO2:** 154-184g/km **MPG**: 36.2-54.3 **Gearbox:** Man/Auto **Boot:** 520/1460 **Seats**: 5

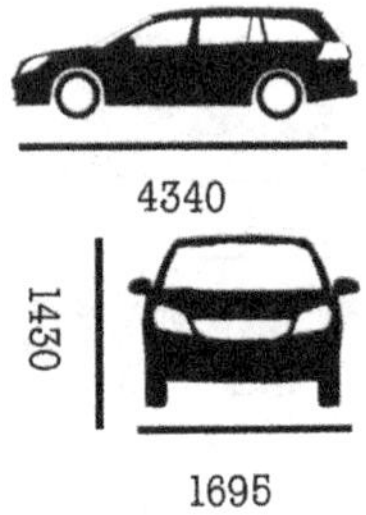

Volkswagen Golf 1997 - 2004

Less room than Mark 3, nice wide flat load area. 60:40 rear seat split, remove the headrests for a flat floor. Classy chrome restraining hooks & stainless steel protecting bumper/tailgate. 12V power output. Standard roof rails. Rust on sills and arches. Engine ECU cause rough running. Brake callipers and water pump failure.

Engines: 1.4 • 1.6 • 1.8 • 2.0 • 1.9D **CO2:** 143-194g/km
MPG: 32.1-57.6 **Gearbox:** Man/Auto **Boot:** 459/1470

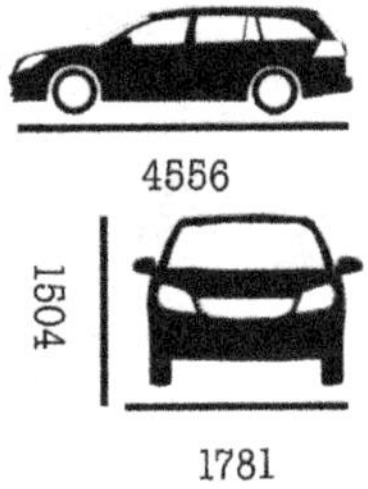

Volkswagen Golf 2007 - 2009

Huge boot gets bigger with 60:40 seats dropped. There are storage holes under the load bay and in the sides. Standard roof rails all ready for work. Mostly fitted with diesels which get a solid 60mpg plus, the 1.9TDi is all most owners need for lugging, or the 2.0TDI. Does the aircon work? Door seal failure, diesel turbo oil seal failure, poor air con.

Engines: 1.6 • 1.9D • 2.0D **CO2:** 137-178g/km **MPG**: 38.2-53.3 **Gearbox:** Man/Auto **Boot:** 505/1550 **Doors:** 5

Volkswagen Golf 2008 - 2013

Golf 6 thankfully qualifies for the Bangerpedia and is rather modern. Has plenty of oddments space Again the 1.4TSI engine is one of the most sensible and economical units in a world that is turning against diesels. It is much more of a computer on wheels that needs reboots and updates. Coil pack failures on petrols, worn suspension bushes.

Engines: 1.4 • 1.6D • 2.0 • 1.9D • 2.0D **CO2:** 119-144g/km

MPG: 45.6-62.8 **Gearbox:** Man/Auto **Boot:** 505/1495

Volkswagen Passat 1997 - 2001

Generously equipped, had fine engines and subtle styling with a large load area, that can obviously be extended with the split/folding seats. Easy to load, too with a conveniently low and flat lip. From the S climate control, electric sunroof, engine immobiliser and ABS brakes. Smoking diesels, damaged suspension, worn brakes and steering

Engines: 1.8 • 2.0 • 2.3 • 2.8 • 1.9D • 2.5D **CO2:** 149-259g/km

MPG: 26.9-51.4 **Gearbox:** Man/Auto **Boot:** 475/1500

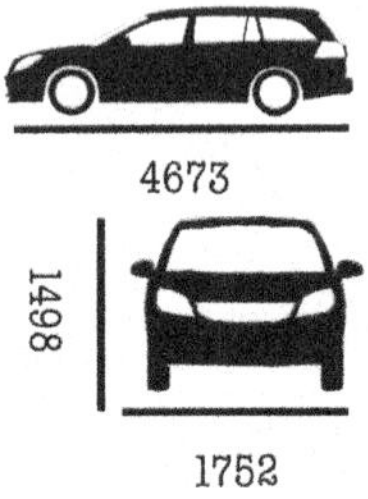

Volkswagen Passat 2000-2005

Version 5.5 is another quality update. Heavy duty load bay carpet with lots of hook and eye lashing points. Good level of standard safety and comfort equipment. All the engines, especially the 1.9 TDI diesel are excellent and there is something for everyone. Worn turbocharger, needs proof of timing belt changes, leaking air con into cabin.

Engines: 1.8 • 2.0 • 2.3 • 2.8 • 4.0 • 1.9D • 2.5D

CO2: 151-317g/km **MPG**: 21.4-50.4 **Boot:** 450/1545

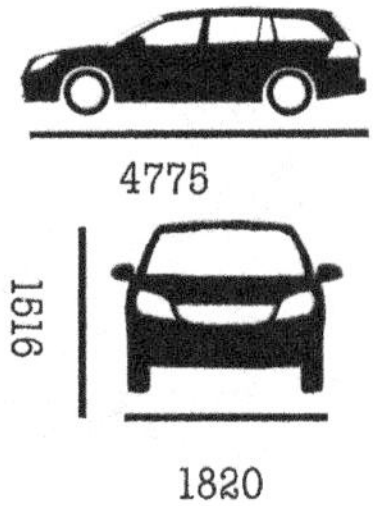

Volkswagen Passat 2005 - 2011

Mark 6 is bigger than before with a bewildering number of engine options. 1.9 Tdi remains the go to engine, otherwise the petrol TSIs are the ULEZ safe ones. Perfect company car, which is why some caution always required when it is time to buy. Oil pump failure, ECU failures especially after water leaks into cabin, electrical faults and flat batteries.

Engines: 1.4 • 1.8 • 2.0 • 3.2 • 2.0D **CO2:** 153-206g/km

MPG: 36-48.7 **Gearbox:** Man/Auto **Boot:** 603/1731

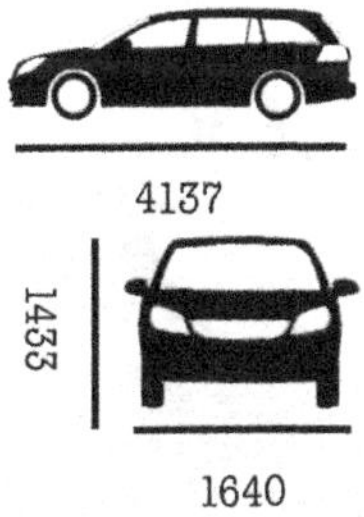

Volkswagen Polo 2000 - 2002

Average looks but a decent amount of room inside. Plenty of engines from the 1.0, 1.4 and 1.6 petrols which are the ones to aim for. There are plenty of 1.4 and particularly 1.9 diesels around. Last of the small estate line, but worth finding. Can chew through tyres and brakes, mostly wear items to look at and rust. Worn gearbox.

Engines: 1.0 • 1.4 • 1.6 • 1.9D **CO2:** 120-173g/km **MPG**: 40.4-62.8 **Gearbox:** Man/Auto **Boot:** 391/1248 **Doors:** 5

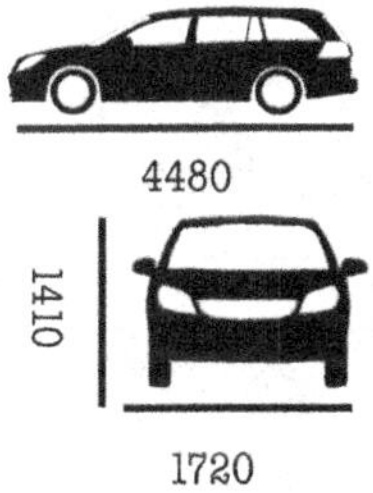

Volvo V40 1996 - 2004

Quite a pokey estate for a Volvo. Successful and most buyers went for the 1.9D which is an excellent Renault Turbo diesel. 1.6 or 2.0 petrols are better options for the modern world. Updated in 2000 with new engines and styling. Survivors can be a bit rusty. Corroded loom causes poor electrics, air con/power steering can fail, brake calipers break.

Engines: 1.6 • 1.9D **CO2:** 142-185g/km **MPG**: 35.8-52.3 **Gearbox:** Man/Auto **Boot:** 368/1422 **Seats**: 5 **Doors:** 5

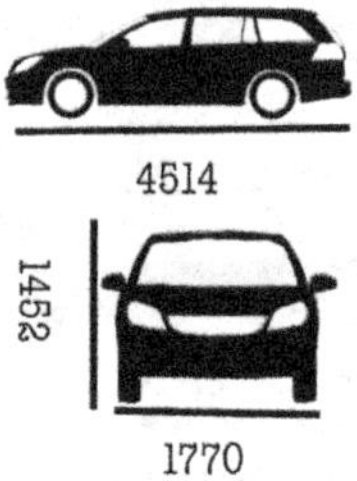

Volvo V50 2004 - 2012

Adorable brick. Related to a few other cars including a Ford Focus. Can come with a folding luggage retainer is useful. Other than petrol/diesel, Flexifuel runs on bioethanol. Facelifted 2007, looks identical, fresh interior & few minor mechanicals. Faulty fuel injectors, ECUs and air mass sensors means rough running. Automatic gearbox issues.

Engines: 1.6 • 1.8 • 2.0 • 2.0F • 2.4D **CO2:** 154-174g/km **MPG**: 38.7-48.7 **Gearbox:** Man/Auto **Boot:** 417/1307

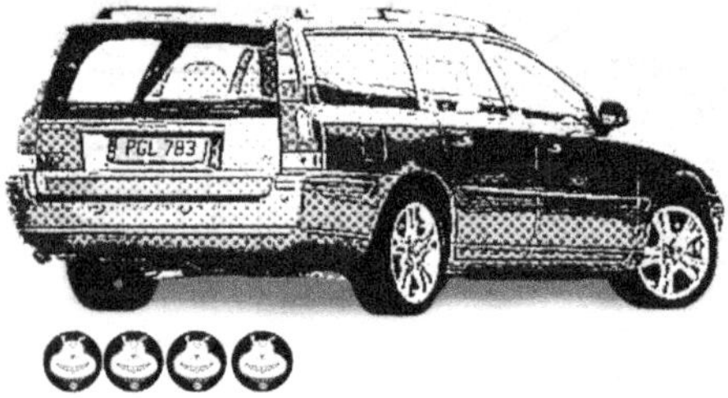

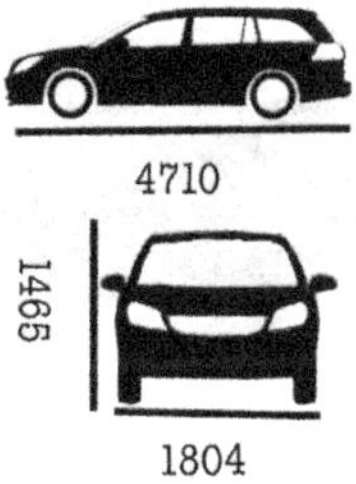

Volvo V70 2000 - 2007

The iconic estate car. Can have kiddie friendly rear seats in the boot. Posh carpeted load lugger. Got a revamp in 2004. Engines are very tough whether diesel or petrol, or even flexifuel, but finding gas is an issue now. Obviously safe, spacious and could not be more solid or stylish. Buy while stocks last. Suspension bushes, broken throttle bodies.

Engines: 2.0 • 2.4 • 2.5 • 2.4D • 2.5D **CO2:** 169-249g/km **MPG**: 31.7-52.3 **Gearbox:** Man/Auto **Boot:** 485/1641 **Seats**: 5/7

Volvo V70 2007 - 2012

Familiar looking flat pack swallowing body. Lots of meaty engines and diesels remain the best if there are miles to be done. Petrols just soldier on forever. Becoming complicated, but on the whole Volvos manage to avoid being too flakey in the mechanicals and electricals departments. Smoking worn turbos, damaged suspension. Mostly wear and tear.

Engines: 1.6 • 2.0 • 2.5 • 3.0 • 1.6D • 2.0D • 2.4D **CO2:** 114-244g/km **MPG**:43.5-62.8 **Boot:** 575/1600 **Doors:** 5

Banger Vans People Shifters + Space Ships

On the face of it the dullest vehicles in all the Bangeruniverse. Useful though. Yes, Space Ships are all about packing people and other stuff inside. That's right, these are vans. With windows. Always have been.

Literally in the case of Citroen Berlingo, Fiat Doblo and Renault Kangoo. Arguably the best working incarnation as they compact and practical. Genuinely innovative creations like the Fiat Multipla lasted the decade whereas the Honda FRV is all but forgotten, but built to incredible standards that makes it Banger material to this day/

For a time these were the most popular family wagons and that explains why there are so many of the things. In many ways the van derived models as mentioned are the most honest and practical of the lot.

Others can be awful (Mercedes Vaneo) , some just seat five (like a capable Mazda Premacy), but allow you to wear a top hat (Peugeot Bipper Tepee). People Carriers remain the most credible alternative to an estate car and the best way of seating more than five people, but not necessarily packing in all of their luggage, unless you have a trailer or a top box.

Only buy a van if you need a van. Otherwise it will always be a waste of space and money. It will have worse mpg and take up most of your drive if you are lucky enough to have one.

The traditional MPVs are mostly the winners because the Ford Galaxy drives like a car and is easy to live with. Despite being the French and slightly unreliable the compact Renault Scenic delivered a flexible family hatch option.

The thing is these vans can be in average to good condition for the year like any used car, or worn out by lots of families, or a bloke who used it as a taxi. More worrying there is the genuine scenario where a jobbing plumber hoofed out all the seats so that he could fit all those bags of rubble, tiles, copper pipe and kitchen sinks inside.

Choose carefully.

Chevrolet Tacuma 2005 - 2008
Basically a Depressingwoo. Anyway it continued as before but with Chevy badges. Good standard kit of course, but it is only a five seater, but oddly does the Captain's Passenger Chair thing which can face the rear as well as rear middle that can fold flat. So it is versatile. Pretty reliable, just electricals to check that they work, especially air con.

Engines: 1.6 • 2.0 **CO2:** 191-212g/km **MPG**: 32.8-35.3
Gearbox: Man/Auto **Boot:** 347/1847 **Seats**: 5 **Doors:** 5

Chrysler PT Cruiser 2000 - 2008
Bugsy Malone five seater in Classic, Touring and Limited trim, yet underneath it is just a stupid Neon. Eventually got a Mercedes 2.2 Diesel in 2002. Quite clever inside with removable seats and a performing parcel shelf and up to 64 cubic feet in old money. Revamp in 2005. Leaking air con, head gasket failure, power steering fluid leak, worn clutch.

Engines: 2.4 • 2.2D **CO2:** 185-212g/km **MPG**: 31.7-40.9
Gearbox: Man/Auto **Boot:** 538/1821 **Seats**: 5 **Doors:** 5

Chrysler Voyager 1997 - 2001

This does seem to be roughly the size of a planet without actually being a van with windows. There was an even grander, Grand model. Everything as standard specification. Not just large American related petrols, also an Italian made VM Turbo diesel. Oil leaks, automatic gearbox, diesel turbo smoke and rust.

Engines: 2.0 • 2.4 • 3.3 • 2.5D **CO2:** 185-297g/km **MPG**: 22.2-36.7 **Gearbox:** Auto **Boot:** 445/4080 **Seats**: 5/7**Doors:** 5

Chrysler Voyager 2001 - 2008

Safer than the '90s one and so ginormous it needed three zone air con. After 2005 the seats folded into the floor, really quick and one handed. With the seats up there is loads of litres of underfloor storage. Lots of things electrical to check. Oil leaks, power steering, ECU failure leading to rough running, tired suspension and brakes.

Engines: 2.4 • 2.5D • 2.8D **CO2:** 203-234g/km **MPG**: 28.5-36.7 **Gearbox:** Man **Boot:** 1000/4100 **Seats**: 5/7

Chrysler Grand Voyager 2008- 2015

Ultra Massive Yank People Shifter. Stow 'n Go seating system means all five rear seats folding simply and quickly into the floor turning it into a panel van. Second row of seats rotate through 180 degrees so you have office space. comes with either a 2.8-litre CRD diesel engine or a 3.8-litre V6 petrol, overall average quality and electricals.

Engines: 3.8 • 2.8D **CO2:** 207-222g/km **MPG**: 23.-351

0-60: 12.2-12.4 **Boot:** 756/3296 **Seats**: 7 **Doors:** 5

Citroen Berlingo 1998 - 2009

Endearing little van. Flexispace models have extra cubbies and also roof shelf areas. Not a lot of standard items on early models. Usually cheap to run whether as a diesel or small petrol. Beware ones actually used as vans. Suspension repairs can be expensive, cabin and under bonnet water leaks, ECU malfunction. Complicated wiring. Rust.

Engines: 1.4 • 1.6D **CO2:** 150-195g/km **MPG**: 34.5-49.6

Gearbox: Man/Auto **Boot:** 624/2800 **Seats**: 5 **Doors:** 5

Citroen Berlingo Multispace 2009 - 2018

Not as van like as the old one, ideal for practical minded normies. Specs vary hugely, certainly safer and probably better built than before. Rear seats can be removed rather than just folded. Split opening tailgate option. Fancier than before, but bigger. More complex electrics can go wrong. Damage to brakes and suspension.

Engines: 1.6 • 1.6D **CO2:** 143-177g/km **MPG**: 38.2-52.3

Gearbox: Man/Auto **Boot:** 675/3000 **Seats**: 5 **Doors:** 5

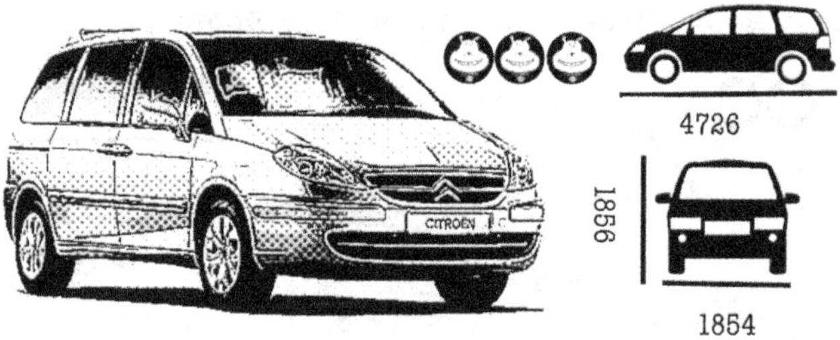

Citroen C8 2003 - 2010

One of three identikit MPVs, this is arguably the 'value' one. Best driven with the 2.0HDi. You couldn't get more versatile seats, which slide, detach and spin to name just a few moves. Four zone air con is just one of the complexities that may well go wrong. Electronics sticky, windows, locks and sliding doors. Power steering failure. Faulty ECUs and air con.

Engines: 2.0 • 2.2 • 2.0D • 2.2D **CO2:** 188-218g/km

MPG: 31.0-39.2 **Boot:** 225/2948 **Seats**: 5/7

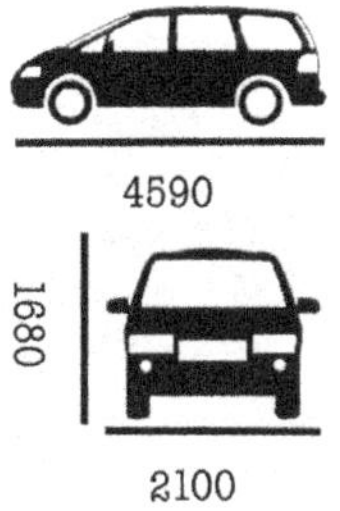

Citroen Grand C4 Picasso 2007 - 2013

The artist formally known as the Xsara. Car sized, but will seat seven. Updated 2010, lots of electrical complication like the handbrake and optional automatic gearbox. Although well built, caution needed when checking one over. Electrics complicated, check windows, dash display and air con. Dual mass flywheels on diesels fail. Suspension and brakes.

Engines: 1.6 • 1.6D • 2.0D **CO2:** 135-159g/km **MPG**:40.4-47.9

Gearbox: Man/Auto **Boot:** 208/1951 **Seats**: 5/7

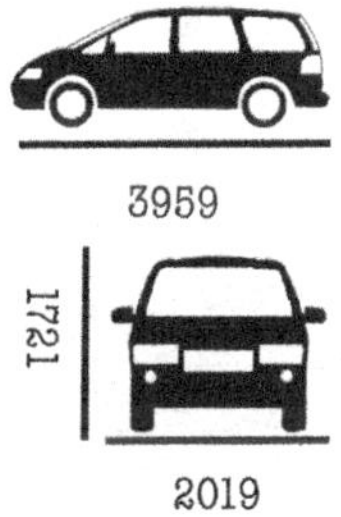

Citroen Nemo Multispace 2009 - 2012

Another one of those vans, a Nemo one. Unapologetically a commercial vehicle with some seats at the back, at least it does have a parcel shelf out back. Not overly refined, but that isn't the point. Also, may not have air con, but surely that doesn't matter. Air con may not work, oil leaks and general damage inside. Electrics.

Engines: 1.4 • 1.4D **CO2:** 119-155g/km **MPG**: 42.2-62.8

Gearbox: Man **Boot:** 360/2500 **Seats**: 5 **Doors:** 5

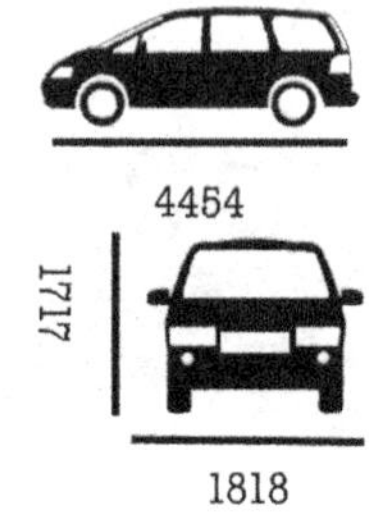

Citroen Synergie 1995 - 2002

One of three, see also Peugeot and Fiat. A box with seats which went from six to seven and eventually eight depending on the model. All upgraded for '98 and overall the specification was fine even though this was 'cheapie' of all the van versions. Leaking coolant, clutch failure, ECU issues and MOT failure level rust.

Engines: 2.0•2.0D **CO2:** 182-232g/km **MPG**: 30.7-42.2
Gearbox: Man/Auto **Boot:** 340/3313 **Seats**: 5/7 **Doors:** 5

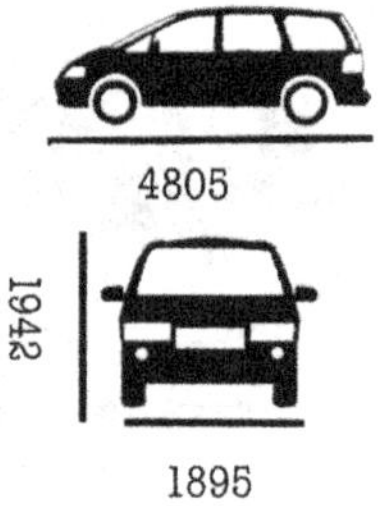

Citroen Dispatch Combi 2007 - 2016

Part of a Fiat and Peugeot, big van project. Depends what badge you want and engine preference. Longer one has more seats. 2.0 HDi with this one. Essentially a colossal people shifter. That's it really, same story with each version. Oil leaks, worn suspension bushes and possible ECU failure.

Engines: 1.6•2.0 **CO2:** 168-194g/km **MPG**: 37.0-42.0
Gearbox: Man/Auto **Boot:** 310/2200 **Seats**: 6/9 **Doors:** 5

Citroen Xsara Picasso 2000 - 2010

Egg shaped small MPV works well for families with five decent seats. It has plenty going on inside from big door pockets to rear picnic tables, under floor storage and a wardrobe sized glovebox. Decent kit overall, a lot have been neglected now and are quite Shabbitat. MOT level rust failures, brakes, ECUs and oil leaks.

Engines: 1.6 • 1.8 • 1.6D • 2.0D **CO2:** 135-187g/km

MPG: 36.7-51.4 **Gearbox:** Man **Boot:** 549/2127

Daewoo Tacuma 2000 - 2005

Compact people carrier was actually a high point for the Korean carmakers. Five good individual seats, armrests up front, storage trays, rear seat fold forward picnic tray and countless cubbyholes. Plus that swiveling passenger seat. See back Chevy. Cheap. Cheap, Cheap. Check electricals, windows, locks, specially air con. Wear and tear.

Engines: 1.6 • 2.0 **CO2:** 208-236g/km **MPG**: 26.2-34.0

Gearbox: Man/Auto **Boot:** 456/1155 **Seats**: 5 **Doors:** 5

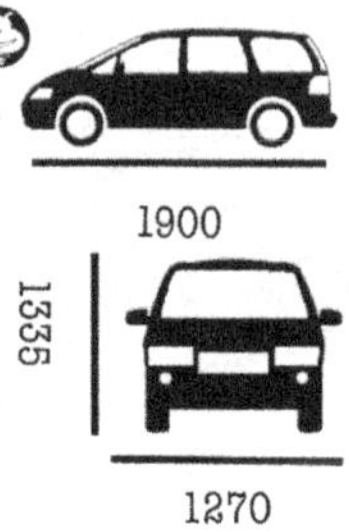

Daihatsu Hijet 1986 - 2004

A real micro van with more seats than are truly comfortable or safe. Considering its size the 1.3 engine is more than enough. Will seat six incredibly. Better enjoyed as a van, tipper or pick-up. Seem to buzz around well beyond their recorded time of death. Enjoy. Rust have killed most. Smoke and oil leaks and general decrepitude.

Engines: 1.3 **CO2:** 159g/km **MPG**: 38.0 **Gearbox:** Man

Boot: 560kg**+** **Seats**: 6 **Doors:** 5

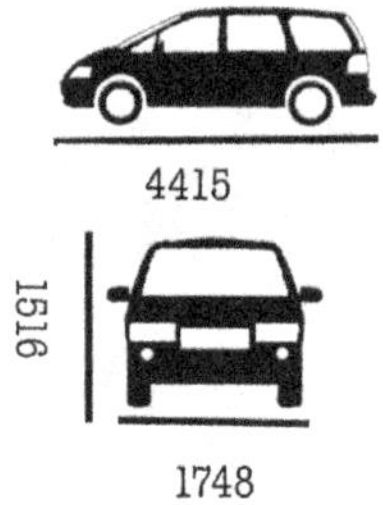

Dodge Caliber Crossover 2006 - 2009

The clue is in the name, it looks like an SUV, but it's actually an MPV. Not that well thrown together but it does have a Volkswagen diesel. Big boot, fold flat passenger seat and you can play banging tunes when parked on the tailgate if it has 'Music Gate'. Who cares? Brake discs, suspension wear, worn clutch, getting parts is a challenge. Check electricals.

Engines: 1.8 • 2.0 • 2.0D **CO2:** 161-192g/km **MPG**: 34.9-46.3

Gearbox: Man/Auto **Boot:** 525/1360 **Seats**: 5 **Doors:** 5

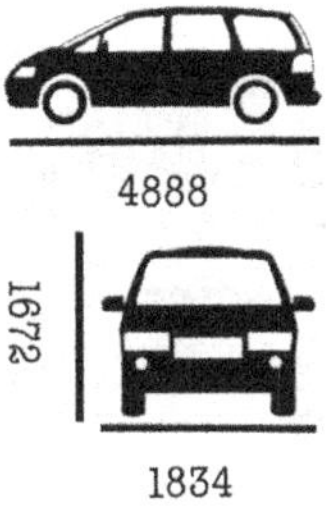

Dodge Journey 2008 - 2010

Chrysler Grand Voyager related, looks a bit more 4 x 4. Choc full of cubbyholes under the floor, air con glovebox, picnic table backs, tilt and slide centre row seats and all that jazz really. Apart from looking more hardcore than the average van, just a fairly noisy box. Must be cheap. Check the brakes, suspension and that the electricals work, air con, locks.

Engines: 2.4 • 2.0D **CO2:** 170-290-g/km **MPG**:32.1-43.5

Gearbox: Man/Auto **Boot:** 167/2117 **Seats**: 5/7 **Doors:** 5

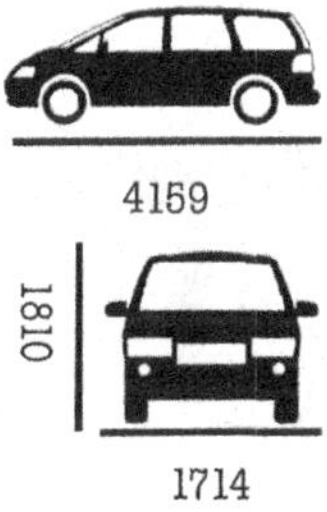

Fiat Doblo 2001 - 2010

So ugly and slow, it is rather lovable. Another real van mostly powered by basic and reliable diesels and one tiny petrol. Has a steel chassis which can rust, but helps make it old fashioned and rock solid if you wanted to muck about with it. Revamped in 2004. Get a JTD for serious work. Engine related ECU issues, worn brakes and suspension.

Engines: 1.2 • 1.4 • 1.3D • 1.9D **CO2:** 119-191g/km **MPG**: 36.7-62.8 **Gearbox:** Man **Boot:** 330/2500 **Seats**: 5 **Doors:** 5

Fiat Multipla 2000 - 2010

MPV designed by a committee is surprisingly successful despite looking like a camel. 3×3 seating is clever and stumpy footprint is useful. Making it look duller in 2004 didn't help anyone. Wide rather than deep boot. Live with it. Rotten exhausts, rotting bodywork, Fiat electrics, windows, central locking. Smoking tired engines.

Engines: 1.6 • 1.9D **CO2:** 205g/km **MPG**: 32.8-44.1 **Gearbox:** Man/Auto **Boot:** 430/1900 **Seats**: 6 **Doors:** 5

Fiat Qubo 2009 - 2020

A.K.A. Fiat Fiorino van. It is though an honest little thing. The 1.3 Multijet diesel is very reliable, enthusiastic and economical delivering a solid 60mpg. Properly compact van plus the seats fold and come out and there's big glove box. It's a modern and very useful box. Seems very tight and reliable, just look for wear and tear, suspension, brakes.

Engines: 1.4 • 1.3D **CO2:** 119-165g/km **MPG**: 40.4-62.8 **Gearbox:** Man/Auto **Boot:** 330/2500 **Seats**: 5 **Doors:** 5

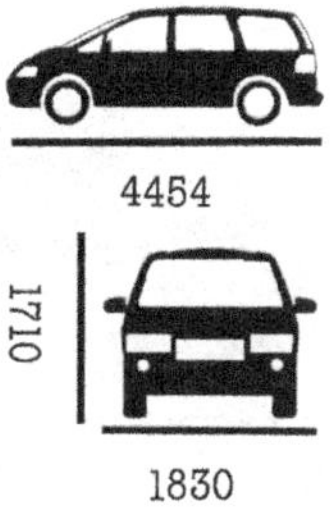

Fiat Ulysse 1995 - 2003

There's a Citroen and Pug version, these were the ones which always ended up as the cheapest in the classifieds. Still the case for any survivors. Seat options are from five to eight and most of them perform lots of swiveling, sliding turning tricks. Remember it is a Fiat. Structural rust, oil leaks, ECU failure, oil leaks and smokey exhaust.

Engines: 2.0•2.0D **CO2:** 182-221g/km **MPG**: 30.7-42.2 **Gearbox:** Man/Auto **Boot:** 340/3313 **Seats**: 5/7 **Doors:** 5

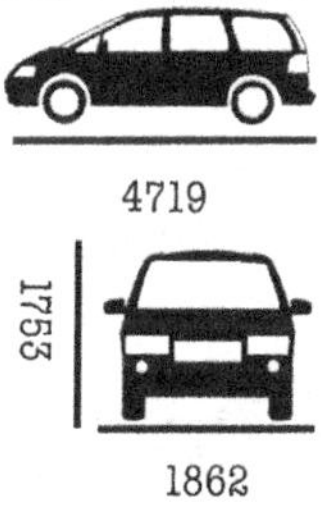

Fiat Ulysse 2003 - 2006

Has a big old air con system to cope with the interior which can go on the blink as might the electric sliding door, if fitted. Removable rear seats and three row ones are all mounted on runners so slide forward and indeed, back. Electrics air con, central locking and ECU issues, also power steering, brakes and suspension.

Engines: 2.0•2.0D•2.2D **CO2:** 186-218g/km **MPG**: 31.0-40.4 **Gearbox:** Man/Auto **Boot:** 324/2948 **Seats**: 5/7

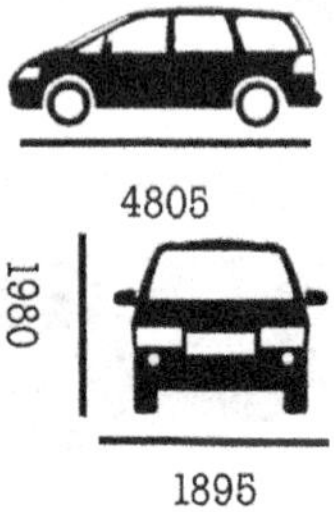

Fiat Scudo Panorama 2007 - 2016

In two Banger minds about featuring a related group of great big vans. This one is fitted with 2.0 JTD multijet. Has five to six seats as standard, but there is an even longer model with eight to nine. These are suitably huge and practical. There's a Citroen and a Peugeot. Oil leaks, worn suspension bushes and possible ECU failure.

Engines: 2.0D **CO2:** 189g/km **MPG**: 39.0

Gearbox: Man/Auto **Boot:** 310/2200 **Seats**: 5-9 **Doors:** 5

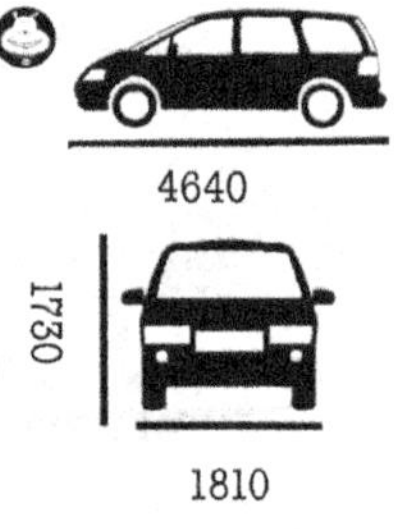

Ford Galaxy 2000 - 2006

The nation's absolutely favourite people shifter. Not the cleverest of all MPVs, just decent to drive, practical and spacious enough. Good engines from the VW diesel to the creamy V6 and Ford 2.3 petrol compromise. Reliability got better. Oil leaks, turbocharger wear. MOT failure structural rust. Corroded brake pipes, air con and gearbox failures.

Engines: 2.3 • 2.8 • 1.9D **CO2:** 176-242g/km **MPG**: 28.0-43.5

Gearbox: Man/Auto **Boot:** 256/2610 **Seats**: 5/7 **Doors:** 5

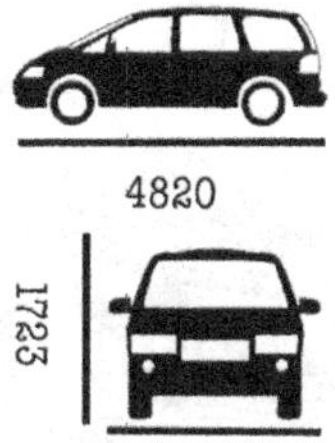

Ford Galaxy 2006 - 2010

Bulkier version of the nearby S-Max. Got the decent TDCI diesels and like previous ones, it is groovy to drive. Inside there was the Ford Fold Flat System and apparently allowed 32 different layouts, both the back rows allow a double bed or load bay. Water leaks in the cabin, air con failure, all electrics: windows, locks and sometimes ECUs.

Engines: 2.0 • 1.8D • 2.0D • 2.2D **CO2:** 165-197g/km **MPG**: 34.4-45.6 **Gearbox:** Man/Auto **Boot:** 308/2325 **Seats**: 5/7

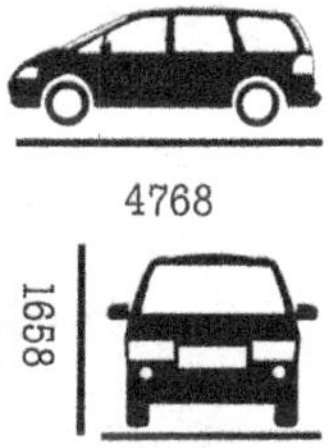

Ford S-Max 2006 - 2014

Underneath is a Mondeo which explains why it is good to drive. Still has seven seats and it still folds flat. There are still something like 26 random cubbyholes. Odd that air con not standard, but most would have paid. Revamped a bit in 2010 with some new engines. water leaks in the cabin, air con failure, all electrics windows, locks and sometimes ECUs.

Engines: 2.0 • 2.5 • 1.8D • 2.0D • 2.2D **CO2:** 164-194g/km **MPG**: 34.8-44.1 **Boot:** 285/2000 **Seats**: 5/7 **Doors:** 5

Honda FRV 2004 - 2009

Front seats three abreast, a real novelty outside of '50s saloons bench seats. Odd certainly and the fact that it would be reliable, didn't convert many buyers. Apparently the folding front seats allow it to do the double bed trick. The only Flexible Recreational Vehicle you really. Not much to worry about, wear and tear, brakes and suspension.

Engines: 1.8 • 2.2 **CO2:** 167-177g/km **MPG**: 37.7-44.8

Gearbox: Man/Auto **Boot:** 439/1049 **Seats**: 6 **Doors:** 5

Honda Stream 2001 - 2005

One of those rare MPVs which is actually OK to drive. Doesn't have proper performing seats but the rear ones do fold flat for the maximum amount of boot. Decent standard kit and a reputation for not having any mechanical sulks. Worth finding. Rust on the old ones, parts are difficult to find, otherwise just wear and tear.

Engines: 1.7 • 2.0 **CO2:** 183-203g/km **MPG**: 36.7-32.8

Gearbox: Man/Auto **Boot:** 158/435 **Seats**: 7 **Doors:** 5

Hyundai i800 2008- 2014

Incredibly no an equivalent van, it was actually designed like this, which is a good thing so it does all the luggage and the people. There are a couple of rear benches that fold and recline and the centre row slides around. That's flexibility on a grand scale. Heavy use by hire companies so lots of wear and tear issues, brakes and suspension.

Engines: 2.5 **CO2:** 225g/km **MPG**: 33.2 **Gearbox:** Man/Auto **Boot:** 851 **Seats**: 8 **Doors:** 5

Hyundai Matrix 2001 - 2010

Designer bodywork, cool model name, doesn't matter if it just takes five. It will though take every shape of cubby as there is a hole for it. Split fold and fold again rear seats, plus they will also slide. Reliable and dead easy to park as it is incredibly compact. Wear and tear issues, just check for oil leaks, the brakes, exhaust and tyres.

Engines: 1.6 • 1.8 • 1.5D **CO2:** 142-205g/km **MPG:** 33.2-52.3 **Gearbox:** Man **Boot:** 350/1284 **Seats**: 5 **Doors:** 5

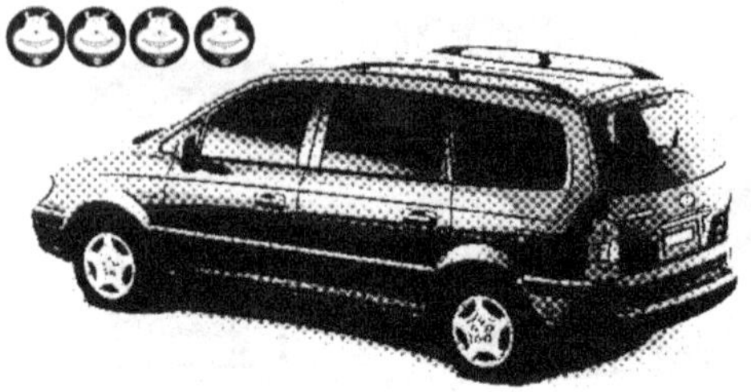

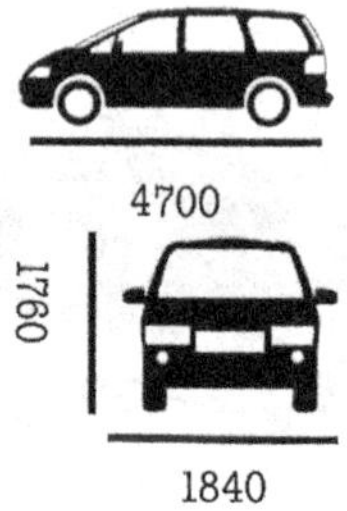

Hyundai Trajet 2000 - 2008

Kia Sedona with a different badge. Has Captain's swiveling chairs, seatback trays and seats can be removed. Cubby boxes by the dozen obviously. Lots of kit whatever the grade and in 2004 it got a new grille and everything. Heavy wear and tear on these, needs brakes and maybe suspension bushes. Look for oil and coolant leaks.

Engines: 2.0 • 2.7 • 2.0D **CO2:** 192-271g/km **MPG**: 25.0-37.2

Gearbox: Man/Auto **Boot:** 524/3237 **Seats**: 7 **Doors:** 5

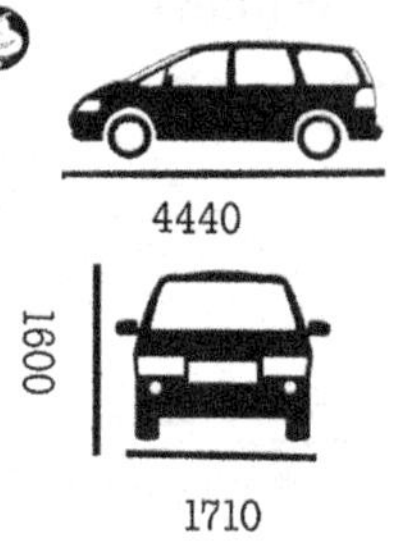

Kia Carens 2000 - 2006

Officially a five, it can squeeze six bodies on board if ordered and they will be sitting in a couple of extra pews in the boot. Rear seat folds, but can't be taken out. Reliability is guaranteed unless really hard used and at the end of its useful life. Another wear and tear issue, may have rust and possibly fluid leaks.

Engines: 2.0 • 2.0D **CO2:** 171-192g/km **MPG**: 34.9-43.5

Gearbox: Man **Boot:** 422/1632 **Seats**: 5 **Doors:** 5

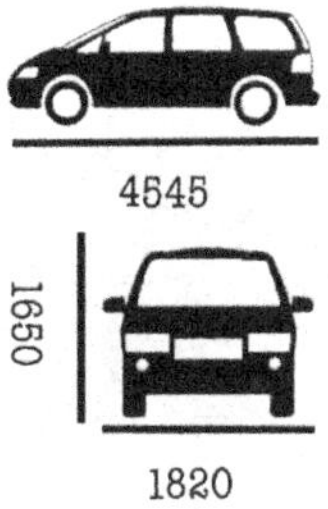

Kia Carens 2006 - 2011

Cutely revamped small people carrier now with added diesel. Available as either a five or seven seater. Big clutch failures . Not a huge amount of recalls but it is a modern 2.0 diesel so needs to be more carefully checked than the old petrol. Blocked particulate filters, worn suspension and brakes otherwise just short haul family journey damage.

Engines: 2.0 • 2.0D **CO2:** 165-201g/km **MPG**: 33.6-45.6

Gearbox: Man **Boot:** 430/2106 **Seats**: 5/7 **Doors:** 5

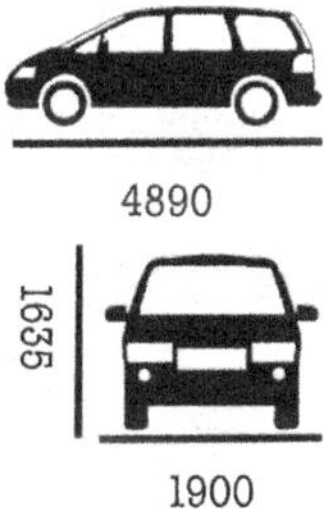

Kia Sedona 1999 - 2006

Effectively the people's Carrier. An awful lot of seats and room for the money. Lots of kit. Captain's swivel chairs in the second row don't just swivel, they slide as well. Bed recliners and double figure cupholders, including rear air con controls. MOT failure level structural rust to floor. Electrics ABD sensors and ECUs.

Engines: 2.5 • 2.9D **CO2:** 217-261g/km **MPG**: 25.7-34.4

Gearbox: Man/Auto **Boot:** 588/3616 **Seats**: 7 **Doors:** 5

Kia Sedona 2006 - 2012

Much more of the same, so not especially clever seats although they still slide and recline like before, plus there is the walk through section in the centre seats, like a plane. Actually smaller than the previous one and the 2.9 diesel does get used as a minicab. Serious wear and tear inside and out, brakes and suspension mostly.

Engines: 2.2D **CO2:** 179g/km **MPG**: 42.2

Gearbox: Man/Auto **Boot:** 851 **Seats**: 7 **Doors:** 5

Mazda 5 2005 - 2010

Compact People Mover that was quite easy to overlook but there was a slight prettification in 2007. Two petrols and diesels and bunch of trim levels. Seats 6 comfortably, and there was room for luggage too, only 7 uncomfortably. Sliding door operation, electrical issues are the only points to check, windows, locks and other gadgets.

Engines: 1.8 • 2.0 • 2.0D **CO2:** 159-187g/km **MPG**:35.8-46.3

Gearbox: Man/Auto **Boot:** 112/857 **Seats**: 7 **Doors:** 5

Mazda MPV 1999 - 2004

Has super clever name. It is just a seven seat bus with sliding doors on both sides of the bodywork, which also contains Mazdareliability, which is more than enough when it is banging about later in life. Lots of standard kit and the usual seatage stuff, sliding folding and removing. Wear and tear now on these, rust on the earliest ones, brake discs and suspension bushes.

Engines: 2.0 • 2.3 • 2.0D **CO2:** 188-237g/km **MPG**: 28-39.8
Gearbox: Man **Boot:** 294/866 **Seats**: 5 **Doors:** 5

Mazda Premacy 1999 - 2004

Takes Five. Those seats do all the usual, recline, fold flat, hinge forward. Good to drive, if that really matters. All that matters is Mazdareliability. The standard specification is pretty good, if you want any more than the GSi delivers. This is a bit like a Ford Galaxy, but less useful. Won't break down though, light rust, wear and tear.

Engines: 1.8 • 2.0 • 2.0D **CO2:** 177-209g/km **MPG**: 33.6-49.6
Gearbox: Man/Auto **Boot:** 371/719 **Seats**: 5 **Doors:** 5

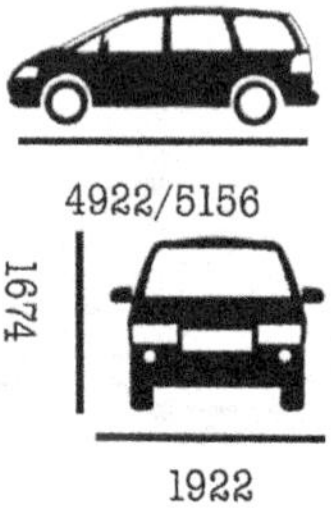

Mercedes R-Class 2006 - 2012

No more than a pregnant E-Class estate. Available as normal and L for longer, the former was gradually phased out as pointless. It is quite a complicated package with all-wheel drive. Costly to keep. Electricals are what slows it down, ECUs, oil and coolant leaks and pot hole damage to suspension plus worn bushes and brakes.

Engines: 5.5•3.0D **CO2:** 246-317g/km **MPG**: 21.2-30.4
Gearbox: Auto **Boot:** 212/2057 **Seats**: 5 **Doors:** 5

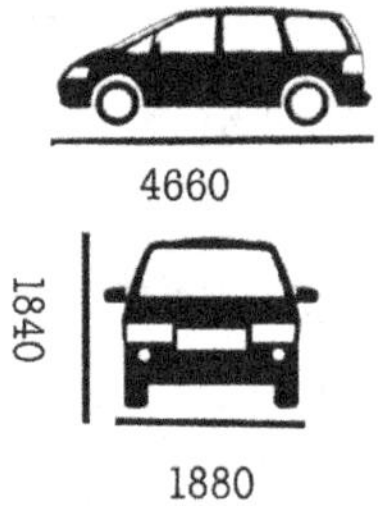

Mercedes V-Class 1996 - 2002

It is one of those jolly useful Vito vans. Whopping sliding doors with S luxury Class touches. Here is a utility vehicle, for the posh people. Diesel or petrol, depends how well off the buyer is. Ones that are used by the hire industry might be tired, also pretty rusty by now. Suspension and brakes have to be checked also tired diesels, leaking fluids etc.

Engines: 2.3•2.8•2.2D **CO2:** 199-330g/km **MPG**: 20.5-35.5
Gearbox: Auto **Boot:** 580/4559 **Seats**: 6 **Doors:** 5

Mercedes Vaneo 2002 - 2005

Van by name. Contains very heavy seats that you may not want to hoof out. Passenger seat comes out for awkward loads but really the fact it will take five and luggage is a good thing, unless you want the seven seat. More useful than the hate mob would have you believe. Electrical faults with windows, locks and air con. Surface to serious rust.

Engines: 1.6 • 1.7D • 1.9D **CO2:** 157-197g/km **MPG**: 34.6-48.1 **Gearbox:** Man/Auto **Boot:** 150/3000 **Seats**: 5

Mercedes Viano 2004 - 2015

Civilised V-Class, still quite commercial. Has a rail system for the seats, so they slide around individually. Not only that, they fold flat or tilt as required. Easy to live with CDis. When it comes to fixing, van parts are genuinely affordable. Plenty of recalls like most modern vehicles, so needs history. Exhaust can crack, diesel ECU can fail, look for fluid leaks.

Engines: 3.5 • 2.0D • 2.2D • 3.0D **CO2:** 254-309g/km **MPG**: 21.9-32.8 **Gearbox:** Auto **Boot:** 730 **Seats**: 6/8 **Doors:** 5

Mitsubishi Grandis 2004 - 2010

An almost handsome MPV, which is long, low and narrow. Never mind, the seats are intelligent so each one can be folded quickly into the floor. Oddly the third row of seats can be swiveled to face out the rear window. Reasonable amount of standard kit. Obviously reliable. Wear and tear only with these rarities, may need suspension bushes and brake discs.

Engines: 2.4 • 2.0D **CO2:** 176-223g/km **MPG**: 30.1-42.8

Gearbox: Man/Auto **Boot:** 320/1545 **Seats**: 7 **Doors:** 5

Mitsubishi Space Wagon 1999 - 2004

Third incarnation is larger, but still more estate than van like. Seven seats in three rows and that second row is an old fashioned bench. Top of the range GLS has 6 separate seats allowing passengers to walk through. Folding and removing are also options. Yes its all about space. Suffers from rust, wear and tear, check exhaust and brakes..

Engines: 2.0 • 2.4 **CO2:**225-229g/km **MPG**: 29.7-28.0

Gearbox: Auto **Boot:** 241/1931 **Seats**: 7 **Doors:** 5

Nissan Almera Tino 2000 - 2005

Pumped up hatch with seemingly not very much to recommend it. Well actually it is something of a space creator. Three individual rear seats which can be removed or folded and they slide too. There's hooks, hammocks and multi-level parcel shelf. Almost interesting. Engine sensors,ECUs causing running issues, some rust.

Engines: 1.8 • 2.0 • 2.2D **CO2:** 168-223g/km **MPG**: 30.4-44.1

Gearbox: Man/Auto **Boot:** 439/850 **Seats**: 5 **Doors:** 5

Nissan Serena 1993 - 2001

It was a narrow Japanese market van with added seats. Quite ancient tech, but incredibly there are 8 seat versions. Seats can be removed and moved about, which is no longer a novelty. Old school Nissaness means that it they are really tough. Rot kills them eventually. Some big door mirror imports around. Rust, worn interior/exterior.

Engines: 1.6 • 2.0 **CO2:** 244g/km **MPG**: 25.9-28.2

Gearbox: Man/Auto **Boot:** 350/1900 **Seats**: 5/8 **Doors:** 5

Peugeot 3008 2009 - 2016

Modern take on the MPV, which is also SUV, hatchback and so possibly a crossover family car. Colossal boot and more than enough room for everyone. Petrol and diesel engines of course, mostly this is a very complex motor to be a Banger. Useful though. Has sticky brake calipers, electrics and ECU issues, needs to be plugged into a diagnostics machine.

Engines: 1.6 • 1.6D **CO2:** 137-176g/km **MPG**: 38.2-55.4

Gearbox: Man/Auto **Boot:** 432/1604 **Seats**: 5 **Doors:** 5

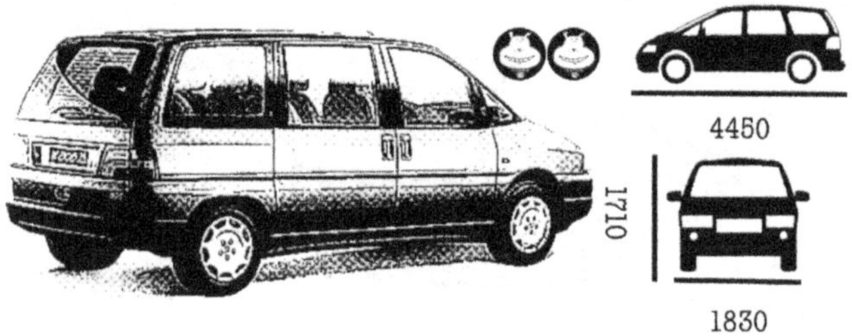

Peugeot 806 1995 - 2002

Familiar because it is also a Citroen and a Fiat. This Pug was also available with 5 or 7 seats. It is just a big old fashioned box, but still pretty useful. The seats do the usual, so that's recline, swivel fold and remove. These aren't looked after and can be quite the worse for wear. Rust, leaky sunroofs, ECUs, smoky diesels, most things really.

Engines: 2.0 • 2.0D **CO2:** 182-221g/km **MPG**: 30.7-42.2

Gearbox: Man/Auto **Boot:** 340/3313 **Seats**: 7 **Doors:** 5

Peugeot 807 2002 - 2010

See also Citroen C8 and Fiat Ulysse. Regularly upgraded which meant it became progressively more complicated with added electronics. From '05 seats more easy to pop out. HDi diesels are the pick of the engines. Electronics sticky, windows, locks and sliding doors. Power steering failure. Faulty ECUs and air con.

Engines: 2.0•2.0D•2.2D **CO2:** 182-213g/km **MPG**: 31.4-40.9 **Gearbox:** Man/Auto **Boot:** 324/2948 **Seats**: 7

Peugeot Bipper Tepee 2009 - 2014

Another one of those clones, also a Fiat Qubo and Citroen Nemo. Compact MPV which is all most families will ever actually need. Goes back to van if you take the rear seats out and the sliding doors are fantastically convenient. Worth a punt on this Pug. Seems very tight and reliable, just look for wear and tear, suspension, brakes.

Engines: 1.4•1.4D **CO2:** 119-155g/km **MPG**: 42.8-62.8 **Gearbox:** Man/Auto **Boot:** 356/884 **Seats**: 5 **Doors:** 5

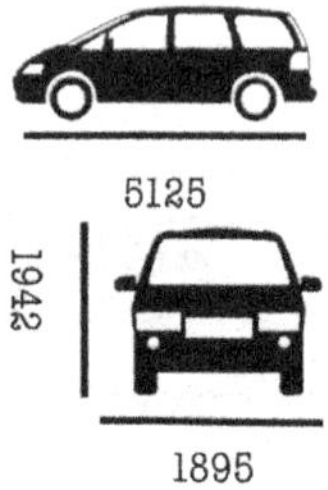

Peugeot Expert Tepee 2007 - 2015

Large scale people mover with a van aesthetic and related to the Citroen Dispatch Combi and Fiat Scudo Panorama. Largely a van rather than luxury spec. Great for the private hire market a proper Saturday Night Special. Up to nine flipping seats. Oil leaks, worn suspension bushes and possible ECU failure.

Engines: 1.6•2.0 **CO2:** 168-194g/km **MPG**: 37.0-42.0
Gearbox: Man/Auto **Boot:** Huge unless seats **Seats**: 6/9

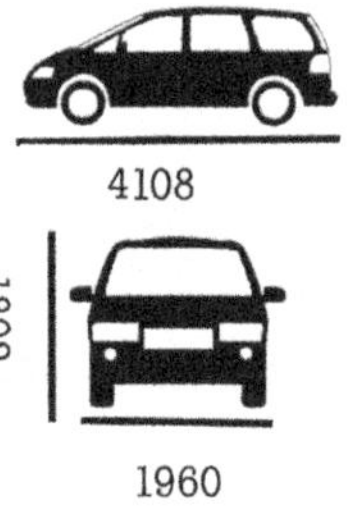

Peugeot Partner Combi 2001 - 2010

Partner van related and also Pug's version of the Multispace. So it is another alternatively badged handily sized versatile vehicle. The seats can be taken out for the full van experience and there is under floor storage as well. That's all there is to it. Bigger Tepee in 2008. Suspension repairs can be expensive, cabin and under bonnet water leaks, ECU malfunction. Complicated wiring. Rust.

Engines: 1.4•1.9D•2.0D **CO2:** 152-181g/km **MPG**: 37.7-49.6
Gearbox: Man **Boot:** 624/2800 **Seats**: 5 **Doors:** 5

Renault Espace 1997 - 2003

MPV royalty. Sophisticated third version spacious and versatile, with decent standard kit. Got the removable and folding seats and the higher spec RXEs have rail system for the seats to slide on. It's a Renault, so the electrics can be flakey especially if an electric handbrake is fitted. Clutches can have a short life, smokey diesels and failing ECUs.

Engines: 2.0 • 3.0 • 2.2D • **CO2:** 189-273g/km **MPG**: 24.4-39.8 **Gearbox:** Man/Auto **Boot:** 600/2600 **Seats**: 7

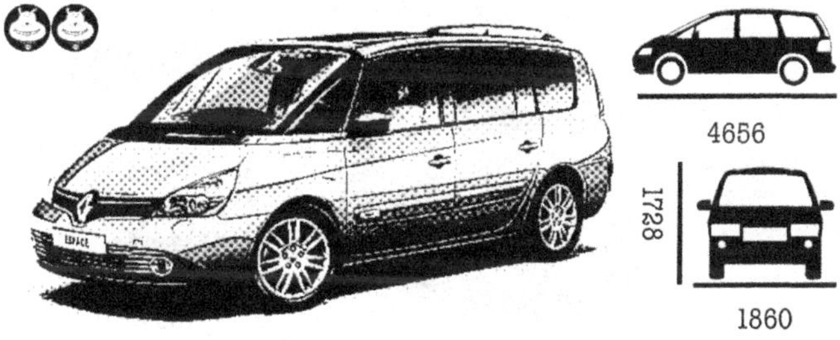

Renault Espace 2003 - 2010

More reassurance for the middle class school runners before they switched to SUVs. Branded as 'New', but not really, still complicated aluminum, plastic and steel construction. Revamped in '05 and '06. Problems usually electrics, meaning all the gadgets and if fitted, clutches which can have a short life, smokey diesels and failing ECUs.

Engines: 2.0 • 3.5 • 2.0D • 3.5D **CO2:** 196-227g/km **MPG**: 29.4-38.2 **Gearbox:** Man/Auto **Boot:** 456/2860 **Seats**: 6/7

Renault Grand Espace 2003 - 2012

Grander version of the hip original so there really is hectares of spaces inside. Quite hard to see the real advantages are apart from a bit more luggage space. Seats may fold, but they don't disappear into the floor. Loadsa bins storage solutions, just more space for used money. Faults as standard issue No Grand Espace, so it's electrics mostly.

Engines: 2.0 • 3.5 • 2.0D • 3.5D **CO2:** 196-227g/km **MPG**: 29.4-38.2 **Gearbox:** Man/Auto **Boot:** 456/3050 **Seats**: 6/7

Renault Scenic 1999 - 2003

Not just a Megane anymore, now its own life form. Essentially an overgrown hatch with seats on rails, air con extended to glove box and posh models had spilt tailgate as well. A ton of trim levels and engines, Renault fragility and electrics can spoil it all. Sensors and various ECUs, coil packs, clutch, heater matrix, rust.

Engines: 1.6 • 1.8 • 2.0 • 1.9D **CO2:** 165-224g/km **MPG**: 29.7-49.6 **Gearbox:** Man/Auto **Boot:** 411/1798 **Seats**: 5

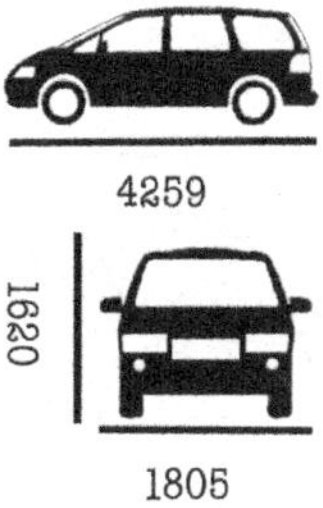

Renault Scenic 2003 - 2009

Looks like a fat Megane and it is much bigger than before. Seats can be moved about, but not as versatile as some. Reputation for safety plus point for buyers. This Scenic II was revamped inside and out in 2006. Seemed a bit better built than before. Air con problems, diesel issues with glow plugs and EGR valves, coil packs and the usual electrics.

Engines: 1.4 • 1.6 • 2.0 • 1.5D • 1.9D • 2.0D **CO2:** 138-182g/km **MPG**: 37.2-54.3 **Gearbox:** Man/Auto **Boot:** 430/580 **Seats**: 5

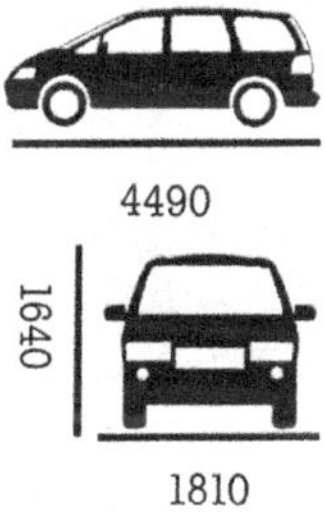

Renault Grand Scenic 2004 - 2009

Baby Espace which seats seven, although a five seater also became available, plus that freed up a load more underfloor storage. Essentially, this is a conventional MPV and the two extra seats folded into the floor, like other carriers. .Air con problems, diesel issues with glow plugs and EGR valves, coil packs and the usual electrics.

Engines: 1.4 • 1.6 • 2.0 • 1.5D • 1.9D • 2.0D **CO2:** 138-182g/km **MPG**: 37.2-54.3 **Boot:** 200/605 **Seats**: 5/7

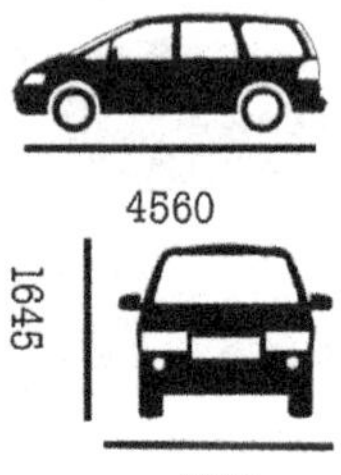

Renault Scenic 2009 - 2016

Usual performing rear seats which slide back and forth, but as before the seats stay bolted to the floor rather than fold into it. Just more of the same old Scenic then. Cubbyholes, fridges and underfloor stowage options. Plus if you want even more space there is still a Grand version. Check all the electrics, windows, locks and air con.

Engines: 1.4 • 1.6 • 2.0 • 1.5D • 1.9D • 2.0D **CO2:** 145-173g/km **MPG**: 38.7-51.4 **Boot:** 437/1837 **Seats**: 7 **Doors:** 5

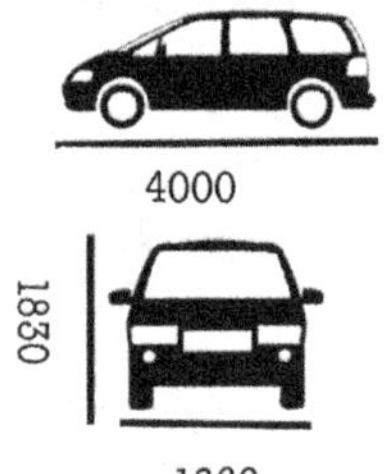

Renault Kangoo 1999 - 2008

Vannish rural France appeal in the mould of ancient Renault 4s. No performing or luxury interior, it's a hose down affair. The rear seat pivots forwards so there is instant load space. Lots of extras available to fancy it up and many buyers did just that. Basic engine options, which is good. Electrics from the really basics like wipers, ignition coil packs, automatics can fail. EGR valves regular Renault soft spot.

Engines: 1.2 • 1.6 • 1.5D **CO2:** 139-190g/km **MPG**: 35.3-53.3 **Gearbox:** Man/Autos **Boot:** 600/2600 **Seats**: 5 **Doors:** 5

Renault Kangoo 2009 - 2012

No van underneath the new one, it's a Scenic in overalls. More engine variety, more modern, not necessarily easier to fix. Seats five with a ton of luggage space and cubbyhole options. Multi height parcel shelf. All the family friendly safety kit. Still problem electrics, so from windows and locks to the cockpit, failing ECUs as well as cosmetic damage.

Engines: 1.6 • 1.5D **CO2:** 140-190g/km **MPG**: 35.3-56.5
Gearbox: Man/Auto **Boot:** 660/2866 **Seats**: 5 **Doors:** 5

SEAT Alhambra 2000 - 2010

Wasn't a Ford or Volkswagen which came with the better VW engines. Folding and removable seats if you can manage to hoof the heavy old things out. Front ones on some models do the Captain swivelling thing. Pretty tough overall. Oil leaks, turbocharger wear. MOT failure structural rust. Corroded brake pipes, air con and gearbox failures.

Engines: 2.0 • 2.8 • 1.9D **CO2:** 176-259g/km **MPG**: 26.2-43.5
Gearbox: Man/Auto **Boot:** 259/2610 **Seats**: 7 **Doors:** 5

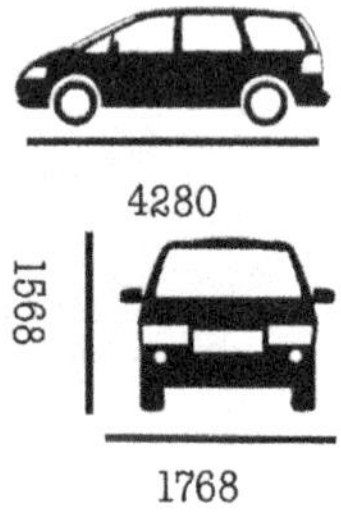

SEAT Altea XL 2004 - 2015

Billed as a sporty MPV, compared to the non XL it has loads of headroom, more than any other similar vehicle, if that actually matters. The rear seats split, fold and slide about. It is also full of bins, cupholders and a massive glovebox. It is sporty though. Timing belt tensioner failure, electrics specifically cruise control, ABS sensors and ECU.

Engines: 1.4 • 1.6 • 2.0 • 1.6D • 1.9D • 2.0D **CO2:** 135-187g/km **MPG**: 36.2-56.5 **Boot:** 409/1320 **Seats**: 5 **Doors:** 5

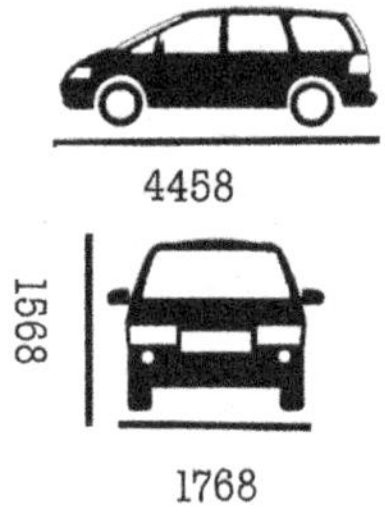

SEAT Toledo 2005 - 2009

Previously a big booted saloon, repurposed into an MPV. It may be a smaller people carrier, but it has a two level boot and the luggage area is huge. You can muck about with the parcel shelf and move it around, which is useful. Electrics can go on the blink, windows, locks, tyre pressure sensors. ABS ECU fail, more serious/expensive dual mass flywheels.

Engines: 1.9D • 2.0D **CO2:** 146-157g/km **MPG**: 47.9-51.4 **Gearbox:** Man/Auto **Boot:** 500 **Seats**: 5 **Doors:** 5

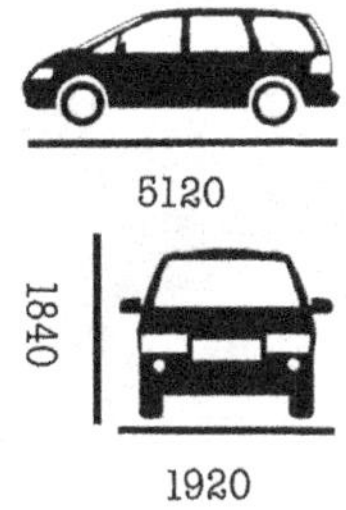

SsangYong Rodius 2005 - 2013

Ugliest MPV in all humanity. Nobody cares because all they want is space and reliability. A ton of standard kit, rear bench also weighs a ton whereas the second row is a fixture. Does the double bed thing, if that is important. Solid bus which will only suffer from wear and tear items brake discs, suspension parts, but also some ECU issues.

Engines: 2.7D **CO2:** 223g/km **MPG**: 32.1 **Gearbox:** Auto **Boot:** 875/3043 **Seats**: 7 **Doors:** 5

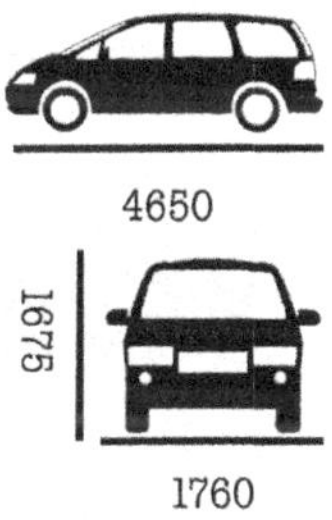

Toyota Avensis Verso 2001 - 2008

Not that related to the hatch/saloon, it is low, long, will seat seven and won't breakdown. There's an identical Corolla version. The third row of seats are fixed whilst the second are on rails and can be folded/removed. In theory a small compact MPV, actually a normal size one that's overlooked. Clutch and gearbox failures over time, brake fluid leaks.

Engines: 1.6 • 1.8 • 2.2D **CO2:** 167-184g/km **MPG**: 36.7-44.8 **Gearbox:** Man/Auto **Boot:** 63/779 **Seats**: 7 **Doors:** 5

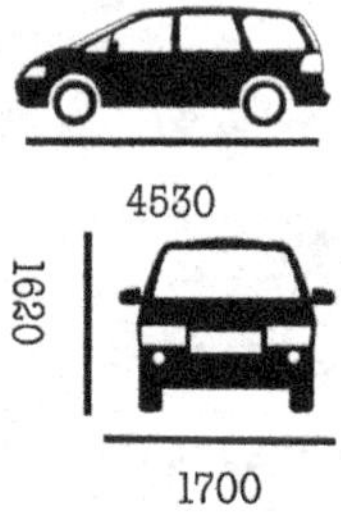

Toyota Picnic 1997 - 2001

Great name, actually called a Family Fun Vehicle by the manufacturer with funky two tone paintwork that seated six inside its compact dimensions. Optional seven seats as was dual sunroofs. Simply buy because it is a Toyota rather than a packaging masterclass. And because of the name. Old age issues, so rust, wear and tear to suspension and brake parts.

Engines: 2.0 • 2.2D **CO2:** 226g/km **MPG**: 29.1-36.1

Gearbox: Man **Boot:** 182/1841 **Seats**: 6 **Doors:** 5

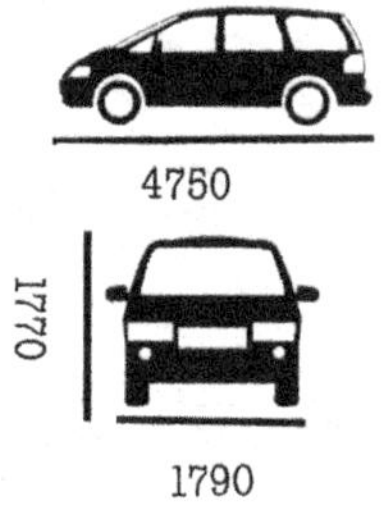

Toyota Previa 2000 - 2007

Supersized MPV which seats up to eight and will just keep on working for the family. Diesel from 2002. Reclining and sliding rear seats. Few take out the seats unless relocated to the lounge. Double bed option. Mostly this is the best big family MPV. Wear and tear mostly, so it will be tyres, suspension bushes and brake parts. Smokey tired diesels.

Engines: 2.4 • 2.0D **CO2:** 191-226g/km **MPG**: 30.1-39.2

Gearbox: Man/Auto **Boot:** 516/2929 **Seats**: 7 **Doors:** 5

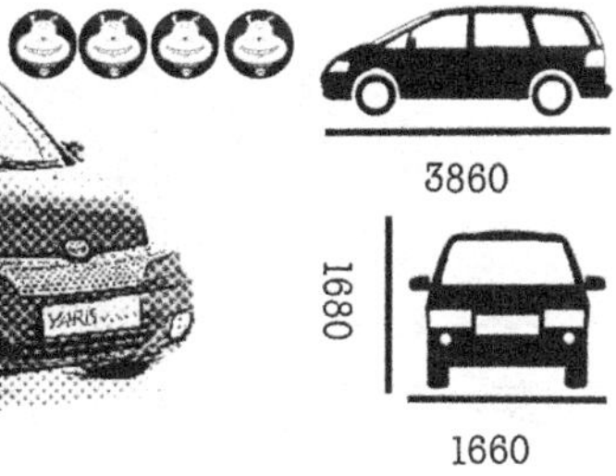

Toyota Yaris Verso 2000 - 2005

Micro MPV van type thing, but still a squeeze when five up, so better as a shopper than people shifter. Has overhead storage plus cupholders and great big glove boxes. As usual, worth buying because it is an unbreakable Toyota with optional roof bars. Automatic gearbox can fail, early model head gaskets. Overall take a lot of abuse.

Engines: 1.3 • 1.4D **CO2:** 133-154g/km **MPG**: 44.1-56.5

Gearbox: Man **Boot:** 390/2160 **Seats**: 5 **Doors:** 5

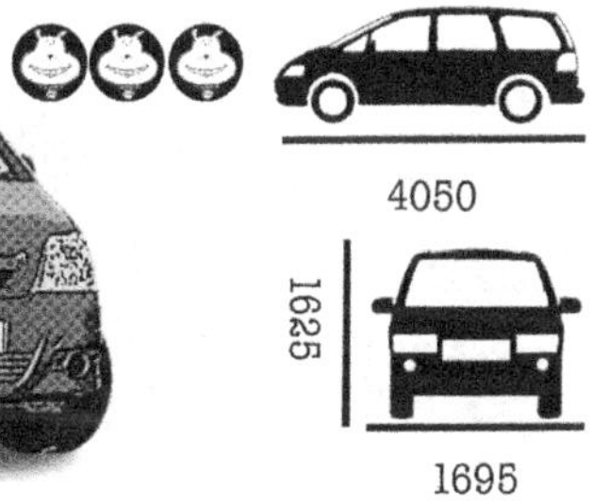

Vauxhall Meriva 2003 - 2010

Corsa based little bus. But with a 60/40 split-folding rear seat FlexSpace centre seat can fold flat into the footwell, leaving the two remaining seats to slide independently, backwards/forwards, also on runners side to side. 2006 facelift reprofiled bumpers some chrome and two new engines. Electric power steering problems, suspension bolts.

Engines: 1.4 • 1.6 • 1.8 • 1.3D • 1.7D **CO2:** 134-161g/km **MPG**: 36.2-53.3 **0-60:** 10.3-13.4 **Boot:** 397/1496 **Seats**: 5 **Doors:** 5

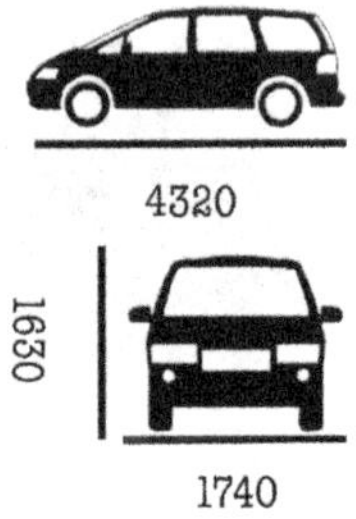

Vauxhall Zafira 1999 - 2005

Compact MPV for the masses which is an Astra with seven seats. Middle seats slide and the rearmost fold into the floor. That does create a massive amount of load space if required. A hard working and flexible people shifter that isn't a pig to drive or park. Water pump failures, seized brakes, water leaks into cabin and ECU irregularity.

Engines: 1.6 • 1.8 • 2.2 • 2.0 • 2.0D **CO2:** 175-243g/km

MPG: 27.9-43.5 **Boot:** 150/1699 **Seats**: 7

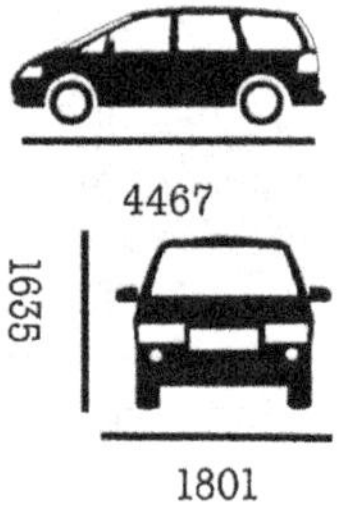

Vauxhall Zafira 2005 - 2014

Seven seats, three in the middle & fold flat rears, +more legroom. Decently bigger, so rather more useful. Loads around and the most serious problem was spontaneous combustion. Lots of recall action around these. Poor electrical performance when it comes to windows, lights and locks. Smokey turbos and dual mass flywheels.

Engines: 1.6 • 1.8 • 2.0 • 2.2 • 1.9D **CO2:** 167-197g/km **MPG**: 34.4-45.5 **Gearbox:** Man/Auto **Boot:** 140/1820 **Seats**: 7

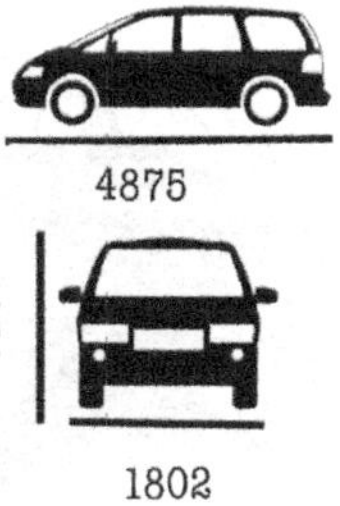

Volkswagen Caddy mx life 2008 - 2010

Looks like a van, because it was, so not that spectacular in the seat department (rearmost ones weigh a ton), but it has loads of space a couple of sliding doors and a VW badge. Van rather than lifestyle ambience. Never less than useful and these seem to be reliable, just don't use diesels for short runs. Check electrics, windows, locks and cockpit dials.

Engines: 1.9D • 2.0D **CO2:** 164-167g/km **MPG**: 44.8

Gearbox: Man/Auto **Boot:** 530-918 **Seats**: 5/7**Doors:** 5

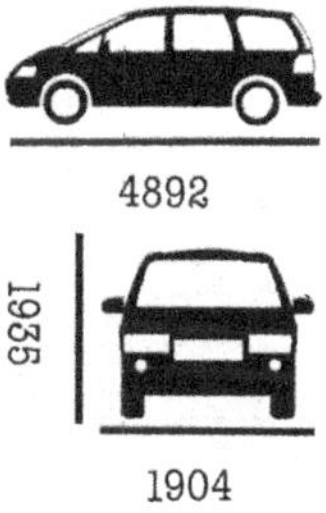

Volkswagen California 2005 - 2015

A motorhome, but cooler. Torn over whether to include a sleeper, it does though have a pop up roof and if times are really hard you could live in it comfortably. A four seater, but a two sleeper. Two folding seats in the tailgate. Might have an awning. Expensive new so should be looked after, ECU issues, smokey diesels and coolant/oil leaks.

Engines: 2.5D **CO2:** 219-232g/km **MPG**: 21.0-34.9 **Gearbox:** Man/Auto **Boot:** 580 **Seats**: 5-6 **Doors:** 5

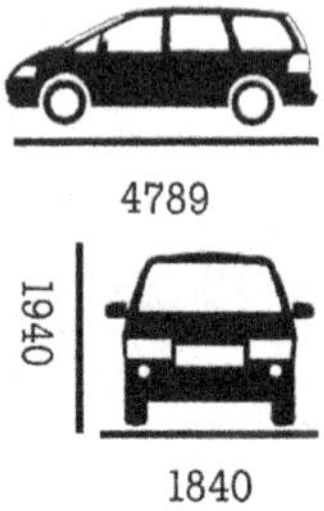

Volkswagen Caravelle 2003 - 2015

The XXL people carrier of choice for a family in search of the most square footage. It isn't all about seatage this is the posh van, Business and Executive specifications with fridges and walnut trimmings. Option of short and long versions. Not cheap. Expensive new so should be looked after, ECU issues, smokey diesels and coolant/oil leaks.

Engines: 2.0D **CO2:** 199-203g/km **MPG**: 36.7-37.2
Gearbox: Man/Auto **Boot:** 5474 **Seats**: 5/7**Doors:** 5

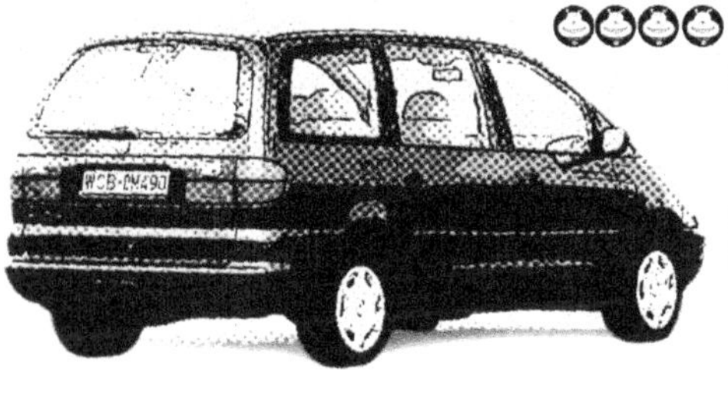

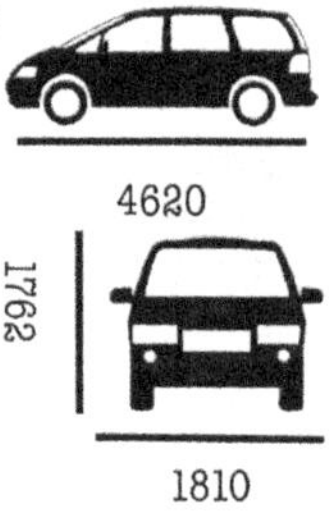

Volkswagen Sharan 1995 - 2000

Identikit MPV (Galaxy/Alhambra) VW has Passat 2.0 engine and usual turbodiesels. What drivers like is that it is like a car to steer. V6 Synchro is a pointless laugh. Seats do the fold down and pull out tricks. Five or seven seats. Often worn out. Oil leaks, turbocharger wear. MOT failure structural rust. Corroded brake pipes, air con and gearbox failures.

Engines: 1.8 • 2.0 • 2.8 • 1.9D **CO2:** 178-278g/km **MPG**: 24.4-42.8 **Gearbox:** Man/Auto **Boot:** 255/2608 **Seats**: 5/7

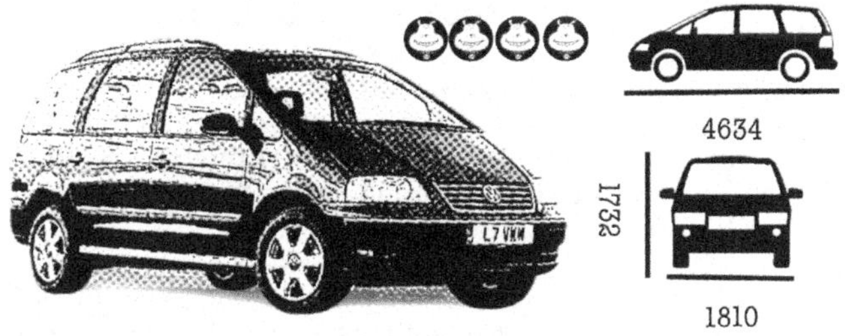

Volkswagen Sharan 2000 - 2010

Seven seats standard. Engine upgrades over the years the Pump Duse diesels are great and the BlueMotion in 2008 delivered the most mpg. Swiveling front seats and endless cubbyholes and cupholders. The go to Family MPV with a badge. Oil leaks, turbocharger wear. Air con and gearbox failures, ECU failures, brake parts may be needed.

Engines: 1.8.•2.0•1.9D•2.0D **CO2:** 181-184g/km **MPG**: 41.5-42.2 **Boot:** 852/2610 **Seats**: 7

Volkswagen Touran 2003 - 2010

Mini MPV, with five or seven seats and pretty decent to drive. Seat 6 and 7 are for nippers, but at least they fold flat into the floor. The middle rear folds away to make a spacious four seater and those also slide individually. So an adaptable little thing with a middle class badge. Electrics from windows, locking and cockpit, engine ECUs and the ABS.

Engines: 1.4•1.6•1.9D•2.0D **CO2:** 162-182g/km

MPG: 37.2-47.1 **Boot:** 695/1989 **Seats**: 5/7

Banger Utility Vehicles - 4x4s & Beyond

Not sure anyone saw this coming, effectively the democratisation of the Range Rover and civilizing the Defender to a stupid degree. Very few drivers need any 4 x 4 ability as so few know how to use it anyway. A Rover Streetwise, or the Mantra Rancho is as far as the whole Sports Futility Vehicles thing needed to go.

But hey, it happened and we need to deal with it. Just look at the colossal number of models here. There's millions of them. The good news is that most have not been used down on the farm, or the building site.

All of a sudden there were dozens of Germans to choose from, wanna be Range Rovers that would more comfy and reliable if not more able. The Quattro Audis went from the estate based Allroad to the ever expanding range of Q badged SUVs. BMW proved the concept with e X5 and then came up with smaller knock offs that were just as popular and almost as impressive. Mercedes were not to be left behind with the M and GL building up a following.

The sensible, not the fashionable money went East. Toyota delivered reliability in abundance, especially the early model Rav 4 and of course anything with the Landcruiser badge will last forever, if not longer. Otherwise there are always the Koreans who supplied practical options in absolute spades. Whether you go for a Huyndai Santa Fe, the Tuscon, or the Kia Sportage version of the latter they deliver a reliable, affordable experience.

Land Rover Discoveries had their place, but really they didn't make a lot of sense if you wanted to complete a journey. Far better to get a Shogun, or any of the overlooked Sport variants.

There is of course a genuine demand for a vehicle that can do a job, whether it is towing, or something else where extra traction is needed. Caution is always needed, but it is worth remembering that it can cost a bomb to sort out a 4×4 that's gone wrong. Otherwise there is lots to enjoy. Potentially if you choose the right sub commercial vehicle, or Toyota, it could last you the rest of your motoring days. Have fun.

Audi Allroad 2000 - 2005

An A6 Audi Quattro on stack heels plus body appendages to make the point that this is an offroader. Equipped with V6 turbo petrol and diesel engines. Not a great deal happened over five years apart from a wonderfully overpowered V8 in 2003. Audi build quality combined with equivalent repair costs especially for the adaptive air suspension.

Engines: 2.7 • 2.5D **CO2:** 240-307g/km **MPG**: 22.1-31.7

Gearbox: Man/Auto **Boot:** 456/1591 **Towing:** 2300

Audi Allroad 2006 - 2011

Gentrified A6 didn't look so much 'let's off all road' anymore. Still did the serious muddy business though and the new one bigger all round. More diesel and petrols to choose from and of course a lot of extra technology and aluminum suspension parts that will cost a fortune to fix. Needs a caring previous owner to make buying sense.

Engines: 3.7 • 4.2 • 2.7 • 3.0 **CO2:** 221-322g/km

MPG: 25.2.33.6 **Boot:** 565/1660 **Towing:** 2100

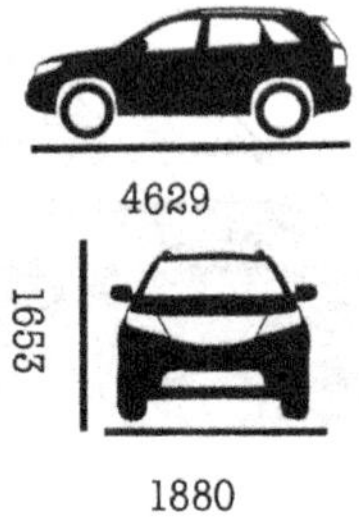

Audi Q5 2008 - 2016

Almost sensibly sized Quattro equipped luxoSUV with a 2.0-litre TDI diesel and 2.0 TFSI petrol as the easy way into ownership although a thumping six-cylinder 3.0-litre TDI diesel was an option. SE added loads of kit and S-Line models some cosmetic style. Fully into the complicated era now. Worn ball joints, water pump failure, engine ECUs.

Engines: 2.0 • 3.2 • 2.0D • 3.0D **CO2:** 175-199g/km

MPG: 33.2-42.1 **Boot:** 540/1560 **Towing:** 1750-2400

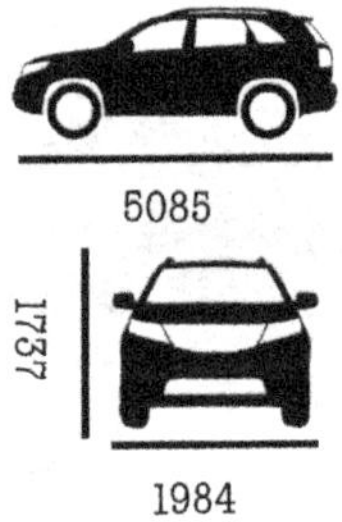

Audi Q7 2006 - 2014

Mammoth dimensioned SUV. Rear seats slide, the back ones fold. Colossal 775 litre boot. Lots of petrol and diesel engine options, but not one that made it less stupid. Upgrades in 2009 meant minor restyle and desperate fuel saving tech, such as kinetic energy recovery and envirofriendly AdBlue injection. So don't bother. Damaged ECUs and electrics.,

Engines: 3.6 • 4.2 • 3.0D • 4.2 • 6.0 **CO2:** 233-350g/km

MPG: 20.8-27.2 **Boot:** 775/2035 **Towing:** 2050-3500

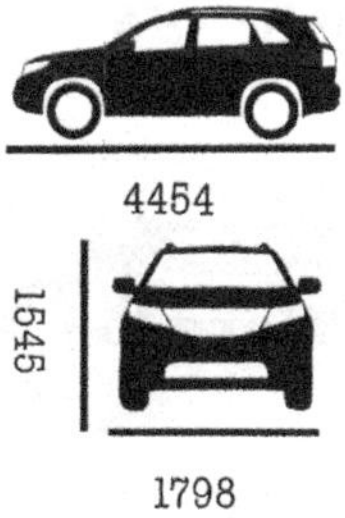

BMW X1 2009 - 2015

Offroad 3 series, unless you got the two wheel drive when it just was a pumped up 3. Even if you did get the xDrive it had marginal off road ability. Pretty decent boot, but otherwise no more roomy for 5 than a saloon and at this point a cheaper Touring looks like the far better option. The 2.0D engine that breaks timing chains still needs to be avoided.

Engines: 1.8D • 2.0D **CO2:** 136-153g/km **MPG**: 48.7-54.3

Gearbox: Man/Auto **Boot:** 420/1350 **Towing:** 1700-2410

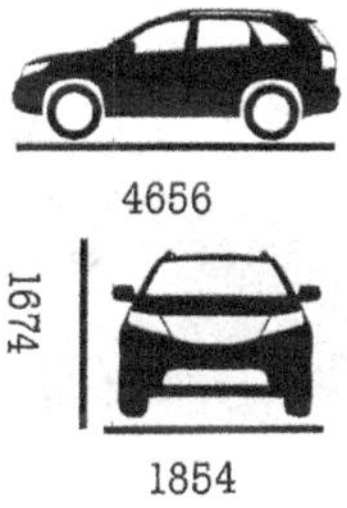

BMW X3 2004 - 2010

Scaled down X5, not as handsome or as good. Had some inappropriate exterior black plastic cladding and seemed to be not as well bolted together. Upgraded with diesel engines and body coloured plastic if anyone really cared. 2006 new engines, EfficientDynamics from 2007 all make them more costly and complicated. Electric upsets and faulty ECUs.

Engines: 2.5 • 3.0 • 2.0D • 3.0D **CO2:** 172-233g/km

MPG: 29.1-43.5 **Boot:** 480/1800 **Towing:** 2000-2600

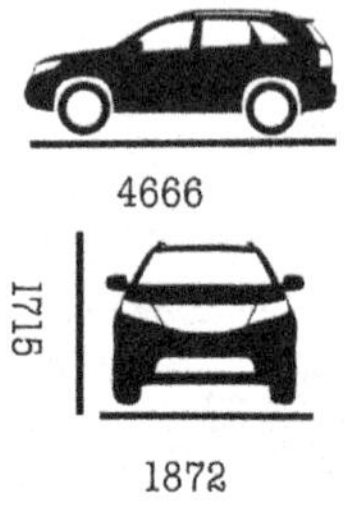

BMW X5 2000 - 2006

A tall 5 Series brilliant from the split rear hatch forward. Engines excellent, the first generation still make a great Bimmer Banger buy. Revamped for 2004 and they benefit from the xDrive system. Normal and Sport models, plus whatever the original buyer specified. Oil and coolant leaks, minor electric upsets, mostly reliable with a 3.0 petrol.

Engines: 3.0 • 4.4 • 4.8 • 3.0D **CO2:** 229-335g/km **MPG**:20.3-32.8 **Gearbox:** Man/Auto **Boot:** 465/1550 **Towing:** 2700+

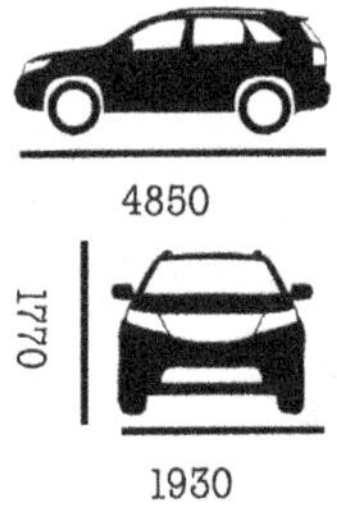

BMW X5 2007 - 2013

Grew very slightly to accommodate an extra row of seats, so a sort of X7 then. Even though the seats are rather marginal. The six cylinder petrols remain as the best option in challenging times and the V8s if you don't care. Owners might be excited by the extra storage options. iDrive is something to be frightened of. Minor electricals.

Engines: 3.0 • 4.8 • 3.0D **CO2:** 214-286g/km **MPG**: 23.5-34.4 **Boot:** 620-1750 **Towing:** 2700

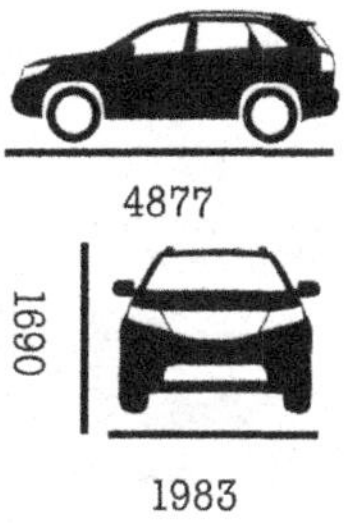

BMW X6 2008 - 2014

A less useful X5. Four seats and despite the bulk, not that spacious, there is a lot of equipment which all goes towards packing it full of things that ultimately will go wrong. Obviously good to drive and great engines, but otherwise over two tonnes of nonsense. Minor electricals, abuse and overmodification.

Engines: 3.0 • 4.4 • 3.0D • 3.5D **CO2:** 217-262g/km

MPG: 25.9-34.4 **Boot:** 580/1530 **Towing:** 3500

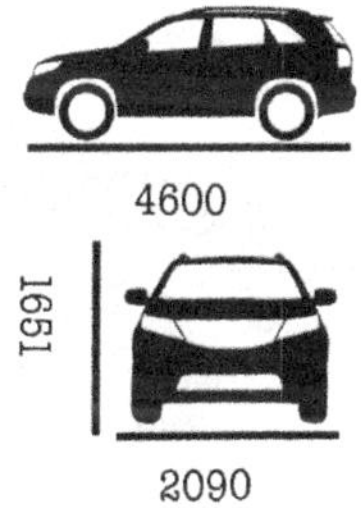

Chevrolet Blazer 1999 - 2002

Very old school 4×4 with quite a soft suspension set up which comes with all the basic luxury spec items which seemed lush in the '90s. Just the five seats and V6 petrol, but you'd buy it as a hack or for towing. Has the proper high and low range gearbox settings. By this stage in its life could be quite corroded. Engine should be fine.

Engines: 4.3 **CO2:** 345g/km **MPG**: 19.1-20.6 **Gearbox:** Man/ Auto **Boot:** 1056 **Towing:** 2041 **Seats**: 5 **Doors:** 5

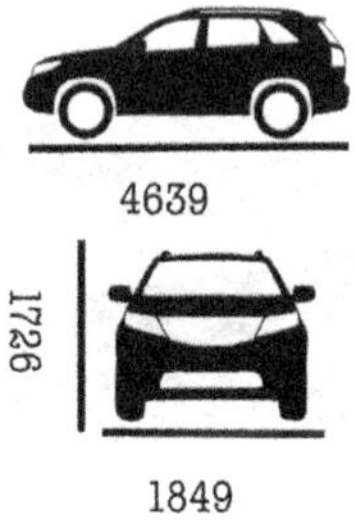

Chevrolet Captiva 2007 - 2014

Mildly more interesting as a Chevy rather than the identical Vauxhall Antara, it is big enough for three rows of seats, but as usual the back ones are best for nippers. The four wheel drive system kick in as and when required and works with the ABS brakes and traction control. Weak clutch, worn suspension and the odd minor electrical.

Engines: 2.4 • 2.0D **CO2:** 195-217g/km **MPG**: 31.7-38.7
Boot: 465/930 **Towing:** 2000 **Seats**: 5/7

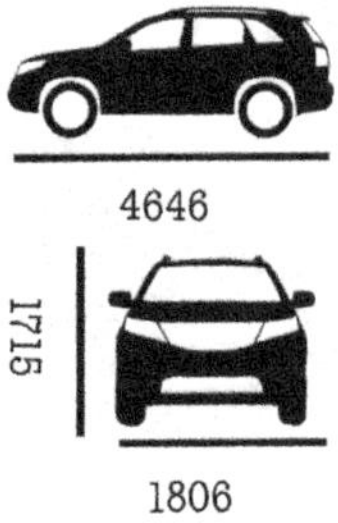

Citroen C-Crosser 2007 - 2012

Cut price Mitsubishi Outlande is a good thing, especially when it comes to reliability and the 2.4 petrol is perfect. Lots of standard items from 2009 including automatic wipers. 5+2 seating attracts many, rearmost seats fold into the floor and the second row slides and reclines. Worth a go. Clutch failure and brake wear.

Engines: 2.2D **CO2:** 191g/km **MPG**:39.2 **Boot:** 184/1686
Towing: 2000 **Seats**: 7 **Doors:** 5

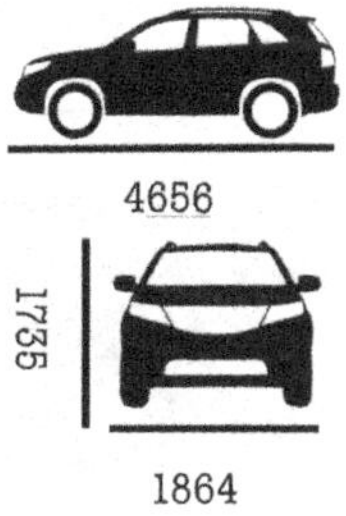

Daewoo Musso 1999 - 2002

Previously home to the tongue twisting SsangYong badge. A quirky Korean which is tough as an everyday tow car with a big boot and more equipment than before. Mercedes related petrol engine and a turbo diesel are both pretty unburstable. Strictly five seats which is fine, just look out for rot and neglect. Fluid leaks and minor electricals.

Engines: 2.3 • 2.9D **CO2:** 245-282g/km **MPG**: 23.3-30.7

Gearbox: Man/Auto **Boot:** 1119-1920 **Towing:** 2800

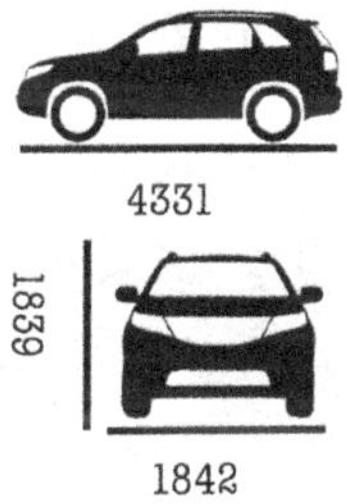

Daewoo Korando 1992 - 2002

Essentially an ugly Jeep which is another SsangYong leftover, this time with a Mercedes five cylinder diesel which was good. Just the three doors, but is decently big inside with a good boot. This is basic 4×4 stuff which isn't bought to pose, but tow and work really hard which is not great when buying used, so rust and worn out mechanicals.

Engines: 2.3 • 2.9D **CO2:** 258-271g/km **MPG**: 27.7-28.8

Gearbox: Man/Auto **Boot:** 479/1464 **Towing:** 2800

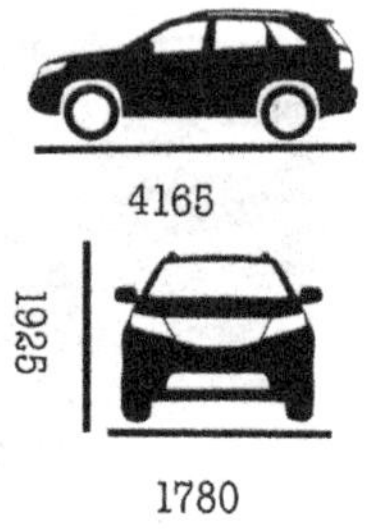

Daihatsu Fourtrak 1984 - 2002

First genuine altDefender 90. Really basic until the '90s with uprated brakes and running gear. Got independent suspension in 1993 and incredibly a seven seat version. Diesels replaced the petrol. Look carefully at what you are actually buying, should be really obvious that it is the Soft Top one. Rust, wear and tear for old farm hacks.

Engines: 2.8 **CO2:** 248g/km **MPG**:30.1 **Gearbox:** Man **Boot:** n/a **Towing:** 3500 **Seats**: 5 **Doors:** 5

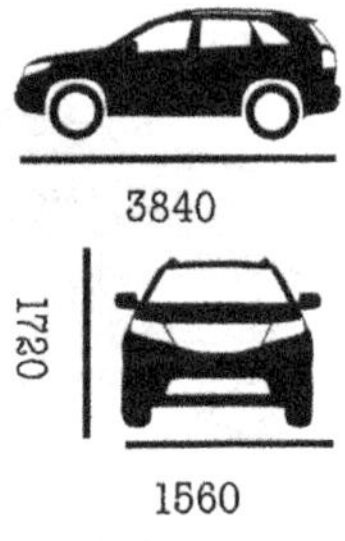

Daihatsu Terios 1997 - 2005

Excellent micro SUV proper off roader as well. Big facelift in 2000 changed the look somewhat and got the 1.3 petrol from a Toyota Yaris, so that will always last. They fiddled with the specifications by realignment of model and specs in 2003, but that's irrelevant. Yes it's really a Starlet/Yaris 4×4. Wear and tear issues only, tyres brakes.

Engines: 1.3 **CO2:** 205g/km **MPG**: 36.7 **Gearbox:** Man/Auto **Boot:** 221/561 **Towing:** 1400 **Seats**: 5 **Doors:** 5

Daihatsu Terios 2006 - 2010

Still cute, a smidge over 14cm longer. Also, underneath this cutie is a full house Toyota Rav 4 which is a fabulous news. Despite all that it sold in tiny numbers. You might struggle to find one these days and the fittings and fixtures means that an SE has everything a tiny family car needs, including air conditioning. Wear and tear and neglect issues only.

Engines: 1.5 **CO2:** 173g/km **MPG**: 39.8 **Gearbox:** Man/Auto
Boot: 380 **Towing:** 1350 **Seats**: 5 **Doors:** 5

Dodge Nitro 2007 - 2009

Breeze Block 4×4 which is unsophisticated and uncomfortable. Not many around, thank goodness and actually a compromised amount of luggage space with seats in place. CRD diesel struggles into the low 30s. Something to be bought on price alone. Actually no things to watch for used, just need service records.

Engines: 3.7 • 2.8D **CO2:** 228-288g/km **MPG**: 23.3-32.8
Gearbox: Man/Auto **Boot:** 369/1994 **Towing:** 2394

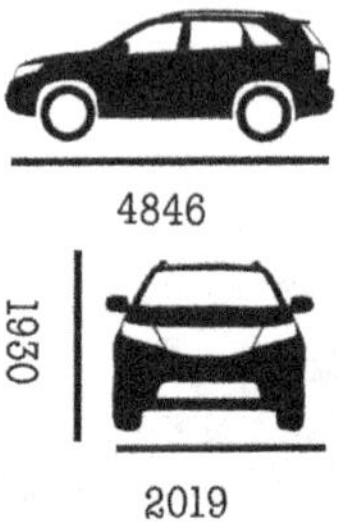

Dodge Ram 2005 - 2007

Yikes, the American dream came to Blighty for a couple of years. Not built to British scale, only available with a V8 petrol. Difficult to find a role for something like this apart from showing off. Not great to drive and makes no real sense at all. That's a shame. Don't pickup. Nothing to look for, parts will be cheap but will have to be air freighted in.

Engines: 5.7 **CO2:** 352g/km **MPG**: 22.1 **Gearbox:** Man/Auto **Boot:** 1741 **Towing: 3878 Seats**: 5 **Doors:** 4

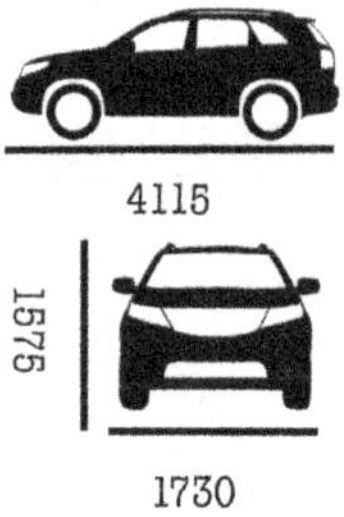

Fiat Sedici 2006 - 2011

See also Suzuki SX4 yet another co-production on the grounds of saving money. With the Fiat badge this little crossover thing is pretty good as a 4 x 4 in the loose stuff. Otherwise a two wheel drive is happy enough to get to shops and back and Multijet diesel does the job the bestest. Wear, tear and neglect issues otherwise, easy to buy.

Engines: 1.6 • 2.0D **CO2:** 173-174g/km **MPG**: 39.8-42.8 **Gearbox:** Man/Auto **Boot:** 270/670 **Towing:** 1200 **Seats**: 5

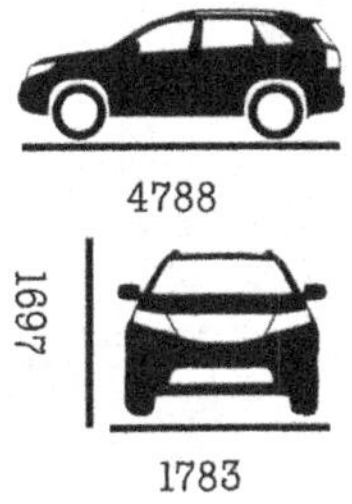

Ford Explorer 1997 - 2001

Reputation for rolling over, packaged with most electrical things as standard. Powered by a 4.0 V6 petrol it provides a comfortable ride and was pretty decent at towing. Easy to live with and getting parts is not impossible on the basis that it is a Ford and a Stateside best seller. Blue engine smoke, worn suspension, broken electrics. Some rust.

Engines: 4.0 **CO_2:** 350g/km **MPG**: 19.3 **Gearbox:** Auto **Boot:** 1209/2308 **Towing:** 2550 **Seats**: 5 **Doors:** 5

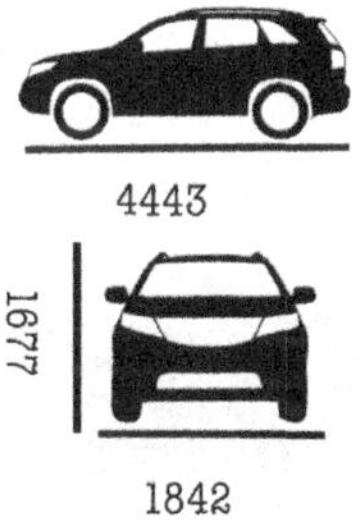

Ford Kuga 2008 - 2012

Fiesta for most eventualities. The 'intelligent' four wheel drive systems was one of those that showed up when the driver needed it. However, a cheaper two wheel drive is all that most needed. Pretty good boot and stowage options. Plenty of minor electrical issues and low level build issues that lets water in the cabin.

Engines: 2.5 • 2.0D **CO_2:** 169-234g/km **MPG**: 28.5-44.1 **Boot:** 360/1355 **Towing:** 1200-2010 2555 **Seats**: 5 **Doors:** 5

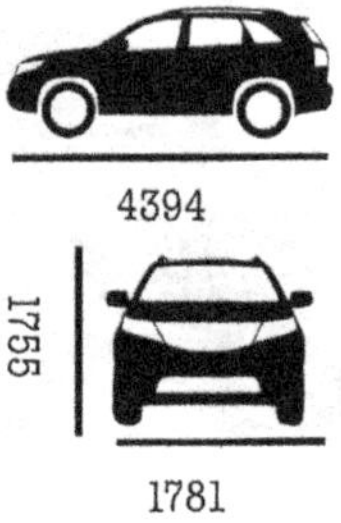

Ford Maverick 2001 - 2003

Also a Mazda Tribute, neither did well in a world that demanded diesel and some sort of credible off road back story. Reliable with a decent boot and interior space. Top spec V6 had lots of kit including leather, loads of airbags and cruise. Worth the effort of finding and won't leave you crying like a Freelander. Just wear and tear issues.

Engines: 2.0 • 3.0 **CO2:** 234-300g/km **MPG**: 22.1-28.8
Gearbox: Man/Auto **Boot:** 490/1830 **Towing:** 1500-1700

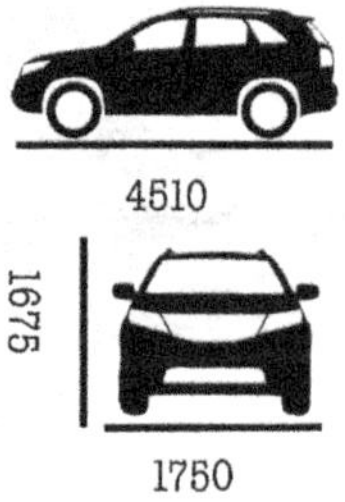

Honda CR-V 1997 - 2001

Impressive standard kit includes the usuals such as electric windows, air conditioning, sunroof and mirrors. Plus there was stuff to keep an adventurer happy such as a removable picnic table, an accessory socket in the boot and a split rear tailgate. The petrol 2.0 VTEC engine is all you need and it won't break down. Wear and tear and rust.

Engines: 2.0 **CO2:** 229g/km **MPG**: 29.1 **Gearbox:** Man/Auto
Boot: 445/668 **Towing:** 1500 **Seats**: 5 **Doors:** 5

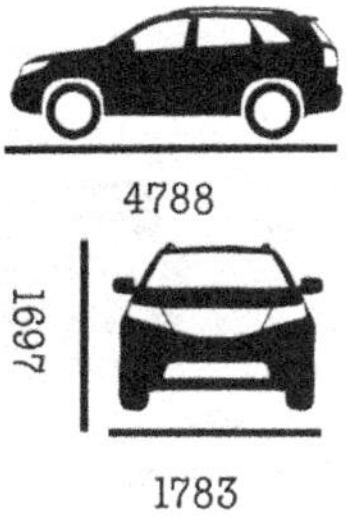

Honda CR-V 2001 - 2006

It's the Brit built one and that means it is well screwed together. Constantly upgraded and although there was a facelift in 2003, suspension tweaks meant that it was even better on road. Cause for celebration for some would be the arrival of diesel. Split rear seats that fold and tumble mean you can get a lot of stuff into the back. Wear and tear only.

Engines: 2.0 **CO2:** 215-216g/km **MPG**: 31.0-31.4 **Gearbox:** Man/Auto **Boot:** 527/952 **Towing:** 1500 **Seats**: 5 **Doors:** 5

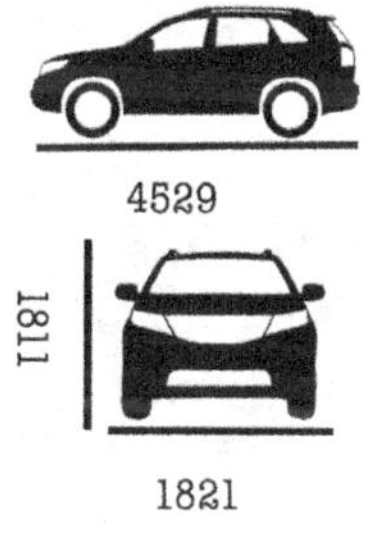

Honda CR-V 2007 - 2012

Less off road chic as the spare goes from the boot to under the floor, tailgate is now conventionally top mounted. Even so, more room inside and still has the 60:40 split rear seat that slides and reclines. Engines are still impressive whether it is the 2.0 petrol which had a power boost or the 2.2 i-CTDi. Wear and tear and smokey neglected diesels.

Engines: 2.0•2.2 **CO2:** 171-190g/km **MPG**: 34.4-43.5 **Gearbox:** Man/Auto **Boot:** 556/955 **Towing:** 577-758

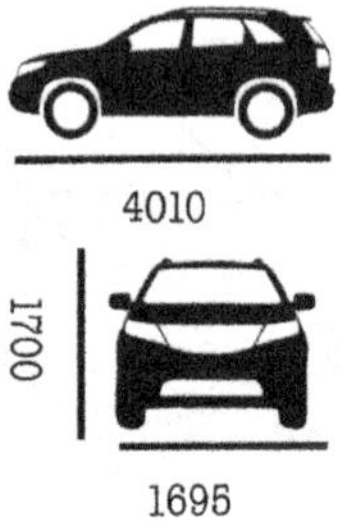

Honda HR-V 1999 - 2005

The world had no idea that it needed a 4 x 4 three door estate styled like a space shuttle. The world was also offered a two wheel drive SE version for a while. Five doors arrived in 2000 which eventually became the sole flavour and there was a complete range which was freshened up in 2001. Very likeable, fun and practical. Suffers from rust and neglect.

Engines: 1.6 **CO_2:** 201-205g/km **MPG**: 32.8-33.6 **Gearbox:** Man/Auto **Boot:** 285/603 **Towing:** 1200 **Seats**: 4 **Doors:** 5

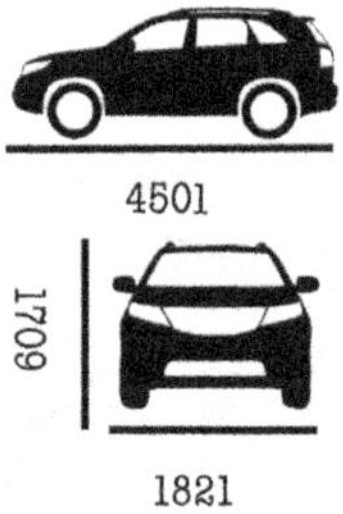

Hyundai Santa Fe 2001 - 2005

Not pretty, has V6 petrols and a diesel CRTD. Split window tailgate is surprisingly handy and overall reasonable amounts of room. Dual power sockets inside were certainly useful and the V6 had loads of standard things from leather to electric most things. These are super reliable, so ignore the looks. Scruffy, low level rust, suspension parts, brakes.

Engines: 2.4 • 2.7 • 2.0D **CO_2:** 202-274g/km **MPG**: 24.8-37.2 **Gearbox:** Man/Auto **Boot:** 850/2100 **Towing:** 700-2300

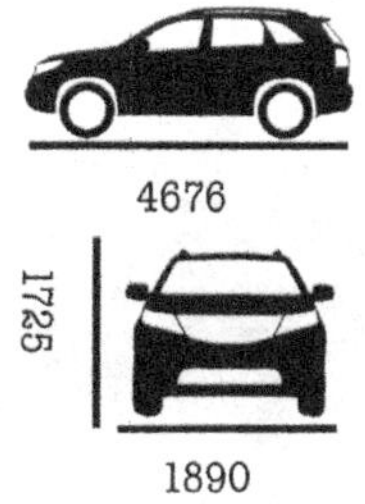

Hyundai Santa Fe 2006 - 2012

Bigger and better, blander. V6 petrol for a while, but the 2.2 diesel best for most applications delivering 40mpg. Seven seat option and that third row folded into the boot floor, then if you fold flat the second row the amount of luggage space is colossal. As is the top spec CDX and CDX+. Air Con and minor electrical issues, locks, windows and cockpit.

Engines: 2.2 **CO_2:** 176g/km **MPG**: 41.5 **Gearbox:** Man/Auto
Boot: 969/2247 **Towing:** 2000-2500 **Seats**: 5/7 **Doors:** 5

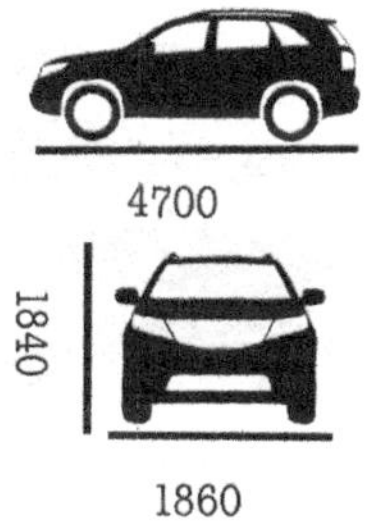

Hyundai Terracan 2003 - 2007

Can't afford a Land Cruiser? It isn't as big or capable. just five seats so inside there is a ton of useful room. The rear seat splits into a 60:40 and then folds. Packed full of the usual equipment plus electric windows and climate control. Underrated, will deliver in spades. Wear and tear issues, tyres, brakes and suspension if used for major towing.

Engines: 2.9 **CO_2:** 222g/km **MPG**: 32.4 **Gearbox:** Man/Auto
Boot: 1170/1955 **Towing:** 2800 **Seats**: 5 **Doors:** 5

Hyundai Tuscon 2004 - 2009

Baffling addition to their 4 x 4 line up as a compact offering. There was also a two wheel drive alternative. Certainly will work hard for a living and much more reliably. Excellent interior flexibility as the rear seats fold flat and even the front seats flatten so there's a bed, meanwhile the split screen tailgate is great. Minor electricals, wear and tear.

Engines: 2.0 • 2.7 • 2.0D **CO2:** 187-237g/km **MPG**: 28.2-39.8
Gearbox: Man/Auto **Boot:** 504/1374 **Towing:** 1400-1900

Infiniti EX 37 2009 - 2015

Nissan with lofty ideas. Indeed, a 370Z with four doors, well built good to drive, a little bit tight inside with a teeny boot. Reliable and well equipped, compared to posh rivals, it will be great value and won't break down. Don't buy for off road work this is a comfy, sporty, rarity which may have a parts supply issue eventually. Needs service history.

Engines: 3.7 **CO2:** 265g/km **MPG**: 25.0 **Gearbox:** Auto
Boot: 340/1175 **Towing:** 2000 **Seats**: 5 **Doors:** 5

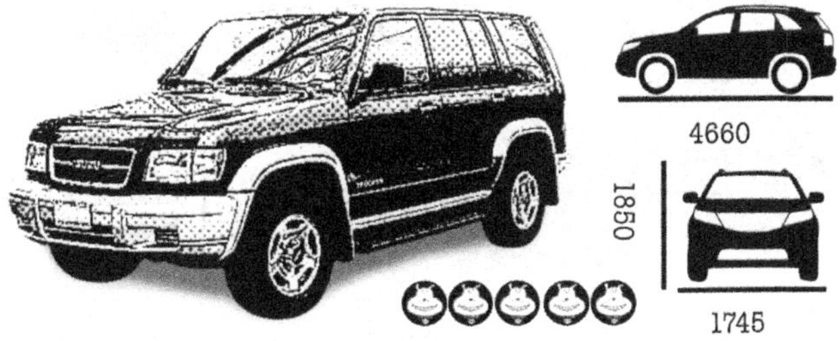

Isuzu Trooper 1992 - 2003

Big Horn back home, third generation from '98 with a V6 petrol or new common rail diesel power. Loved by farmers and builders who just wanted to work and tow. Three or five doors. Clever 4 x 4 system which constantly changed power delivery to the front or rear wheels. Not sophisticated but really tough. Legendary. May have been worked to deat.

Engines: 3.5 • 3.0D **CO2:** 287-326g/km **MPG**: 20.6-26.9

Gearbox: Man/Auto **Boot:** 909/1730 **Towing:** 3300-3500

Jeep Cherokee 1993 - 2001

Proper ancient off roader that copes with the rough and can tow all day long. Go for Limited specs that at least give the passenger an airbag and the driver gets ABS. Not massive inside but adequate. 2.5 Turbo diesel noisy and the 4.0 petrols not economical but pretty reliable and tough overall. Rust, worn suspension, tired smokey engines and gearboxes.

Engines: 2.5 • 4.0 • 2.5D **CO2:** 296g/km **MPG**: 18.2-31.1

Gearbox: Man/Auto **Boot:** 412/1265 **Towing:** 3150

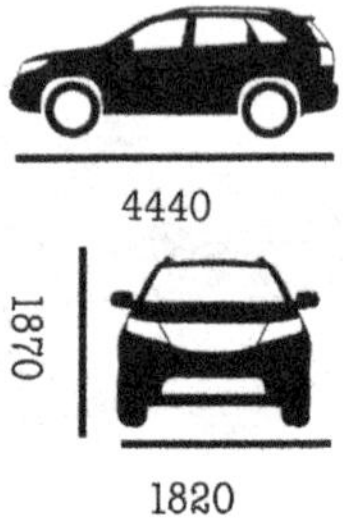

Jeep Cherokee 2001 - 2007

Fairly all new, bug eyed model which got a refresh in 2005, followed by uprated 21st century technology in the shape of sharper traction control. Fairly plastic experience inside making it something you could actually hose down. 2.8 CRD engines seem strong enough while the petrols really will last as long as your overdraft. The odd ECU and electrical.

Engines: 2.4 • 3.7 • 2.8D **CO2:** 260-333g/km **MPG**: 20.2-28.5
Gearbox: Man/Auto **Boot:** 821/1954 **Towing:** 2000

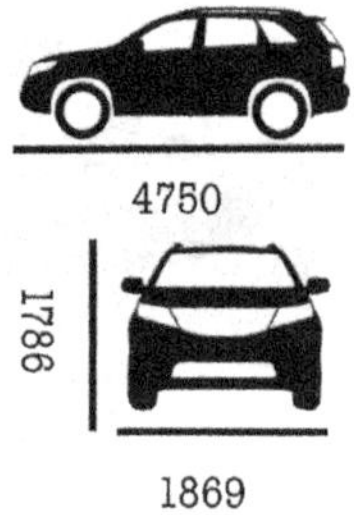

Jeep Cherokee 2008 - 2010

Third generation victim to EU diesel regs. Just the one 2.8 CRD engine then which delivered low 30s mpg. Proper off road tech with high/low range box and a clever hill descent gadget. In practical terms it is a bit bigger than the last one and the designers still love hard wearing plastics, which is fine. Check automatic gearbox and brakes/suspension.

Engines: 2.8 **CO2:** 228g/km **MPG**: 32.8 **Gearbox:** Man/Auto
Boot: 419/1404 **Towing:** 2800 **Seats**: 5 **Doors:** 5

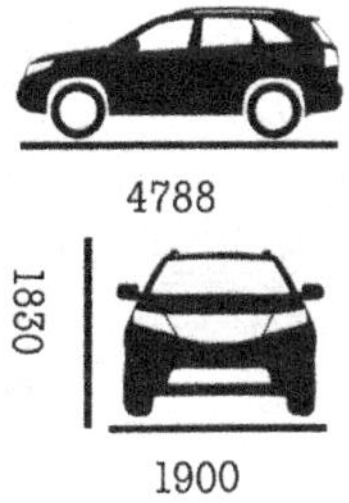

Jeep Commander 2006 - 2009

Cheaper and less likely to break down than a Discovery. V8 petrol for a year so mostly 3.0 diesel. Everyone gets a seat and a view because of the stepped floor, as usual rearmost ones for kiddies. Fold them into the floor for more luggage space. Not that many imported. Not rated highly by owners, spread of mainly electrical and ECU issues.

Engines: 3.0 **CO2:** 284g/km **MPG**: 26.2 **Gearbox:** Auto **Boot:** 212/1940 **Towing:** 3500 **Seats**: 7 **Doors:** 5

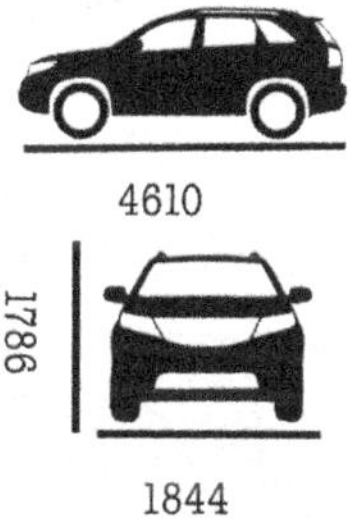

Jeep Grand Cherokee 1999 - 2004

LuxoSUV they invented. New generation Grand had just the one smaller V8 petrol. 3.1 Turbo diesel replaced by the better 2.7 CRD. Reasonable amount of room inside, with clever touches like the electrical retractable split tailgate glass. Dated but useful. Automatic gearbox, electrical niggles like locks and windows, general post towing scruffiness.

Engines: 4.0 • 4.7 • 2.7D **CO2:** 255-380g/km **MPG**: 17.7-29.1 **Gearbox:** Auto **Boot:** 1110/2059 **Towing:** 3360-3500

Jeep Grand Cherokee 2005 - 2010

Wholesale revamp to bring the Grand into the current year. 3.0 V6 CRD Mercedes related and the usual V8 petrols. Much more refined and comfortable than before. Still came with an absolute ton of equipment whether you chose the Predator, Limited or Overland. Some were parked in fields for years unsold. Electrics, some rust and related issues.

Engines: 6.1 • 3.0D **CO2:** 270-388g/km **MPG**: 17.2-27.7
Gearbox: Man/Auto **Boot:** 978/1909 **Towing:** 3500

Jeep Patriot 2007 - 2011

Remember this? Course not. Compact sized SUV alternative to a Freelander, or RAV 4. Has a VW sourced diesel 2.0 CRD as well as a 2.4 petrol. Well equipped when it comes to safety kit and also from the Sport upwards there is air conditioning and electric windows. Limited adds lots more, worth finding. Suspension ball joint wear, shock absorbers and electrics.

Engines: 2.4 • 2.0D **CO2:** 180-206g/km **MPG**: 32.5-42.2
Gearbox: Auto **Boot:** 536/1277 **Towing:** 1500 **Seats**: 5

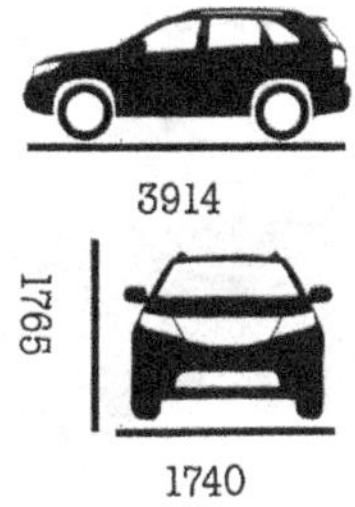

Jeep Wrangler 1993 - 2006

The 24 carat classic. Two or three doors, hard or softop. Proper trad styling from 1996 with the TJ when it got airbags and the Sahara model had a removable hardtop. Flappy plastic side windows, but really not worth bothering with. Should be used and enjoyed as a convertible 4 x 4. Back to basics tough Banger for the end of most days.

Engines: 4.0 **CO2:** 302g/km **MPG**: 22.1 **Gearbox:** Auto

Boot: 320/1577 **Towing:** 1001-2000 **Seats**: 2/3 **Doors:** 2

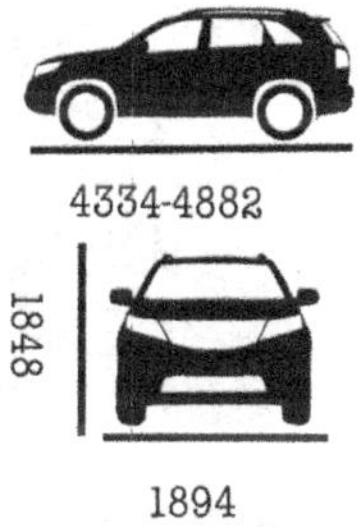

Jeep Wrangler 2007 - 2014

The 24 carat classic but now with extra doors. Gives off a Defender vibe but without the 'retro-tax'. Does off roading really well so you actually can use this for proper work. Powerful and tough engines, so you can tow with ease. Feels a bit cheap, but that suits the character and should mean lower prices. Slightly more complicated than the old one.

Engines: 3.8 • 2.8D **CO2:** 268-275g/km **MPG**: 24.4-29.1

Gearbox: Man/Auto **Boot:** 320/1577 **Towing:** 3500

Kia Sorento 2003 - 2009

King of the budget big 4×4s, but at least these are tough and reliable. 2.5 diesel or 3.5 petrol, depending on the day to day running allowance of the buyer. A revamp in 2006 of equipment and engines. The 4×4 system varies between the specs, going from part time to full time. Wear and tear issues, suspension/brakes.

Engines: 2.5D **CO2:** 204g/km **MPG**:36.7 **Gearbox:** Man/Auto **Boot:** 898/1849 **Towing:** 2300 **Seats**: 5 **Doors:** 5

Kia Sportage 1995 - 2003

Ahead of the curve, well finished, handy sized SUV action. A Korean, built in Germany by Karmann, so decently finished. Just the one 2.0 petrol engine. Loads of different specifications over the years and restyled a bit in 2000. It is a no more than a slightly more capable estate car, but it is reliable. Wear +tear, rust and general old age and overuse.

Engines: 2.0 **CO2:** 275g/km **MPG**: 25.7 **Gearbox:** Man/Auto **Boot:** 483/1319 **Towing:** 1250 **Seats**: 5 **Doors:** 5

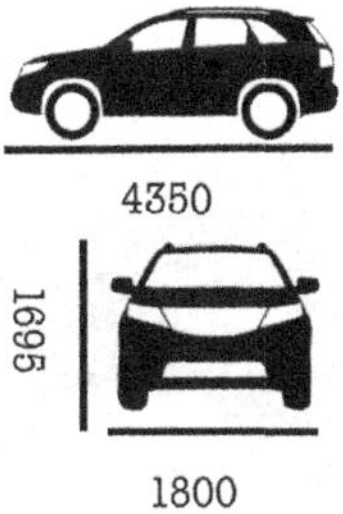

Kia Sportage 2005 - 2010

Can also be bought as a Hyundai Tuscon and returned with 2.0 petrol and diesel engines plus a 2.7 litre petrol option. In 2008 it got a restyle and some mechanical upgrades that made it better to drive. There was also a two wheel drive version because so few owners did anything seriously dirty. Electrics, wear and tear.

Engines: 2.0 • 2.7 • 2.0D **CO2:** 187-237g/km **MPG**: 28.2-39.8 **Gearbox:** Man/Auto **Boot:** 667/1887 **Towing:** 1600

Land Rover Defender 1990 - 2012

The shape never changed much, just short 90 or long 110, with progressively better diesels. in 1999 it got the 5 cylinder unit from the Discovery. Then from 2007 a 2.4 common rail diesel and overall more refined than previous generations. It remained an unsophisticated lorry. Never a cheap option. Rust, electrics. Once sorted it's a long termer.

Engines: 2.4D **CO2:** 299g/km **MPG**:28.2 **Gearbox:** Man **Boot:** 1510 **Towing:** 3500 **Seats**: 2/7 **Doors:** 3/5

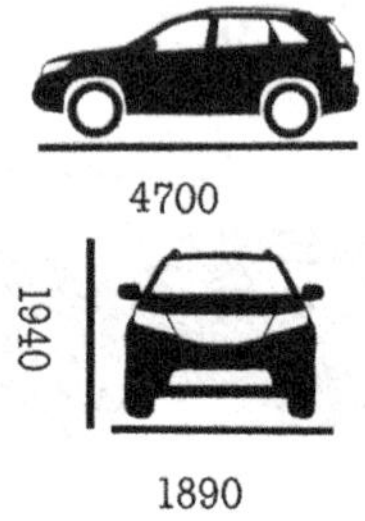

Land Rover Discovery 1998 - 2004

Discovery II looked the same but only the tailgate shared with the cast iron original. Owners swear by and at them. Too much complicated technology meant the air suspension could fail and the electronics would have a mental breakdown. 2002 changes included a centre diff lock option, Td5 diesel preferred. Hard to find an upside.

Engines: 4.0 • 2.5D **CO2:** 262-397g/km **MPG**:17-30.1 **Gearbox:** Man/Auto **Boot:** 881-1699 **Towing:** 3500

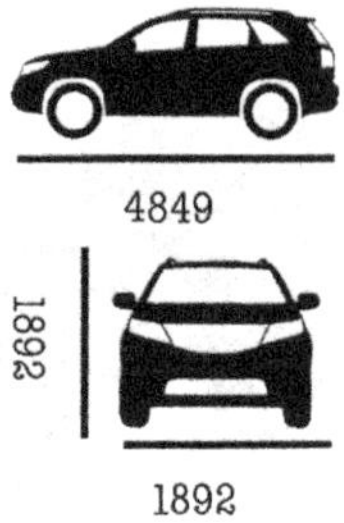

Land Rover Discovery 2004 - 2017

Disco 3 was bigger, whether it was truly better is open to question. Certainly there is a lot more useable space except when the seven seat model is fully occupied. 2.7 diesel and 4.4 V8 petrol provide the power. Plenty of recalls and that air suspension did not get any better causing all sorts of issues. There are reliable ones, but why take the risk?

Engines: 2.7D • 3.0D **CO2:** 244g/km **MPG**: 30.7 **Gearbox:** Man/Auto **Boot:** 1260/2558 **Towing:** 3500 **Seats**: 5/7

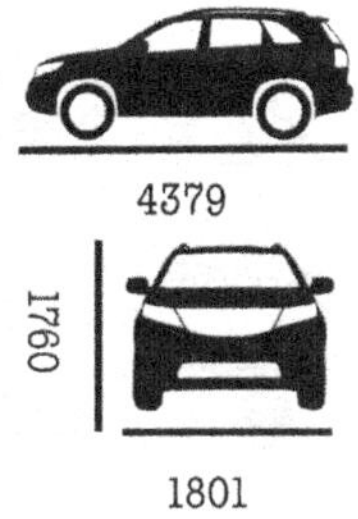

Land Rover Freelander 1997 - 2003

The firstly truly new Land Rover a pint sized off roader. Three door Softback/Hardback and a five door estate petrols and diesels, but beware head gasket issues. Revised model in 2000 with a Td4 engine and specification shuffles. More important is to keep an eye, or rather ear on the wheel bearings. Revised in 2003. Sort of acceptable Road Rover.

Engines: 1.8 • 2.5 • 2.0D **CO2:** 205-298g/km **MPG**: 22.7-37.2
Gearbox: Man/Auto **Boot:** 371/1311 **Towing:** 2000

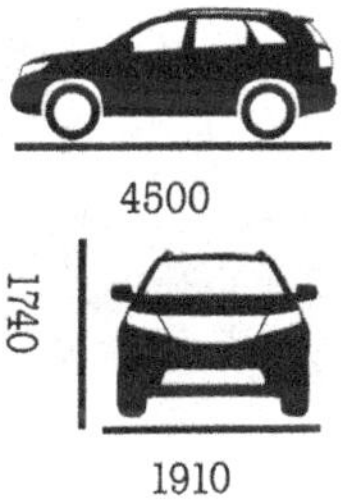

Land Rover Freelander 2006 - 2014

More powerful engines, rearranged specs plus a complex Terrain Response 4 x 4 system. More of a quality build than before. 2.2 Td4 diesel is the popular choice over the 3.2 petrol, but ultra long term it is the fuel hungry winner. Stop/ Start model from 2009 to save fuel. Can still have electrical issues. Arguably the only recent Landie which is half decent.

Engines: 3.2 • 2.2D **CO2:** 194-265g/km **MPG**: 25.2-37.7
Gearbox: Man/Auto **Boot:** 755/1670 **Towing:** 2000 **Seats**: 5

Lexus RX300 2000 - 2004

High rise Lexus with very limited off road ability. Most things standard, most sales were SE with rain sensing wipers, leather electric tailgate, adding the L meant a reversing camera, in colour. Yes it really is all about the spec rather than anything else important. One of the most car like 4 x 4s. Avoid scruffpots. Suspension, brakes, neglect.

Engines: 3.0 **CO2:** 311g/km **MPG**: 21.7 **Gearbox:** Auto

Boot: 1073 **Towing:** 2000 **Seats**: 5 **Doors:** 5

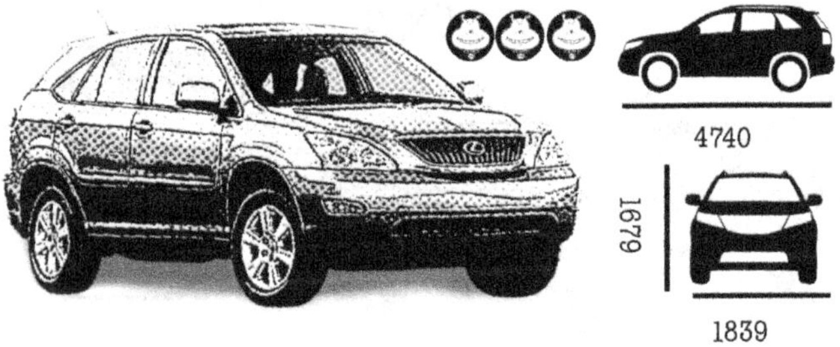

Lexus RX 2004 - 2008

Bigger and redesigned everywhere for a more luxurious experience. Lots of complication from 2005 with the hybrid RX400h. Bigger engines RX 350 and other changes meant that most things are standard by 2006. Not the biggest boot and the rear seats don't fold flat, but this isn't a lugger, its a Lexus. Best to get it plugged into a diagnostic tool.

Engines: 3.0 • 3.3H **CO2:** 192-288g/km **MPG**: 23.2-34.9

Gearbox: Auto **Boot:** 439/1500 **Towing:** 1585 **Seats**: 5

Lexus RX 2008 - 2012

Now hybrid with a Guppy grille. In practical terms there is a 40/20/40 split rear seat which is different and one tug of the boot lever means they fold flat. Average amount of room the rear seats also slide and recline for fun. Still complicated, so plug it in to the proper diagnostics kit. Balance the buying budget to the amount useful life you believe is left.

Engines: 3.5H **CO2:** 148g/km **MPG**:44.8 **Gearbox:** Auto **Boot:** 474/1656 **Towing:** 2000**Seats**: 5 **Doors:** 5

Mazda CX7 2007 - 2011

Doesn't look SUVlike, but it is, seemingly glued together from other products in the range. Petrol engine only at the start a 2.3 Turbo and diesel in 2009 with a revised model and more specification. Rear seats fold flat, not much ground clearance for any 4×4 action. Dual mass flywheels have proved to be troublesome items.

Engines: 2.3•2.2D **CO2:** 199-243g/km **MPG**: 27.7-37.7 **Gearbox:** Man/Auto **Boot:** 455-1348 **Towing:** 1450 **Seats**: 5

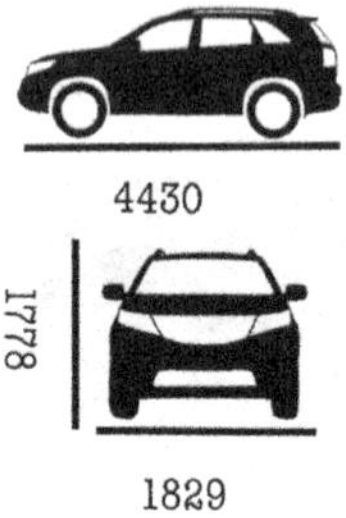

Mazda Tribute 2001 - 2004

Previously spotted wearing Ford Maverick badges. Mazda means reliability and it was offered as both a 4 x 4 and front wheel drive, just powered by petrols. No fancy tricks inside only dull upholstery and a decent sized boot. That sort of simplicity should make it a prime Banger if you can find one. Possibly some rust and wear, but solid.

Engines: 2.0 • 3.0 **CO2:** 231-305g/km **MPG**: 22.1-30.1
Gearbox: Man/Auto **Boot:** 368/793 **Towing:** 1500-1700

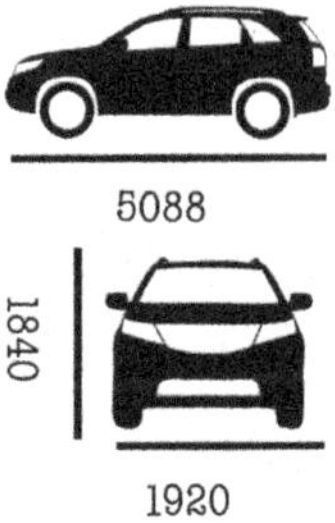

Mercedes GL 2006 - 2012

Gargantuan, more 4×4 than any human being needs. Obviously big on the inside. Very adjustable rear seats that we won't bore you with here, apart from the fact that it all folds flat and will take seven bodies if required, plus luggage. Moved by a big diesel or petrol engine. A big mistake? Wear and tear, electrics and ECUs, so Plug it in.

Engines: 5.5 • 3.0 • 4.0 **CO2:** 261-331g/km **MPG**: 20-28.2
Gearbox: Auto **Boot:** 200/2300 **Towing:** 3500 **Seats**: 7

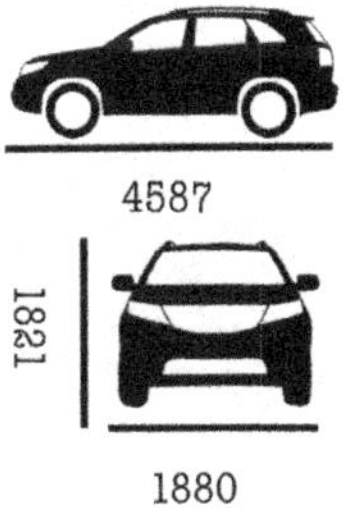

Mercedes M-Class 1998 - 2004

Yet another huge 4 x 4, but not a great time for quality control. That was sort of addressed in 2000 with a range revamp and then again in 2001, but lots of problems with the early examples and many are best left to rot and expire where they are parked. Decent petrol and diesel engines and optional third row of seats. Very scruffy ones around.

Engines: 3.2 • 5.0 • 5.4 • 2.7D **CO2:** 248-354g/km **MPG**: 19.6-30.0 **Boot:** 1061/2020 **Towing:** 2650-3365

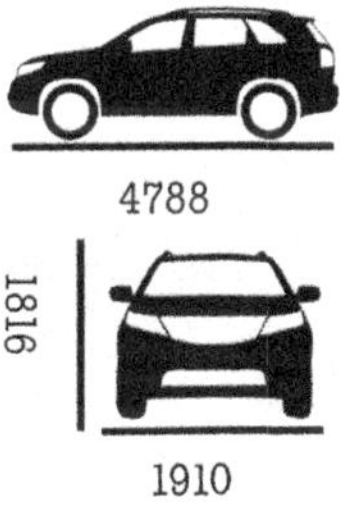

Mercedes M-Class 2005 - 2011

Better built with more emphasis on actual quality. Wider, longer and a tad lower. That resulted in more room to stretch out in the back and for luggage. Removable seat cushions were an option so that there was a flat load bay. Can be cost effective to run with the economical diesel, or an unburstable petrol. Electrics and wear and tear issues.

Engines: 5.5 • 6.2 • 3.0D • 4.0D **CO2:** 246-304g/km **MPG**: 20.1-30.1 **Gearbox:** Auto **Boot:** 551/1830 **Towing:** 3500

Mitsubishi Outlander 2004 - 2007

It's a compact Shogun, more of a crossovery thing. Beyond the nosey styling there is a lot of standard equipment including climate control and electric most things. Powered by a 2.4 petrol and being a Mitsubishi, no chance that it will break down. Never cool, loved or coveted, so cheap. Wear and tear, especially the suspension.

Engines: 2.4 **CO2:** 240g/km **MPG**:28.0 **Gearbox:** Man/Auto **Boot:** 402/1049 **Towing:** 1600 **Seats**: 5 **Doors:** 5

Mitsubishi Outlander 2007 - 2013

Also a Citroen and Peugeot, but a Mitzi with the 2.2 diesel engine is the best value version. Roomy inside and the split tailgate a nice touch along with the automatic folding seat which effectively collapsed the seats so they folded out of the way. Decent boot became massive load bay. Contains three 12V sockets. Wear and tear main issues.

Engines: 2.4•2.0D•2.2D **CO2:** 174-194g/km **MPG**: 38.7-42.8 **Gearbox:** Man/Auto **Boot:** 402/1049 **Towing:** 1500

Mitsubishi Shogun 2000 - 2006

Third Gen models, but so many grey imports over the years, what serious 4×4s users went for instead of flakey Land Rovers. Petrol V6 and Turbo diesels did the hard work. Optional longer 7 seater with forward facing back seats which at least folded into the floor. Facelift for 2003. This is the real deal. Wear and tear suspension and brakes.

Engines: 3.5 • 3.2D **CO2:** 251-339g/km **MPG**: 21.4-29.7 **Gearbox:** Man/Auto **Boot:** 205/1495 **Towing:** 2800-3300

Mitsubishi Shogun 2006 - 2011

As previously you can get a longer 7 seater with seats that fold conveniently away. We liked their 'Hide and Seat' gag, but in the real world it works very well indeed. Continues the company's reputation for reliability and fuss free operation. Facelifted in 2009. Nothing to look out for just check gearbox operation and the brakes/suspension.

Engines: 3.2D **CO2:** 244g/km **MPG**: 30.7 **Gearbox:** Man/Auto **Boot:** 417/1789 **Towing:** 3300 **Seats**: 5/7

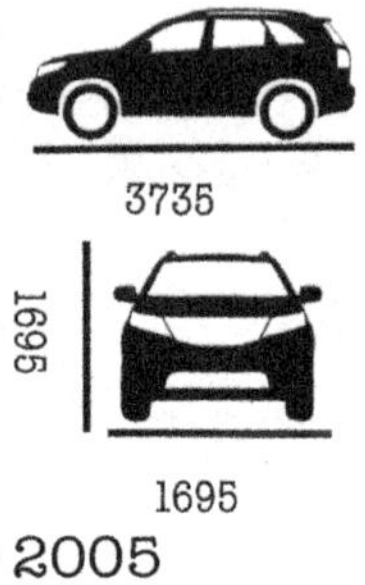

Mitsubishi Shogun Pinin 2000 - 2005

Pocket sized Shogun. Designed by Italians and engineered by the Japanese so the best of all worlds. It is small, but as big as a decent supermini inside, so day to day practical. The GDI engine is very efficient and returns early 30s mpg. Dives in and out of potholes, but pretty good in the dirt. Very reliable, body parts a problem as well as rust.

Engines: 1.8 • 2.0 **CO2:** 217-227g/km **MPG**: 29.7 **Gearbox:** Man/Auto **Boot:** 358/1158 **Towing:** 1500

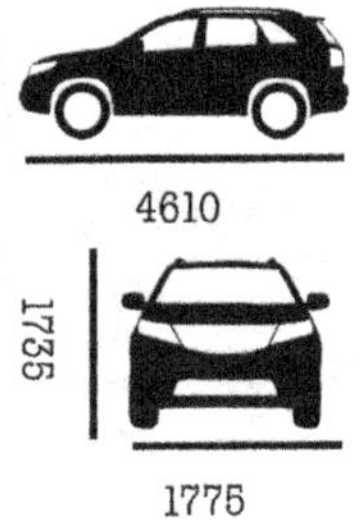

Mitsubishi Shogun Sport 1998 - 2006

Pick-up based Shogun also known as a Challenger. There are compromises with no back seat tricks to pack stuff in, or fold anything flat. It is very basic business transportation. There's a decent 2.5 Diesel and thirsty 3.0 V6. Proper worker so many are worn out, or being sold because there are big bills to fix suspension or engines.

Engines: 3.0 • 2.5D **CO2:** 266-294g/km **MPG**: 22.6-27.2 **Gearbox:** Man/Auto **Boot:** 500/1590 **Towing:** 3100 **Seats**: 5

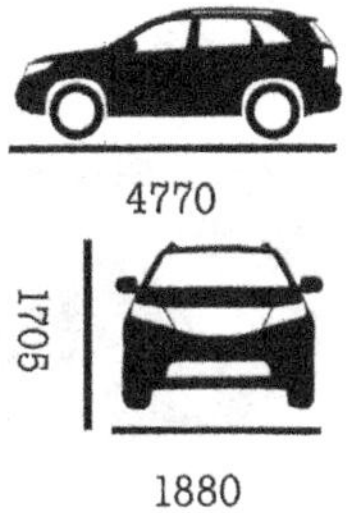

Nissan Murano 2005 - 2008

Pushed along by a motor used in a drifting 350Z, so a proper 'Sports' utility then. Everything standard including leather, satnav and parking sensors and plenty of room in the back. Never designed to lug any loads, just charge about. That's good for used buyers. Simply wear and tear on the suspension, brakes but check electrics.

Engines: 3.5 **CO2:** 295g/km **MPG**: 23.0 **Gearbox:** Auto **Boot:** 438/877 **Towing:** 1500 **Seats**: 5 **Doors:** 5

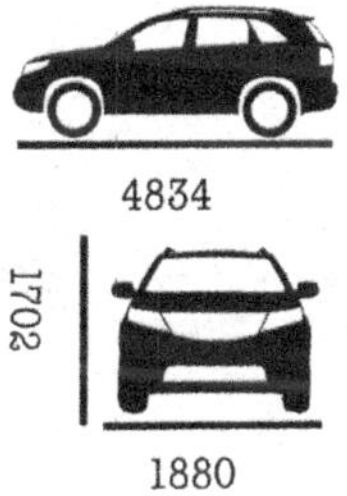

Nissan Murano 2008 - 2011

New undercarriage and revamped interior. Most importantly it was fitted with a 2.5 diesel from 2009 which made no difference to buyer indifference. Quite complicated now with a power tailgate, quite useful kerb camera to help the driver at low speeds and rear electric seats. Electrics worth checking. Diesel condition crucial.

Engines: 3.5 • 2.5D **CO2:** 210-261g/km **MPG**: 25.9-35.3 **Gearbox:** Auto **Boot:** 402/838 **Towing:** 1500 **Seats**: 5

Nissan Pathfinder 2005 - 2014

Underneath is a Navara pickup not great news, rusty chassis broke them in half. Situation ongoing. Caution required. 2.5 diesel and big 4.0 petrol. Seven seats is an option and lots of specs to choose from. Despite some reservations still a safer reliability bet than a Discovery. Check for rust, diesel smoke and ECU failures, otherwise it is towing damage.

Engines: 4.0•2.5D **CO2:** 238-327g/km **MPG**: 19.1-31.4 **Gearbox:** Man/Auto **Boot:** 190/1874 **Towing:** 3000

Nissan Patrol 1998 - 2009

Like a Defender, Jeep or Landcruiser, apparently around since the dawn of time. Big old bus and identified as GR models with three (until 2001) or five doors. Restyled in 2000 and in 2004 with thumpingly good diesel engines. Always better value than the more fashionable alternatives, yet these are rock solid. Watch out for overuse and abuse.

Engines: 3.0D **CO2:** 288g/km **MPG**: 26.2 **Gearbox:** Man/ Auto **Boot:** 183/2226 **Towing:** 3500 **Seats**: 5/7

Nissan Terrano II 1993 - 2005

Also available as a Ford Maverick, but that is just badge this is a proper Datsun. Three and five door versions and a range of diesel and petrol engines. Facelifted in 2000 but what you have is a pretty ordinary compact 4 x 4 was OK for the time (central locking, electric sunroof). At least it is resistant to breaking down. Wear and tear checks.

Engines: 2.4•2.7D **CO2:** 260-292g/km **MPG**: 23.0-28.8
Gearbox: Man/Auto **Boot:** 1642 **Towing:** 2800

Nissan X-Trail 2001 - 2007

Compact 4 x 4 and probably one of the best soft roaders for the busy family buyer. Lots of space and very practical decent 2.0 petrol and 2.2 diesel see later. By 2005 there was a 4 x 2 offering because not everyone needs the full traction option. These have flaws, Renault diesel. Timing chains, blown turbos. Also mild rust and electricals.

Engines: 2.0•2.5•2.2D **CO2:** 190-226g/km **MPG**: 30.1-39.2
Gearbox: Man/Auto **Boot:** 350/410 **Towing:** 2000 **Seats**: 5

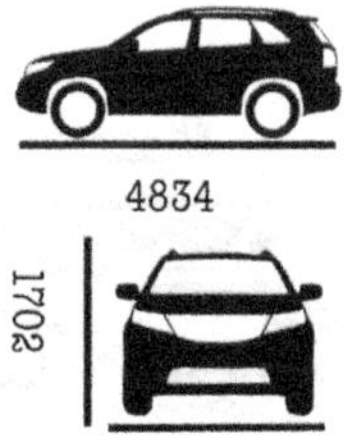

Nissan X-Trail 2007 - 2014

4 x 4 only diesel in different levels of tune and two petrol options, plus specifications, where the lower ones have the all mode system which provides traction when needed and the top level Aventura spec had more sophisticated hill descent and start assist. By 2008 the range was cut back to diesel only. Better than previous one.

Engines: 2.0 • 2.0D **CO2:** 190-208g/km **MPG**: 36.4-39.8 **Gearbox:** Man/Auto **Boot: Towing:** 1350 **Seats**: 5 **Doors:** 5

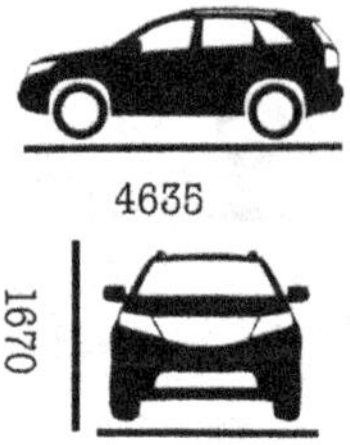

Peugeot 4007 2007 - 2012

2.2 Hdi diesel engines only, simple line up of SE and GT. Has their 5 + 2 seating arrangement, the second row of seats will fold flat at the touch of button for access or loadspace. They also slide and recline while the back row fold into the floor. Lots to like along with the split tailgate. DPF valve can be blocked if low mileage. Electrics can play up and ECUs fail.

Engines: 2.2D **CO2:** 191g/km **MPG**: 39.2 **Gearbox:** Man/Auto **Boot:** 184/1686 **Towing:** 2000 **Seats**: 7 **Doors:** 5

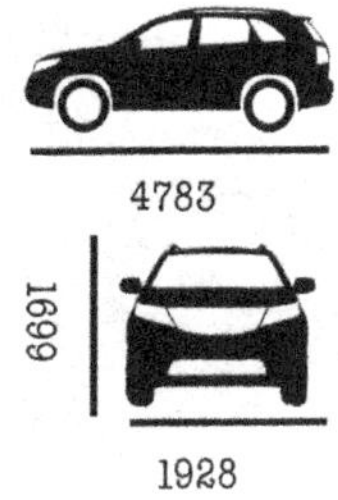

Porsche Cayenne 2003 - 2009

The 'cheap' Porker. Not to fix or service, but an awful lot of high performance 4 x 4 for not much money. Avoid the diesel, the turbo and the air suspension for fixability and durability reasons. Cheaper V6 petrol from the Golf is a consideration though. Facelifted in 2007. Overall, be careful. Oil and coolant leaks, suspension issues.

Engines: 3.2 • 4.5 • 3.0D **CO2:** 244-358g/km **MPG**: 18.0-30.4
Gearbox: Auto **Boot:** 540/1770 **Towing:** 3500

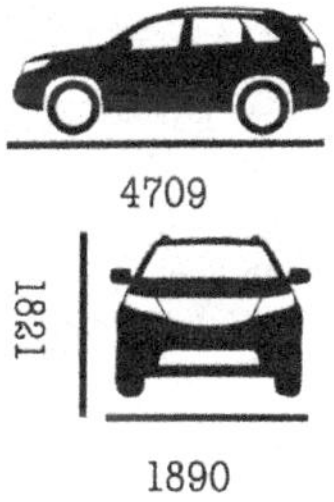

Range Rover 1994 - 2002

Powered by V8 petrols and 2.5TD diesel, the latter boosted for output in 1999 and there was a cosmetic tidy up in 2000. Still flawed when it came to the basics, like reliability. Neglected and older ones are a liability, not the engines, mostly electrics. Certainly does the on and off road better than anything else when it works but not worth the risk.

Engines: 4.0 • 4.6 • 2.5D **CO2:** 304-398g/km **MPG**: 17.2-24.9
Gearbox: Auto **Boot:** 524/1642 **Towing:** 4000 **Seats**: 5

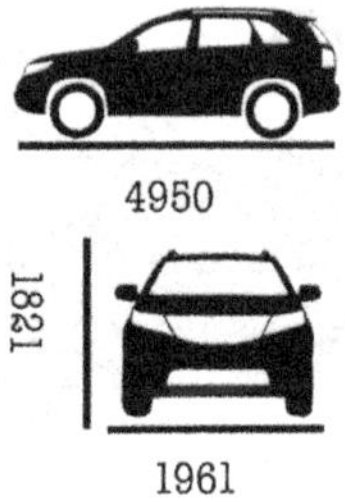

Range Rover 2002 - 2012

Getting better the L322 for you designation nerds. One petrol, but mostly TD diesels, all engines decent of course. Facelift in 2006 then more tech in '07 in the digital shape of Terrain Response from the Disco. A ton to go wrong, cooling leaks, blowing turbos, 'sealed for life' gearbox and air suspension issues, just to name a few. Still big trouble.

Engines: 4.2•4.4•2.7D **CO2:** 299-389g/km **MPG**: 17.4-25

Gearbox: Auto **Boot:** 535/1756 **Towing:** 3500 **Seats**: 5

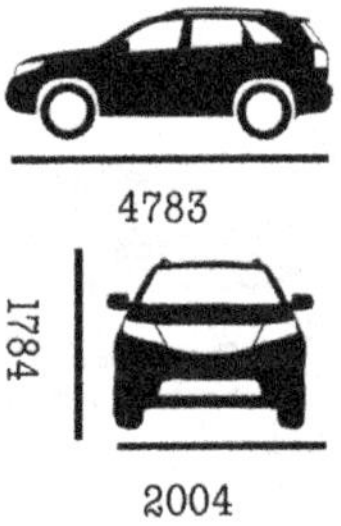

Range Rover Sport 2005 - 2013

The go faster one. Shockingly popular. RR as sports car appealed to many. If you must have a superfast SUV there are better ones which won't explode on you. Everything costs a bomb to fix. Fuel pumps stop doing their thing suddenly and the engine system dashboard mass illuminations reminds you that these are still big trouble.

Engines: 4.2•2.7D **CO2:** 271-374g/km **MPG**: 17.8-27.6

Gearbox: Auto **Boot:** 450/2013 **Towing:** 3500 **Seats**: 5

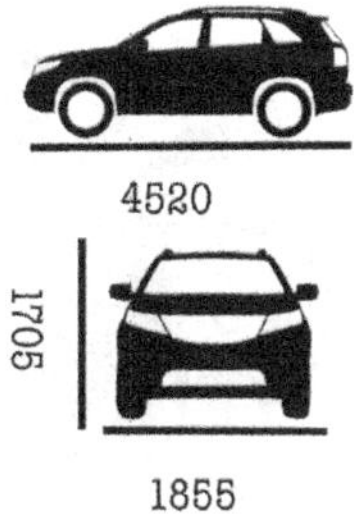

Renault Koleos 2008 - 2010

French interpretation of the Nissan X-Trail. Inside it was like a people carrier with a dashboard bin that pulled out and loads of other cubbyholes. Split tailgate is always welcome, the keycard to start it, less so. 2.0 diesel is good and the high specification models designed to sell on price. Owners happy with it and based on improved Nissan X-Trail.

Engines: 2.0D **CO2:** 191g/km **MPG**: 38.7 **Gearbox:** Man/Auto **Boot:** 450/1380 **Towing:** 2000 **Seats**: 5 **Doors:** 5

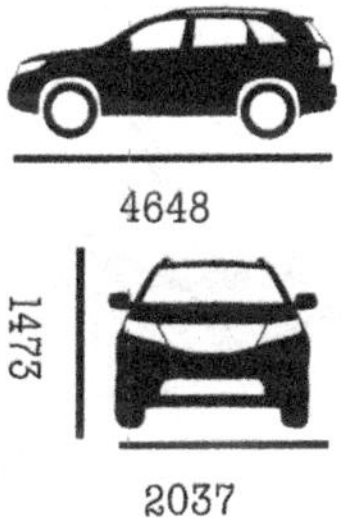

Saab 9-3X 2009 - 2011

Super stylish all wheel drive estate, inside there is loads of useful space with or without the rear seats folded. Always worth bothering to find spare parts to keep one running. Has quite complex electronic traction assistance, but mainly an on road motor. Electrics, windows and locking not great and engine ECU means rough running. Check aircon.

Engines: 2.0 • 1.9D **CO2:** 135-229g/km **MPG**: 28.2-55.4 **Gearbox:** Man/Auto **Boot:** 477/1331 **Towing:** 1600 **Seats**: 5

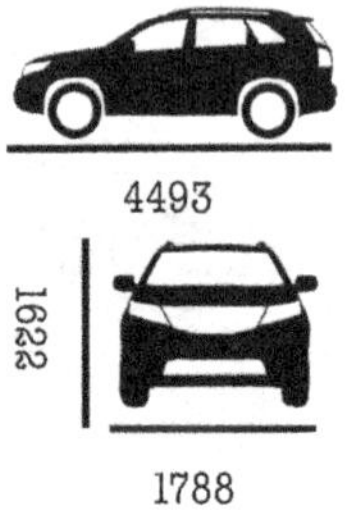

SEAT Altea Freetrack 4 2007 - 2009

People carrier, introduced in 2004 so it has cupholders and cubbyholes. Because it is a 4×4 there is fairly daft black plastic cladding, extra ground clearance and traction system that wakes up when nudged. Rear seats fold and slide, meaning lots of luggage room. Timing belt tensioner failure, electrics specifically cruise control, ABS sensors and ECU.

Engines: 2.0 • 2.0D **CO2:** 179-223g/km **MPG**: 30.1-41.5

Gearbox: Man/Auto **Boot:** 490/1562 **Towing:** 1400 **Seats**: 5

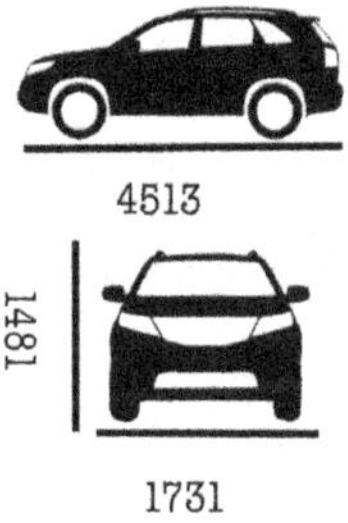

Skoda Octavia 4 x 4 1998 - 2005

Working blokes Quattro, with the emphasis on being able to lug stuff and do a real world job rather than being lifestyle. Plus it has metal guide rails in the floor so you can slide stuff in and it is a metre wide. Diesel ECU failure, water pumps, air mass sensors and ignition coils across the engine range. Electrical failures in the cockpit.

Engines: 1.8 • 1.9D **CO2:**167- 211g/km **MPG**: 31.7 - 41.0

Gearbox: Man/Auto **Boot:** 560/1620 **Towing**: 1500 **Seats**: 5

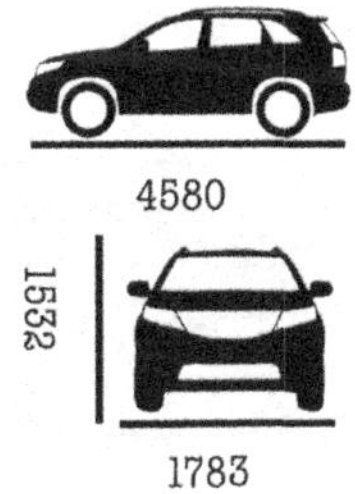

Skoda Octavia Scout 2007 - 2012

Octavia 4 x 4 taken to the show off level with ride and cladding which was turned up a notch in 2009. Estate basis and despite the extra off road plumbing there is plenty of loadspace that is flat and wide with rails inset for slide-ability. Most in circulation are diesel rather than petrol. Very effective full time 4 x 4 system. Better than any SUV.

Engines: 2.0 • 2.0D **CO2:** 173-207g/km **MPG**: 32.0-44.0

Gearbox: Man/Auto **Boot:** 605/1655 **Towing:** 2000 **Seats**: 5

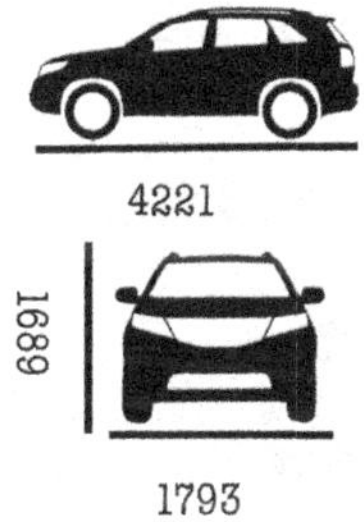

Skoda Yeti 2009 - 2017

Cute, compact crossover fashion accessory with a daft name. Lovable though, with diesel and petrol options, the smallest ones are just front wheel drive. Deservedly popular as it is actually useful, but does not come with all the baggage of an overweight SUV. Buy a simple, low spec one and enjoy. Oil leaks, smoke and electrical issues.

Engines: 1.2 • 1.8 • 2.0D **CO2:** 128-149g/km **MPG**:44.1-52.3

Gearbox: Man/Auto **Boot:** 416/1760 **Towing:** 1400 **Seats**: 5

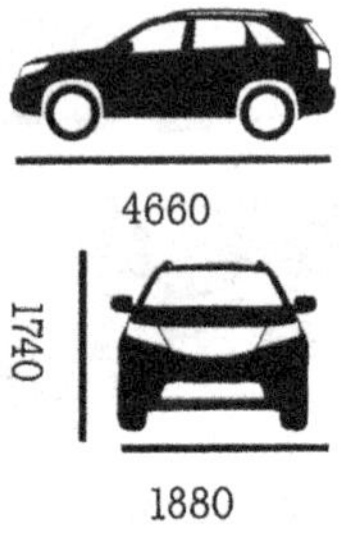

SsangYong Kyron 2006 - 2010

Baby 4×4 also available as just a two wheel drive model. Bought to do hard work, but seems to be up to it hence the to 'infinity and beyond' warranty. The compromises are rear seats that fold, but not flat, yet you have clever tech that gets you down a hill without rolling over. No downsides. Wear and tear issues to suspension. Fluid leaks.

Engines: 2.0•2.7D **CO_2:** 1971206g/km **MPG**: 32.5-37.7 **Gearbox:** Man/Auto **Boot:** 625/1222 **Towing:** 2300 **Seats**: 5

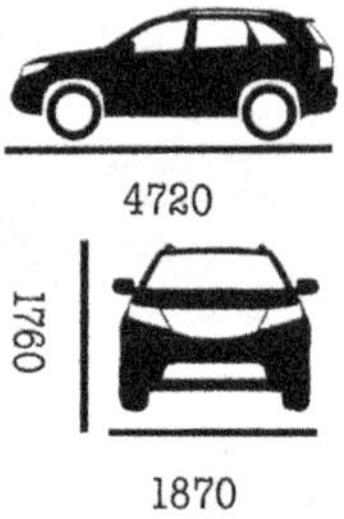

SsangYong Rexton 2003 - 2013

Full fat Merc-looking SUV with none of the style or fragility baggage that most people want. These are bought for towing and working and no one is short changed. Inside the rear seats flip and fold, so there is plenty of luggage depth. Better to drive from 2006 revisions when Rexton II arrived. Excellent all rounder, just wear and tear from towing.

Engines: 2.7D **CO_2:** 228g/km **MPG**: 31.4 **Gearbox:** Auto **Boot:** 935/1524 **Towing:** 3500 **Seats**: 5/7 **Doors:** 5

Subaru Forester 1997 - 2003

Hard nut version of the Legacy and Impreza. Extra ride height, but mostly a brutal estate car with a 2.0 or from '98 Turbo petrol power. All Weather Pack installed air conditioning, contradictory sunroof and heated seats. Facelift from 2000 with more roomy rear seats... Wear and tear and neglect, everything costs a fortune to fix.

Engines: 2.0 **CO2:** 214-235g/km **MPG**: 28.5-31.4 **Gearbox:** Man/Auto **Boot:** 408/1526 **Towing:** 2555 **Seats**: 5 **Doors:** 5

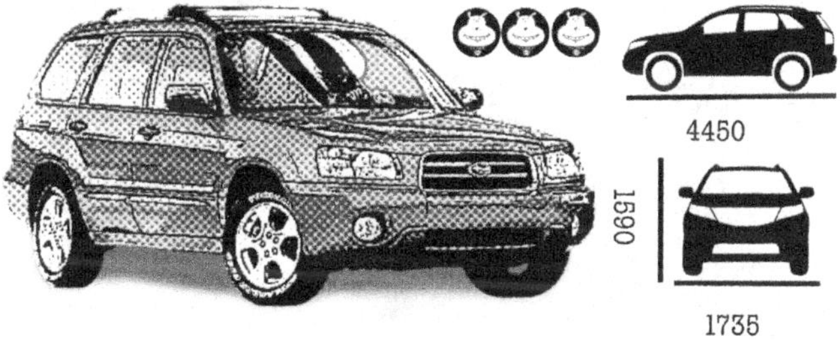

Subaru Forester 2002 - 2008

Restyled, just a tad wider so rear seats have a bit more wiggle room. Also went on a diet with aluminum paneled roof, roof rails, sunroof frame and bumper mounts. The 2.5 Turbo from 2004 in the XT was something quite mad. Lots of standard kit, the reality of running all these models is the sheer expense. Worn transmission, blowing turbo.

Engines: 2.0•2.5 **CO2:** 220-254g/km **MPG**: 26.4-30.4 **Gearbox:** Man/Auto **Boot:** 387/1595 **Towing:** 2000

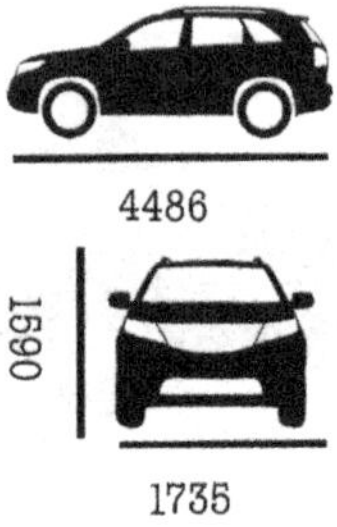

Subaru Forester 2008 - 2012

Finally got diesel power and the following to make it better on and off road, Subaru Vehicle Dynamics Control, self -leveling rear suspension and electric power steering. Different 4 x 4 systems as the manual version has a trad centre diff and the automatic an auto split torque system. Wear and tear mostly especially if used for towing.

Engines: 2.0 • 2.0D **CO2:** 167-198g/km **MPG**: 33.6-44.8
Gearbox: Man/Auto **Boot:** 450/1660 **Towing:** 2000 **Seats**: 5

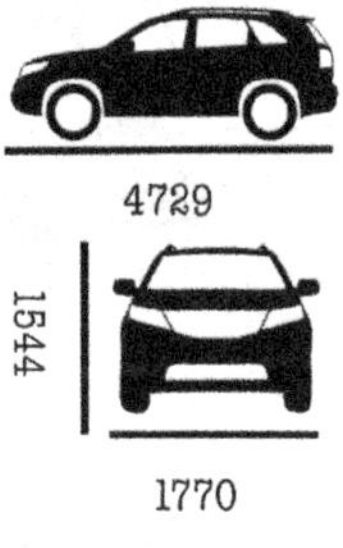

Subaru Outback 2003 - 2009

Pumped up Legacy, a less angry Forester, effectively an ambitious estate car. 2.5 and 3.0 petrol power until 2008 and by then it was too late. Tough as nuts and easy to drive, but never that cheap to own and run. From 2007 a reversible cargo floor so you could choose carpet or lino. Nothing to get too excited about. Check for damage.

Engines: 2.5 • 3.0 • 2.0D **CO2:** 153-233g/km
MPG: 33.0-49.00 **Boot:** 459/1628 **Towing:** 2000

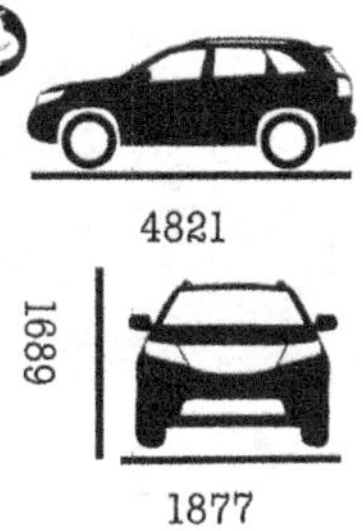

Subaru Tribeca 2006 - 2007

Not pretty, neither is the petrol bill. Two reasons people steered clear when new, as a Banger buy though it will do a tough job. Plenty of room inside, less so if you fill the seven seat model SE7 up with bodies. Subarus repair costs are on the high side which could be offset by a marginal purchase price Wear and tear, towing damage, that's it.

Engines: 3.0 **CO2:** 291g/km **MPG**: 23.0 **Gearbox:** Man/Auto **Boot:** 525/1671 **Towing:** 2000 **Seats**: 5 **Doors:** 5

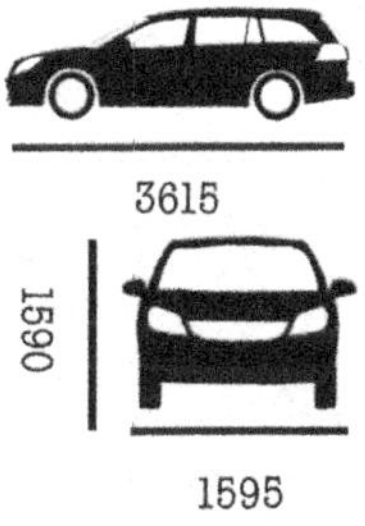

Suzuki Ignis 2003 - 2008

We've transplanted this here for fun, the rather larger five door answer with a 60/40 split rear folding seat which also recline. There is a 1.5 petrol used in the automatic and from 2003 in the 4Grip four wheel drive model which is why it is really here. Always going to be worth having on the basis that it has exceptional reliability and build quality.

Engines: 1.3 **CO2:** 150g/km **MPG**: 44.1 **Gearbox:** Man/Auto **Boot:** 181/419 **Seats**: 5 **Doors:** 5

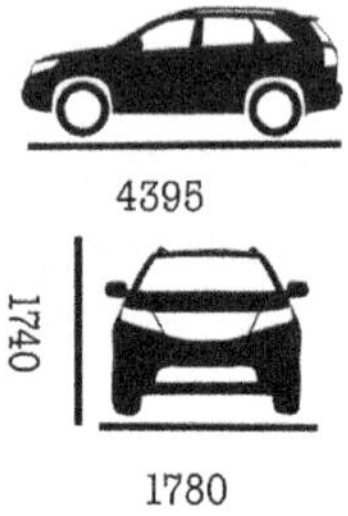

Suzuki Grand Vitara 1998 - 2005

A proper job when going off road, thanks to old fashioned chassis. Three and five door models, plus a soft top, so there is something for everyone there and the bigger XL7 in 2001 meant you could have more seats. Reliable and mostly petrol rather than diesel, so sort of future proofed. Just minor electrical issues, look for coolant and oil leaks.

Engines: 1.6 • 2.0 • 2.5 • 2.7 • 2.0D **CO2:** 260g/km **MPG**: 26.2

Gearbox: Man/Auto **Boot:** 572/984 **Towing:** 2000

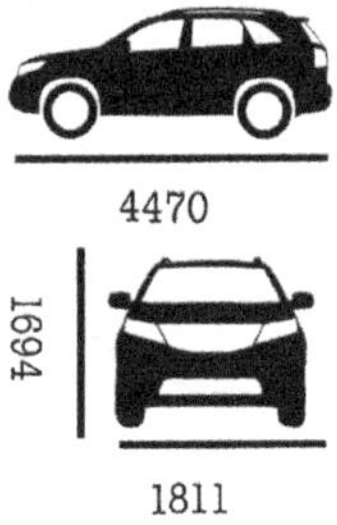

Suzuki Grand Vitara 2005 - 2014

Softened up to be more road friendly as out went the chassis, but still tough. Three and five door models, but never bothered with a multi seat option. Needed 2009 electronic road holding upgrades and perkier engines for a more sophisticated drive and will have the SZ badge. A reliable buy, just look at at the brakes and suspension.

Engines: 1.6 • 2.4 • 1.9D **CO2:** 183-195g/km **MPG**: 34.5-40.4

Gearbox: Man/Auto **Boot:** 398/1386 **Towing:** 1600

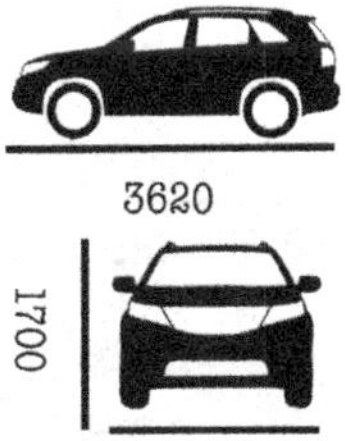

Suzuki Jimny 1998 - 2005

Pint sized Jeepney thing that is brilliant off road. Hard wearing with proper hose down plastic interior. Convertible model is a huge amount of fun, if that roof is missing or damaged, difficult and expensive to fix. The 1.3 petrol engine is keen and relatively simple to look after, just needing a cambelt every 70K miles. Can be killed by rust of course.

Engines: 1.3 **CO2:** 174g/km **MPG**: 38.7 **Gearbox:** Man/Auto
Boot: 113/286 **Towing:** 450 **Seats**: 4 **Doors:** 3

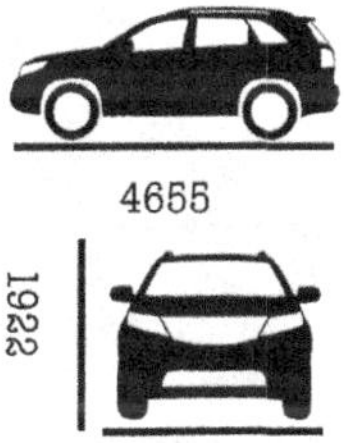

Tata Safari 2000 - 2005

Terrible.

Engines: 2.0D **CO2:** 300g/km **MPG**: 30.4 **Gearbox:** Man
Boot: 530/2050 **Towing:** 2000 **Seats**: 7 **Doors:** 5

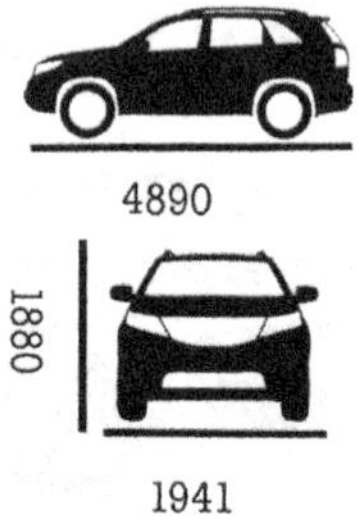

Toyota Land Cruiser Amazon 1998 - 2008

Legendcruiser. effectively the 80 Series from 1990 marks the greatest 4×4 known to humankind, followed by the 100 series in '98. VX V8s petrols, but the diesels always better buys. Never cheap and rightly so. Great at towing, lugging and just going where others struggle. Won't break down if looked after. Lasts forever, no such thing as a cheap one.

Engines: 4.7•4.2D **CO2:** 292-387g/km **MPG**: 17.3-25.4
Gearbox: Man/Auto **Boot:** 252/1114 **Towing: 3500**

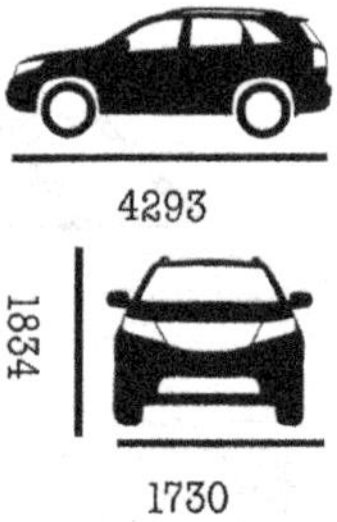

Toyota Land Cruiser Colorado 1996 - 2002

The so-called light duty six cylinder models or LJ90. There's a 3.0 diesel and 3.4 petrol. Everything that applies to the Amazon can be directed at the lickle school run one. Facelift in 1999, air conditioning and ABS became standard. Eight seaters plus power sockets. Check for leaks and damage.

Engines: 3.4•3.0D **CO2:** 253-341g/km **MPG**: 20-29.7
Gearbox: Auto **Boot:** 169/1100 **Towing:** 2800 **Seats**: 5/8

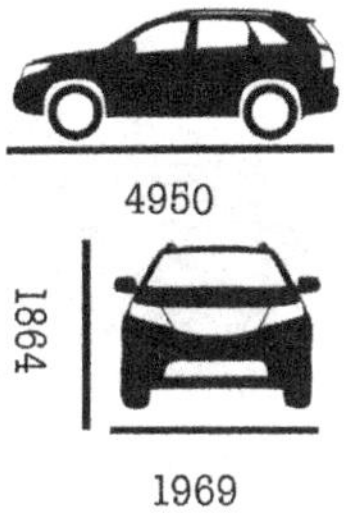

Toyota Land Cruiser V8 2008 - 2011

J200 is not a Range Rover, one of the best reasons to buy. Proper Luxury conveyance moved around by a 4.5 V8 diesel. Revamped fairly considerably but still retained the good old chassis layout as the best way to get around the rough stuff. Everything is standard, proving 4 zone climate control does actually exist. Sublimely special. Never cheap.

Engines: 4.5D **CO2:** 270g/km **MPG**: 27.7 **Gearbox:** Auto **Boot:** 259/1267 **Towing:** 3500 **Seats**: 5 **Doors:** 5

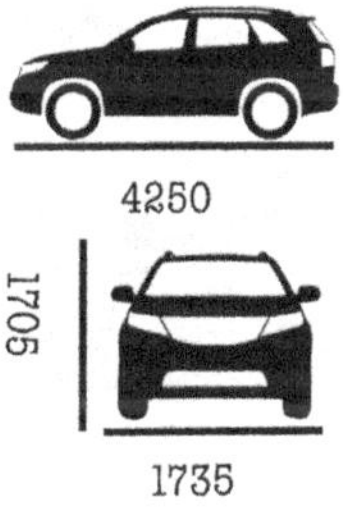

Toyota RAV4 2000 - 2005

Not the fun, funky thing it used to be, rather more grown up, bigger and better, nicer to drive and lots more engines and specs. Still available as a three door, plus there was a diesel from 2001. The rear seats slide back and forth, fold, tumble, plus you can pull them out completely. Two wheel drive on the smallest 1.8 petrol. Just multi-owner servicing neglect.

Engines: 1.8•2.0•2.0D **CO2:** 175-211g/km **MPG**: 30.1-39.8 **Gearbox:** Man/Auto **Boot:** 170/510 **Towing:** 1500 **Seats**: 5

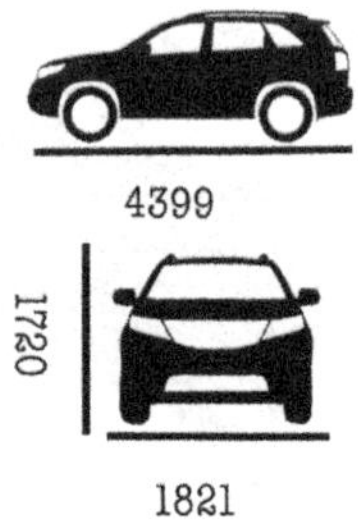

Toyota RAV4 2006 - 2012

Inevitably growing bigger and more of an MPV mash up . Apparently 25% more luggage room as the rear seats folded flat at the pull of a lever very quickly. Those seats also slid around for comfort and space. The 2.2 turbodiesel was the popular choice but the 2.0 petrol is the longer term option. Check the electricals and plug into a diagnostic kit.

Engines: 2.0•2.2D **CO2:** 154-174g/km **MPG**: 38.7-48.7
Gearbox: Man/Auto **Boot:** 586/1469 **Towing:** 1900 **Doora**: 5

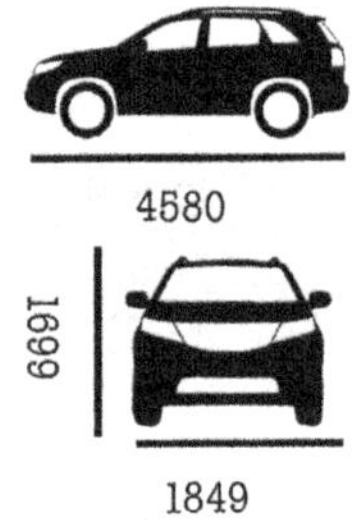

Vauxhall Antara 2007 - 2015

A.K.A. Chevrolet Captiva, but with less (5)seats. At least buyers could pack a load of luggage in the back, especially with the rear seats folded, even if the tailgate is a bit narrow. Few of the petrol 2.4s around, mostly 2.0 diesel. Slightly pointless, just a hatchback with part time 4×4. Weak clutch, worn suspension and the odd minor electrical.

Engines: 2.4•2.0D **CO2:** 198-229g/km **MPG**: 29.4-37.2
Gearbox: Man/Auto **Boot:** 370/1420 **Towing:** 2000 **Seats**: 5

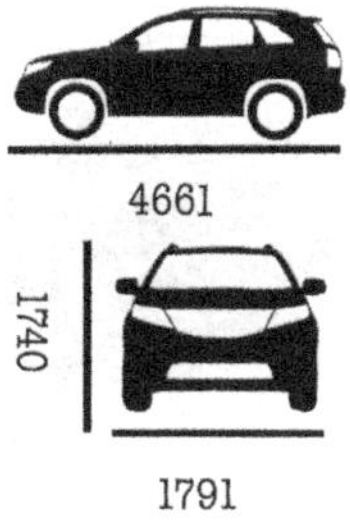

Vauxhall Frontera 1991 - 2003

The Isuzu Trooper connection didn't help much and built in Luton. Aimed at the high street rather than hard work. Revamped in '98 with new engines and styling which did not make it any more interesting. Part time four wheel drive, but essentially this is a big dumb vehicle with not much real charm. Towing damage, worn suspension, leaky engines.

Engines: 2.2 • 3.2 • 2.2D **CO2:** 263-326g/km **MPG**: 22.4-31.0 **Gearbox:** Man/Auto **Boot:** 510/1784 **Towing:** 2400-2800

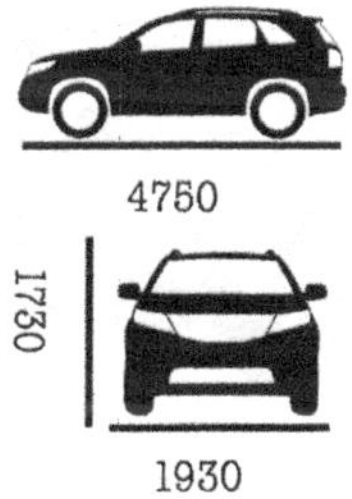

Volkswagen Touareg 2003 - 2009

Porsche Cayenne without the stupidly big engines or poncy badge, although did have the former for a while. It is more off road focused than the Porker which may be one good reason to buy. Uprated in 2007 and it was possible to get a big diesel V10 and even a mildly economical BlueMotion. Check for oil and coolant leaks, suspension wear,

Engines: 3.6 • 2.5D • 3.0D • 5.0D **CO2:** 265-331g/km **MPG**: 20.5-28.8 **Gearbox:** Auto **Boot:** 555/1570 **Towing:** 3500

Volkswagen Tiguan 2008 - 2016

Golf crossover with/without all wheel drive. Escape spec upped the faux 4 x 4 look whilst inside there was a big boot and a sliding rear seat. There is some complex equipment and screens which could eventually go wrong. Most buyers went for the diesel, but the TSIs are our sort of petrols. Air con, electrics seem to be a weak point and power steering.

Engines: 1.4 • 2.0D **CO2:** 189-199g/km **MPG**: 33.6-39.2

Gearbox: Man/Auto **Boot:** 505/1510 **Towing:** 2200-2500

Volvo XC70 2000 - 2007

Middle class country dwelling Labrador owners needed, ride height, matt black plastic, plus sophisticated Haldex diff will get drivers out of serious trouble. Rear seats were versatile and could be folded right down for tricky loads and it would 40/20/40 split too and a luggage restraint that folds out of the roof. Suspension bushes, broken throttle bodies.

Engines: 2.4 • 2.4D **CO2:** 201-244g/km **MPG**: 27.7-37.2

Gearbox: Man/Auto **Boot:** 485/1641 **Towing:** 2100 **Seats**: 5

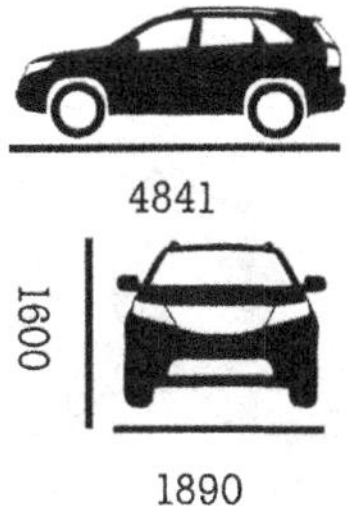

Volvo XC70 2007 - 2016

Lots more specification and clever gadgets like Active Bending Lights and Load Compensating Suspension, the downside is that the complication can eventually cause trouble. For those that didn't need all wheel drive a DRIVe was part of an emissions saving initiative. 2009 mild restyle and engine realignment. Electrical failures, ECU, windows.

Engines: 2.5 • 2.4D **CO2:** 186-244g/km **MPG**: 27.2-40.4
Gearbox: Man/Auto **Boot:** 575/1600 **Towing:** 2100

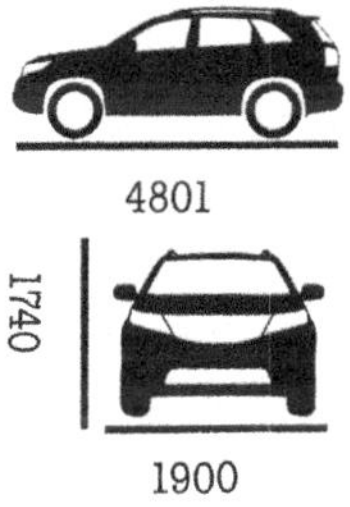

Volvo XC90 2002 - 2014

Acceptable school run fun as a standard seven seater from the off. Those seats lived in the boot floor and are easy to flip up. Otherwise there's a huge luggage area and the rear seats slide independently as well. Most are the D5 Diesel rather than the T6 turbocharged petrol. Hasn't scored that well for reliability over the years, electrical + ECU issues mainly.

Engines: 3.2 • 2.4D **CO2:** 219-289g/km **MPG**: 23.3-34.0
Gearbox: Man/Auto **Boot:** 249/1837 **Towing:** 2250 **Seats**: 7

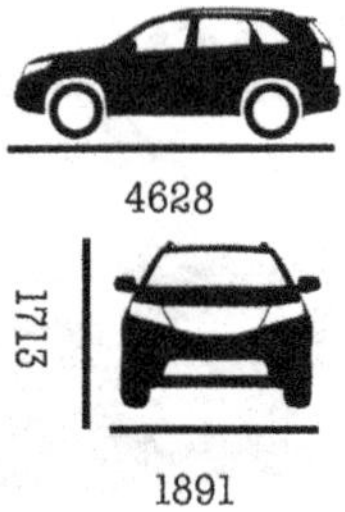

Volvo XC60 2008 - 2013

Playtime off roader with 4WD, or not. Jolly well built inside and out. Actually very roomy and all XC most normal families will ever need. Has the latest don't crash tech, the City Safety system stops you colliding below 20mph. That sort of stuff has to work to pass the MOT which is worth remembering. Electrical issues and general wear and tear.

Engines: 3.0 • 2.4D **CO2:** 183-274g/km **MPG**: 24.1-40.9
Gearbox: Man/Auto **Boot:** 495/1455 **Towing:** 2100-2400

Bangerxecutives - Middle Order Motoring

The whole point of executive motors is to make the owner feel good about themselves. Correspondingly every one else who doesn't have one and makes do with a van, or humble small hatchback, feels not quite as 'special'. Indeed, this was crucial to the whole company car park hierarchy back when there was a functioning economy.

It grew like topsy in the '80s, firmed up in the '90s and in the 2000s there was a massive choice of desirable saloons, yes these are mostly classy boxes. The Alfa Romeos looked great but never sold in big numbers [precisely because they are Alfas. The real bedrock of this market was Audi As and Series BMWs which became as good as they were ever going to get.

Jaguars were an option, except that a Lexus would be the much safer option. More reliable anyway. Last chance to get a Saab, or a Vauxhall Omega as things started to change. Not always for the better.

Depreciation is a huge help in making ownership a much more democratic affair. If you always promised yourself a Mercedes, even though the world is full of Sprinter vans, you still like the idea of a down at heel, slightly rusty E-Class, then that dream can come true.

The good news is that some of these are just an upmarket twist on a much more modest model. That means running costs don't need to be prohibitive at all. Simply do your research to buy parts for less and not worry about getting bits from a breaker's yard rather than a main dealer. Indeed, get the hang of motoring here and you can easily move up to the Luxobarge level in a few pages time.

Caution is always required when buying hard used examples which did indeed work hard in the hands of the regional marketing director. Others may have been bought by a well off retired middle to upper manager who can afford to keep one in the manner to which it should be accustomed. A lot though are several dozen owners down the rabbit hole of abject neglect.

Be careful.

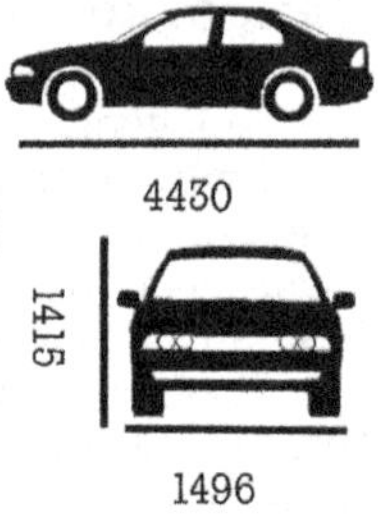

Alfa Romeo 156 1997 - 2005

Inside and out this has always been a good looking car and great to drive. The level of standard equipment was pretty much kitchen sink and all models have climate control, central locking and electric windows. Failing oil pump and air flow sensors. Air con problems, Selespeed gearbox malfunctions.

Engines: 1.6 • 1.8 • 2.0 • 3.2 • 1.9D • 2.4D **CO2:** 161-212g/km
MPG: 32.1-46.3 **Gearbox:** Man/Auto **Boot:** 374

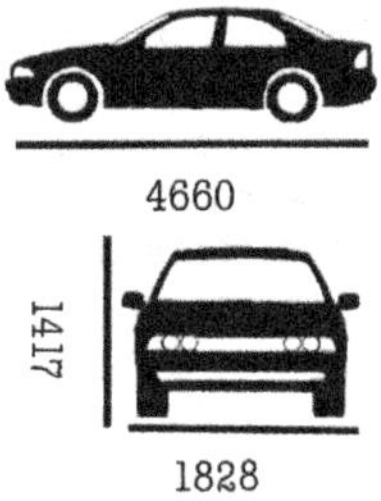

Alfa Romeo 159 2006 - 2011

Powerful and refined diesels which make the most sense. The proper petrols though deliver more of what you want, in particular the 3.2 V6. Fun to drive and it came with plenty of equipment as standard from climate control to alloy wheels, lots more spec levels from 2009. Water pump failures, dual mass flywheels, steering rack breakages.

Engines: 1.8 • 2.2 • 3.2 • 1.9D • 2.4D **CO2:** 138-260g/km
MPG: 25.7-53.3 **Gearbox:** Man/Auto **Boot:** 405

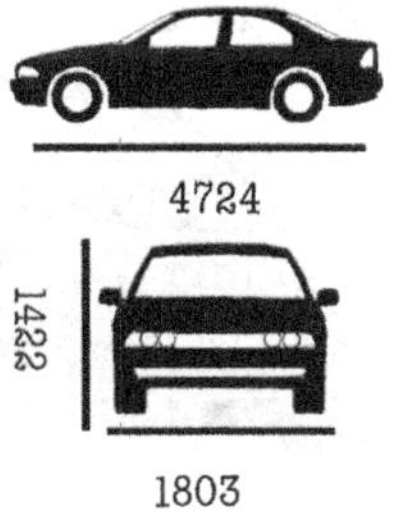

Alfa Romeo 166 1999 - 2005

Inside the dashboard looks lovely, all overlapping dials, timber rimmed steering wheel and plush leather on the Lusso models. It is a big car that likes going quickly and you can take corners very smartly indeed. Suspension bushes, oil pump failure, the usual electricals and needs timing belts sharpish.

Engines: 2.0 • 3.0 • 3.2 **CO2:** 230-297g/km **MPG**: 22.6-29.1 **Gearbox:** Man/Auto **Boot: 490 Seats**: 5 **Doors:** 4

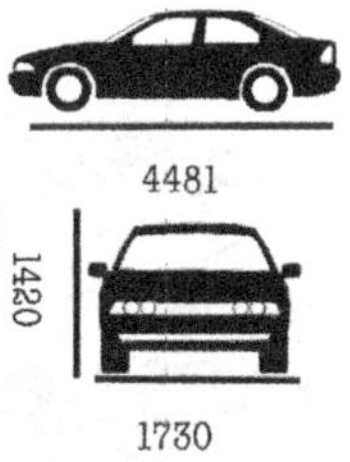

Audi A4 1995 - 2001

Default business saloon Still not a 3 series. From '99 a revamp tweaked the suspension and steering for the better. Rear seat passengers remained cramped and endured rock hard suspension. earlier models had some sparse specs, especially below SE. It is the wear and tear issues of brakes and suspension, otherwise it is just service history.

Engines: 1.6 • 1.8 • 2.4 • 2.6 • 1.9D • 2.5D **CO2:** 142-288g/km **MPG**: 23.0-53.0 **Gearbox:** Man/Auto **Boot:** 439 **Seats**: 5

Audi A4 2001 - 2005

Better to drive and a more socially acceptable level of standard equipment. A bigger car, not by much, but it made a difference The arrival of the 2.0 FSI meant economical and perky petrol engines were a real alternative to those dirty old diesels. A great buy. Early model ignition coils, water pumps, automatic gearbox unreliable.

Engines: 1.6 • 1.8 • 2.0 • 2.6 • 3.0 • 1.9D • 2.5D **CO2:** 149-252g/km **MPG**: 26.9-51.4 **Gearbox:** Man/Auto **Boot:** 445 **Seats**: 5

Audi A4 2005 - 2007

These mid period A4s are dull but decently finished with the usual mind blowing and wallet saving range of engines. Obviously the four cylinder TDi diesels are universally excellent and the fresh Fyi petrol units are very economical too. If mpg is not paramount then the multivalve 2.6, or 2.8 V6s are great. Clutch failing and central locking.

Engines: 1.6 • 1.8 • 2.0 • 3.2 • 1.9D • 2.0D • 2.5D • 3.0D **CO2:** 153-252g/km **MPG**: 34.0-49.7 **Boot:** 460/1200

Audi A4 2008 - 2015

Late period A4 is prime diesel technology. The 2.0-litre TDI engines are great otherwise it is the TFSIs petrols as the obvious go anywhere for less choices. Decent spec includes stability control, electric windows all round, climate control and remote central locking. Blocked Particulate Filters, otherwise just worn brakes and suspension.

Engines: 1.8 • 2.0 • 3.2 • 2.0D • 2.7D • 3.0D **CO2:** 120-252g/km
MPG: 34.0-61.4 **Gearbox:** Man/Auto **Boot:** 480/962

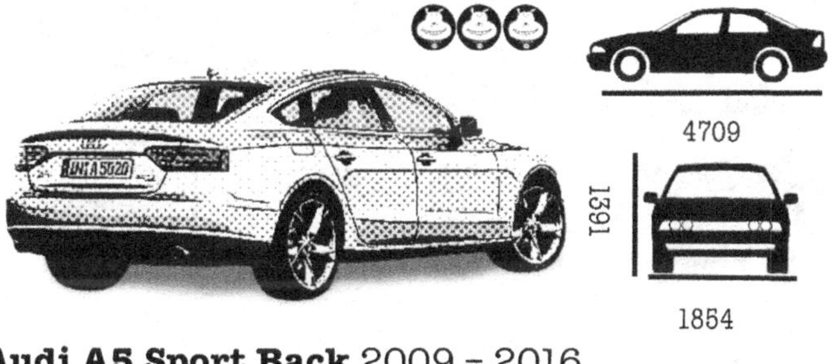

Audi A5 Sport Back 2009 - 2016

Coupe with 4 doors, hatch related to the A4 with bigger boot. Perfect for awkward loads, except for sloping roof, so hardly an Avant. There is a ton of the usual engine options and five trim levels all with climate control and cruise control as standard. Oddly irrelevant. Mostly electrical issues, central locking and in car entertainment problems.

Engines: 2.0 • 3.2 • 2.0D • 2.7D • 3.0D **CO2:** 137-252g/km
MPG: 34.0-54.3 **Gearbox:** Man/Auto **Boot:** 480/980 **Doors**:5

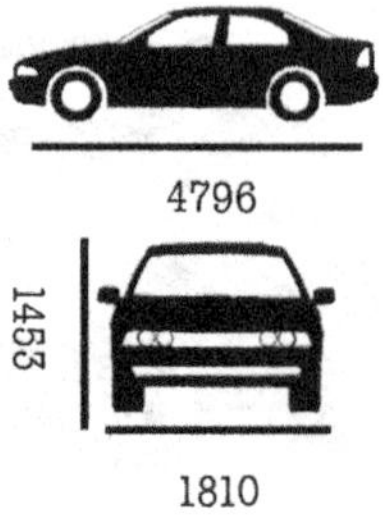

Audi A6 1997 - 2004

Previously a revamped 100, this was an all new safer and more spacious saloon. Best to avoid the Tiptronic clutchless manual on the grounds of complication. They all have twin airbags and climate control. Fuel pumps and air flow meters can fail, wheel bearings, plus with old age electric windows and central locking.

Engines: 1.8 • 2.0 • 2.0 • 2.7 • 3.0 • 4.2 • 1.9D • 2.5D

CO2: 154-312g/km **MPG**: 21.7-49.6 **Boot:** 549

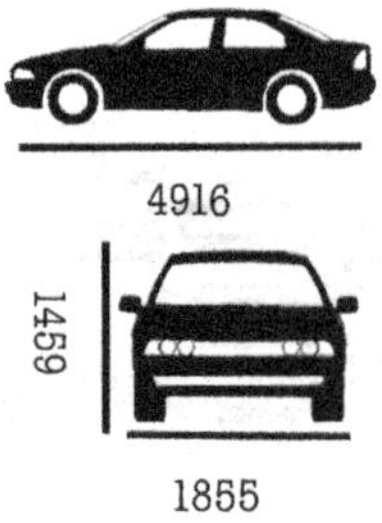

Audi A6 2004 - 2011

More A6 goodness, then in 2008 it got a major freshen up and you might spot the LED tail lights and a bunch of new engines. The 2.0 FSI makes a lot of sense to get into ULEZ areas. Quite a lot of aluminum in the bodywork so can be expensive to fix. Engine ECU failure means rough running. Air conditioning on the blink, brakes and suspension.

Engines: 2.0 • 2.8 • 3.0 • 2.7D • 3.0D **CO2:** 149-219g/km **MPG**: 30.1-49.6 **Gearbox:** Man/Auto **Boot:** 546

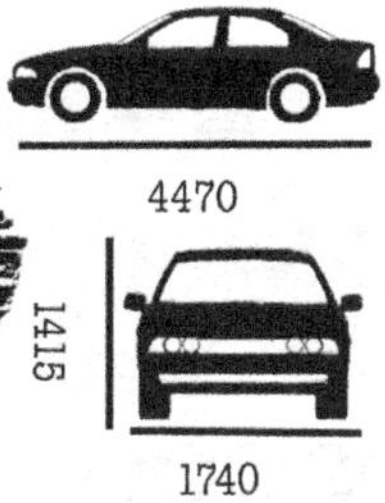

BMW 3 Series 1998 - 2005

Nerds know it as the E46. Good range of engines both petrol and diesel. Hard to go very wrong, just depends on what you can find and afford. Available in several exciting model flavours, but the saloon is almost the perfect small executive. Power loss and smoke, diesel issues, broken air mass meters and cracked inlet manifolds. Worn balljoints.

Engines: 1.8 • 2.0 • 2.2 • 2.5 • 3.0 • 1.8D • 2.0D • 3.0D **CO2:** 151-218g/km **MPG**: 31.0-50.4 **Gearbox:** Man/Auto **Boot:** 439

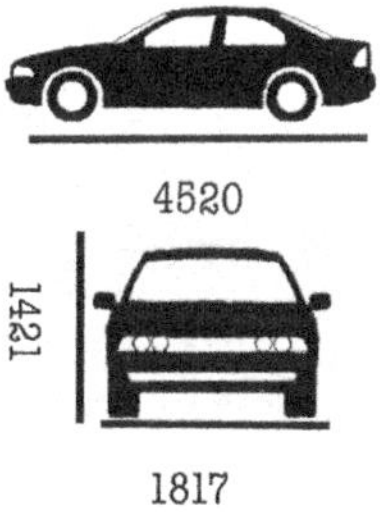

BMW 3 Series 2005 - 2011

Generation 5 saloon. Build quality does not seem to be as excellent as the old days. You do need to choose your engine carefully as the 2.0 diesel from '07 caused a huge amount of grief. Mostly fixed now, but dare you take the risk? There was a revamp in 2008. Efficient Dynamics, stop/start, brake regeneration things to go wrong. Avoid the 2.0d.

Engines: 2.0 • 2.5 • 3.0 • 3.5 • 2.0D • 3.0D **CO2:** 128-232g/km **MPG**: 31.0-58.9 **Gearbox:** Man/Auto **Boot:** 460 **Seats**: 5

BMW 5 Series 1996 - 2003

The last of the fixable BMWs, (E39, anoraks) and greatest. The diesels are great especially the bigger sixes which offer thumping performance combined with regular 40mpg+ economy. Otherwise the petrol sixes will last forever, if not longer. Check electronics on old ones, suspension bushes. Diesels may need new injectors.

Engines: 2.0 • 2.2 • 2.5 • 3.0 • 3.5 • 4.0 • 2.5D • 3.0D **CO2:** 179-295g/km **MPG**: 23.0-42.2 **Boot:** 460

BMW 5 Series 2003 - 2010

So here is the beginning of the unfixable generation of BMWs, or at least ruinously expensive to run models. Looks better now than it first did. Big facelift in 2007 saw the economical upgrades including regenerative braking and the TV may work as it is digital. Looked less scary. Expensive suspension, brake and ECU implications. 2.0d problems.

Engines: 2.5 • 3.0 • 2.0D • 3.0D **CO2:** 165-176g/km **MPG**: 38.2-45.6 **Gearbox:** Man/Auto **Boot:** 520 **Seats**: 5 **Doors:** 4

BMW Gran Turismo 2009 - 2017

This just slips under the Bangerpedia cut off radar, klaxon thing. A great big pointless and not that pretty four door coupe. Based on the 7 it is also 300Kg heavier for some bad reasons and only seats four comfortably. Boot not massive. This really is a completely pointless exercise in bad car design. Faulty water pumps, brake servo failure.

Engines: 3.0 • 4.4 • 3.0D **CO2:** 173-263g/km **MPG**: 25.2-43.5
Gearbox: Man/Auto **Boot:** 440/1700 **Seats**: 5 **Doors:** 5

Cadillac BLS 2006 - 2010

Built from Saab 9-3 underparts in Sweden and 1.9 diesel also found in Vectra, but marginally better. Designed to be comfy and cruiserish and no more. Decent standard equipment, (the Elegance has everything) offers value and slab sided styling. Few sold and the downsides as a Saab 9-3 and has electrical issues and irritating warning lights.

Engines: 2.0 • 2.8 • 1.9D **CO2:** 151-259g/km **MPG**: 26-49
Gearbox: Man/Auto **Boot:** 425/1273 **Seats**: 5 **Doors:** 4

Chrysler 300C 2005 - 2010

Huge saloon improved with the addition of Bentley badge? They may well be correct. Here's a motor with character, plus a ton of standard equipment. Mercedes E Class sourced diesel is the sensible choice, V6 petrol is only OK, whilst the Hemi is outstandingly silly. Diesel electronic throttle sensor failure, oil leaks, starter motors break and engine oil leaks.

Engines: 3.5 • 5.7 • 6.1 • 3.0D **CO2:** 215-288g/km **MPG**: 23.3-34.9 **Gearbox:** Man/Auto **Boot:** 504 **Seats**: 5 **Doors:** 4

Hyundai Sonata 2001 - 2004

Pretend executive, with sub Jag styling and a V6 is pretty much it (not pretty). Big boot and split fold rear seat which will never get used. Well built and reliable, you can drive this quite comfortably into the ground if you wish. Hard to argue against a Sonata provided you can find one. These just get old and tired suspension and brakes, nothing else.

Engines: 2.0 • 2.7 **CO2:** 215-256g/km **MPG**: 26.4-31.4 **Gearbox:** Man/Auto **Boot:** 394 **Seats**: 5 **Doors:** 4

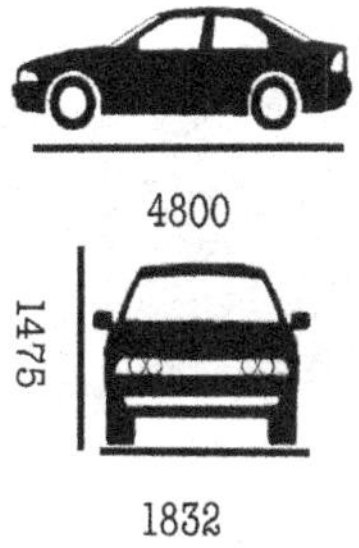

Hyundai Sonata 2005 - 2010

Return of the faux-exec. It is well-built, has good engines, especially the diesel. As before versatile interior and most things are standard. That is all you really need went it comes to driving a biggish saloon. Cancelled in the UK due to lack of interest. Will only have wear and tear items to worry about such as brakes and pot hole damaged suspension.

Engines: 2.0 • 3.3 • 2.0D **CO2:** 163-190g/km **MPG**: 35.3-46.3 **Gearbox:** Man/Auto **Boot:** 462 **Seats**: 5 **Doors:** 4

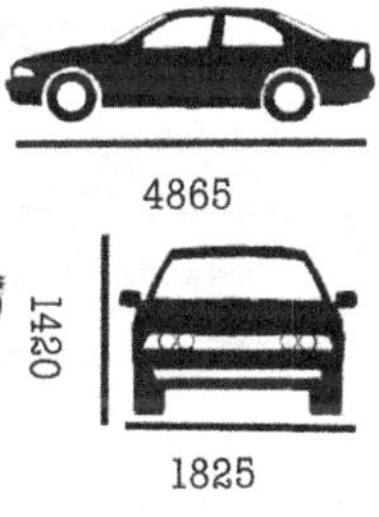

Hyundai XG 2000 - 2003

Like a slightly underwhelming Jaguar XF but better. Absolutely everything is standard, electric most things including a sunroof, plus cruise, plus it has an awful lot of safety kit. Be careful as there is a ton of stuff to go wrong. Even though it might not. Actually it doesn't maybe the odd electrical, look out for rust now.

Engines: 3.0 **CO2:** 256g/km **MPG**: 26.4 **Gearbox:** Auto **Boot:** 411 **Seats**: 5 **Doors:** 4

Jaguar S-Type 1999 - 2007

First of all it came with a 3.0 V6 and 4.0 V8, V6 diesel arrived in 2004 following on from a mildish facelift so it made company car commonsense. Comfortable, easy to operate and pretty quick when required. Take advantage of its unlovedness. V8 timing chain tensioners. . Rust, expensive suspension parts, electrics like windows.

Engines: 3.0•4.2•2.7D **CO2:** 179-259g/km **MPG**: 24.5-41.5 **Gearbox:** Man/Auto **Boot:** 400/810 **Seats**: 5 **Doors:** 4

Jaguar X-Type 2001 - 2010

Some owners have sworn by it, others at it. Underneath is a Ford just not Jaggy enough. Four wheel drive useful, but no Quattro. Not well made. Really not very Jaguar at all, even though we quite like the look of the estate version. Avoid. Automatic gearboxes can fail, four wheel drive transfer box expensive. Otherwise clutch failures. Oil and coolant leaks.

Engines: 2.5•3.0•2.0D•2.2D **CO2:** 149-234g/km **MPG**: 29.5-49.1 **Gearbox:** Man/Auto **Boot:** 452 **Seats**: 5 **Doors:** 4

Jaguar XF 2008 - 2015

21st Century finally arrived. Four cylinder diesel designed to be the bread and butter model. Revised in March 2009 with a V8 petrol engine plus a new 3.0-litre diesel. A bigger facelift was carried out in the middle of 2011, but really that's all you need to know. Electrical issues, particulate filters blocked, modern car stuff, no biggies.

Engines: 3.0 • 5.0 • 3.0D **CO2:** 179-249g/km **MPG**: 26.8-42.0
Gearbox: Man/Auto **Boot:** 500/920 **Seats**: 5 **Doors:** 4

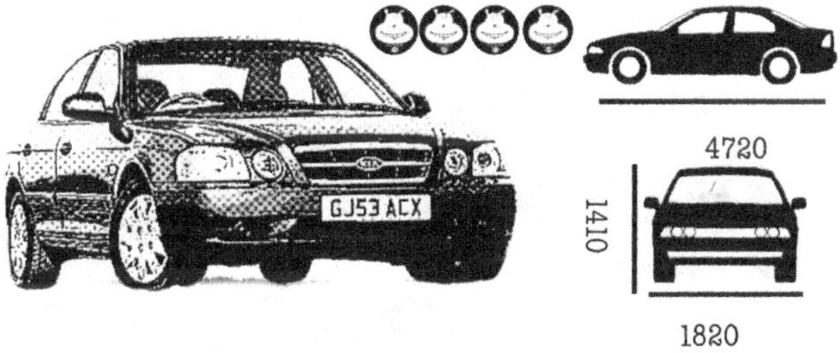

Kia Magentis 2001 - 2005

Pretend executive and sub Mercedes E styling with a V6. That's it pretty much and it certainly isn't pretty. Big boot and split fold rear seat which will never get used. 2002 the V6 2.0 replaced by a four cylinder version. Well built and reliable, you can drive this quite comfortably into the ground if you wish. Wear and tear, so brakes and suspension.

Engines: 2.0 **CO2:** 204g/km **MPG**: 32.8 **Gearbox:** Man/Auto
Boot: 385 **Seats**: 5 **Doors:** 4

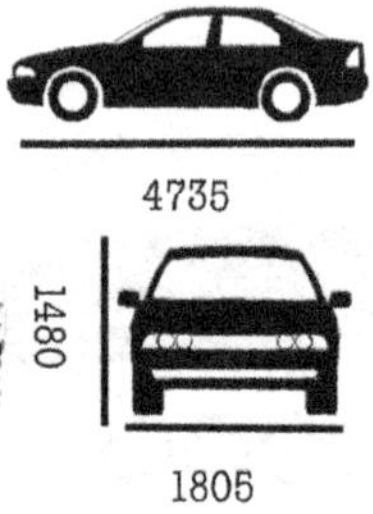

Kia Magentis 2006 - 2010

Return of the faux-exec looks no more special than a Passat. A '93 one. It is though well-built, has good engines, especially the diesel. Despite looking OK cancelled in the UK due to lack of interest, but obviously a jolly good buy because there isn't too much to worry about, just check electrics, suspension and brakes.

Engines: 2.0•2.0D **CO2:** 162-184g/km **MPG**: 36.7-47.1
Gearbox: Man/Auto **Boot:** 420 **Seats**: 5 **Doors:** 4

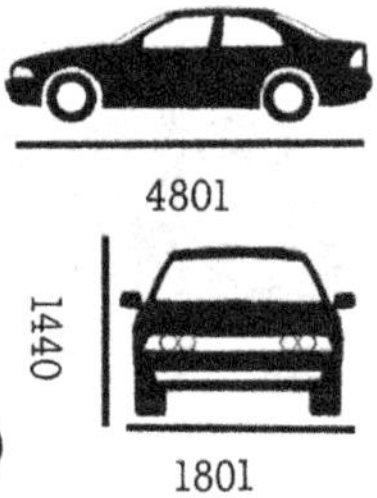

Lexus GS 1993 - 2004

What a mid size Jag should be, ultra reliable with the highest of technical technology. Most things standard, except diesel, they didn't exist back then. V8 or 6 cylinder, what a brilliant choice of engine which won't leave you stranded and waiting for a flatbed. Check suspension and brakes. indeed all parts are pricey despite being a Toyota. Can be rusty.

Engines: 3.0•4.3 **CO2:** 281-296g/km **MPG**: 23-24.4
Gearbox: Auto **Boot:** 515 **Seats**: 5 **Doors:** 4

Lexus GS 2005 - 2011

Smoothed out styled and just the one petrol engine offering which is fashionably hybrid as well. Made by Toyota, so it won't blow up. If the electric motor goes, or you need some chip work done it may cost you a small fortune, but that is the only reason not to buy, that and you don't like the styling. Coil pack failure otherwise just wear and tear,

Engines: 3.0 • 3.5H **CO2:** 180-232g/km **MPG**: 29-37.2
Gearbox: Auto **Boot:** 430 **Seats**: 5 **Doors:** 4

Lexus IS 1999 - 2005

Started life as a Toyota Altezza and that is reason enough to buy one. Copycat 3 Series saloon. Just the one 2.0 6 cylinder petrol engine until the IS300 arrived. SE has most fixtures and fittings and there is the LE too. Best not to buy a wide boy gold plated one. No issues, just wear and tear as usual to fix the brakes, suspension and any missing parts.

Engines: 2.0 • 3.0 **CO2:** 233-256g/km **MPG**: 26.2-29.1
Gearbox: Man/Auto **Boot:** 399 **Seats**: 5 **Doors:** 4

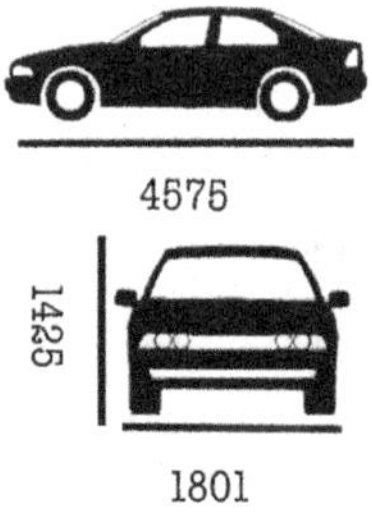

Lexus IS 2005 - 2012

Becoming rather complicated and the arrival of the diesel not the best for peace of mind. The EGR valve can cause all sorts of grief so you are best off sticking to the petrols. There were five levels of specification, but really whatever you chose you get more than enough kit. Anything that needs fixing is pricey, mostly cosmetic bodywork or wear and tear.

Engines: 2.5•2.2D **CO2:** 168-231g/km **MPG**: 28.8-44.8

Gearbox: Man/Auto **Boot:** 378 **Seats**: 5 **Doors:** 4

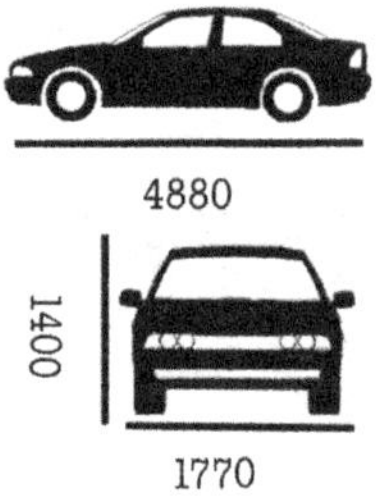

Mazda Xedos 9 1994 - 2001

Not remotely desirable, but quality kit. Great engines, a smooth V6 then from '98 clever 'Miller Cycle' (smaller size, behaves like something bigger) along with more anonymous restyle. Well put together. Good specs, SE had even more. Interestingly dull oddity. Rust is now an issue. Can look quite rough, it is tyres, exhaust and brakes that cost.

Engines: 2.3 **CO2:** 245g/km **MPG**: 28.2 **Gearbox:** Auto

Boot: 416 **Seats**: 5 **Doors:** 4

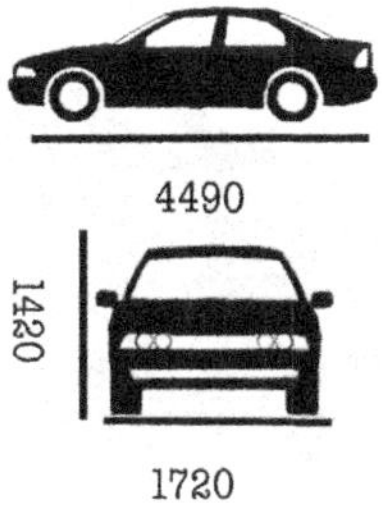

Mercedes C-Class 1993 - 2000

Halfbaked 190. Officially more wonderful, realistically less old school solid Merc more style than classy substance. 1997 revamp was considerable new engines and silly subtle styling things like dark tail lights, if that was your bag. The diesels are great of course. Still rusted though. Indeed rusty suspension parts, wipers cost a bomb to fix, engine oil leaks.

Engines: 1.8 • 2.0 • 2.3 • 2.4 • 2.8 • 2.2D • 2.5D **CO2:** 163-225g/km **MPG**: 26.6-45.6 **Gearbox:** Man/Auto **Boot:** 430

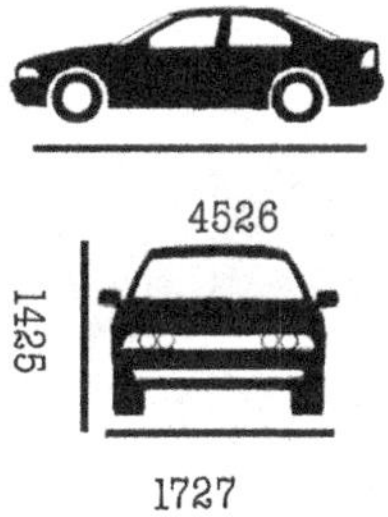

Mercedes C-Class 2000 - 2007

Big changes and it finally looked like a baby S. 'Twinpulse' engines would save the planet and your wallet. 2004 revamp nonsense with a new grille and clear headlamps as clearly they liked messing with light lenses. Still some rust. Electrics patchy so check windows and central locking, turbo oil seals and condition/age of glow plugs.

Engines: 1.8 • 2.5 • 3.0 • 3.5 • 2.1D • 3.0D **CO2:** 169-183g/km

MPG: 35.3-44.1 **Gearbox:** Man/Auto **Boot:** 455 **Seats**: 5

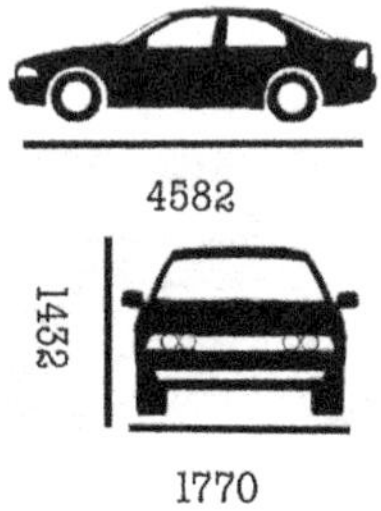

Mercedes C-Class 2007 - 2014

W207 with it a lot of depressing complication. Looks solidly serious and a C220 is all that anyone needs, but the petrols are properly economical and future proof. Changes outside of the Bangerpedia remit include 2011 and 2013 revamps. Instruments can go blank. Diesel wear, suspension + brakes.

Engines: 1.6 • 1.8 • 2.5 • 3.0 • 3.5 • 2.1D • 3.0D **CO2:** 127-187g/km **MPG**: 40.9-58.9 **Gearbox:** Man/Auto **Boot:** 475

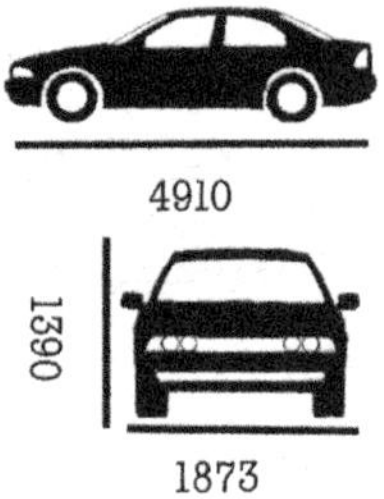

Mercedes CLS 2005 - 2010

Trendy coupe luxobarge saloon which is all E Class underneath and rather handsomely swoopy on top. Frameless windows could become an issue in the future, but otherwise it is a more interesting E that needs attentive previous owners. Petrol engines are great. Electrics will play up, diesels can be tired and leaking oil and coolant.

Engines: 3.5 • 5.0 • 6.2 • 3.0D **CO2:** 202-275g/km **MPG**: 24.4-37.2 **Gearbox:** Auto **Boot:** 505 **Seats**: 4 **Doors:** 4

Mercedes E-Class 1995 - 2002

Es were good, but these were disappointingly rusty. New school saloons not as solid or special as the old W124. Usual diesels and petrols and substantially revamped in 1999, so better to buy those if you must. The common rail diesels are good and the petrols reliable. Oil leaks, rust to brake lines, suspension bushes, engine mountings. More rust.

Engines: 2.0 • 2.4 • 2.8 • 3.2 • 4.3 • 2.1D • 2.7D • 3.2D **CO2:** 164-271g/km **MPG**: 25.0-45.6 **Boot:** 520

Mercedes E-Class 2002 - 2008

Bigger and aluminium panels, bonnet, front wings and bootlid plus a few other bits, worth remembering when it comes to repairs. Mid life quality upgrade from 2006. The E220 CDI is the bread and butter and V6 petrols better if you don't do the miles. Fuel pump failures, tired fuel injectors, catalytic converter issues, broken suspension springs.

Engines: 1.8 • 3.0 • 3.5 • 5.0 • 2.2D • 3.0D **CO2:** 169-202g/km **MPG**: 33.6-42.8 **Gearbox:** Man/Auto **Boot:** 540 **Seats**: 5

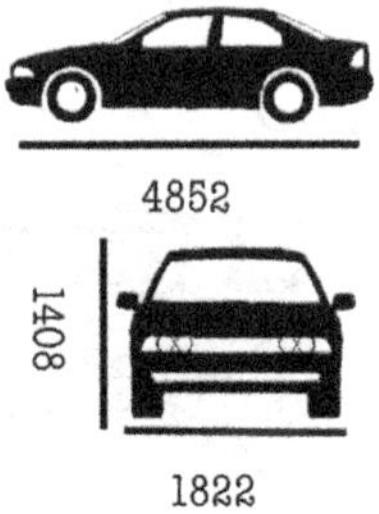

Mercedes E-Class 2009 - 2016

Ultra modern saloon it remained the diesel exec of choice and that's the vast majority of used options. Colossal boot of course and with an 80 litre tank, you will only have to top up with the black stuff every six months. Four cylinder petrol probably the easiest and dullest to own. Worn diesel injectors that smoke, also oil leaks.

Engines: 5.5 • 2.2D • 3.0D **CO2:** 139-256g/km **MPG**: 25.9-53.3 **Gearbox:** Auto **Boot:** 540 **Seats**: 5 **Doors:** 4

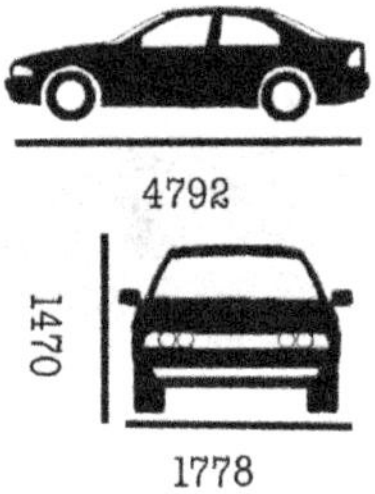

MG ZT 2001 - 2004

Rover 75 with added plastic V6 2.5 litre engine in two states of tune, a 160 and 180bhp. BMW diesel version was pretty good too, oh and a handful of V8s with a Mustang engine. So a complete personality change which seemed to work. Worth it for fun. Rotten sills and wheelarches, blown head gasket, diesel pump failure is major grief. Corroded suspension.

Engines: 1.8 • 2.5 • 4.6 • 2.0D **CO2:** 163-225g/km **MPG**: 30.0-48.8 **Gearbox:** Man/Auto **Boot:** 432 **Seats**: 5 **Doors:** 4

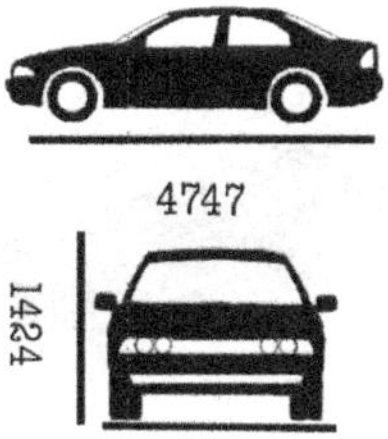

MG ZT 2004 - 2005

Odd death knell facelift with zero tolerance approach to plasticised chrome. Find an enthusiast owner who has truly cherished and kept the body on the straight and narrow, rather than a 23 owner example. Just as rusty as the original. Rotten sills and wheelarches, blown head gasket, diesel pump failure is major grief. Corroded suspension. **Engines:** 1.8•2.5•4.6• 2.0D **CO2:** 163-225g/km **MPG**: 30.0-48.8 **Gearbox:** Man/Auto **Boot:** 432 **Seats**: 5 **Doors:** 4

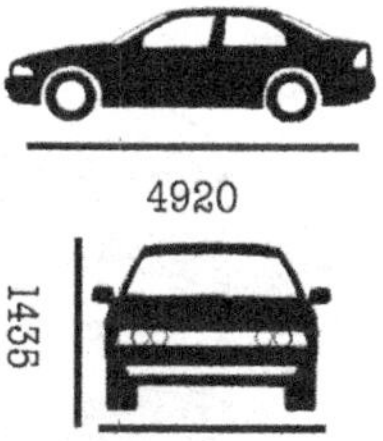

Nissan Maxima QX 2000 - 2002

Big saloons were never duller and anyway everyone else bought a Vauxhall Omega. So the 2000 reinvention, by adding the QX suffix, made it marginally larger and the 3.0 V6 got bhp boost, plus extra specification. Won't break down, but could sap your will to live. Wear and tear issues, or maybe worn out if it has a million miles on the clock.

Engines: 2.0•3.0 **CO2:** 220-252/km **MPG**: 26.9-30.7 **Gearbox:** Man/Auto **Boot:** 440 **Seats**: 5 **Doors:** 4

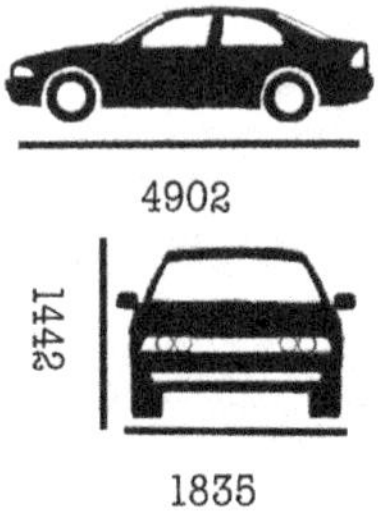

Peugeot 607 2000 - 2009

Great big soft French sponge, with zero business appeal. Those snobs missed out on supreme comfortableness. Petrols fine right up to the V6, but the diesel 2.0 and 2.2HDi made more sense. Revamped a tiny bit in 2004 with massive spec levels to include sat nav and mobiles. Electrics can have nightmares, worn smokey diesels, doors can drop.

Engines: 3.0 • 2.0D • 2.2D **CO2:** 162-242g/km **MPG**: 27.7-46.3 **Gearbox:** Man/Auto **Boot:** 509 **Seats**: 5 **Doors:** 4

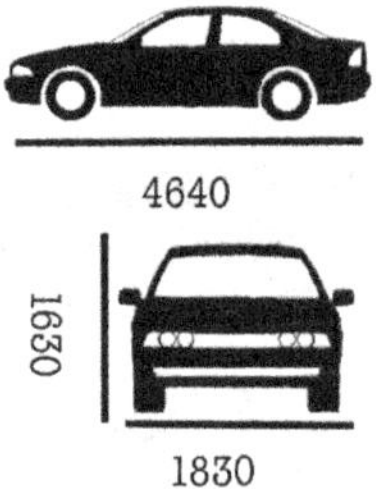

Renault Avantime 2002 - 2003

3 door Espace, perfect for pretentious executives Nobody, pretentious or otherwise bought this oddity. There is a colossal amount of electrics to go wrong. Lots of warning light bingo. Parts not cheap, good luck getting a wing this century. Objectively genius. Subjectively impractical the power steering may fail along with the automatic gearbox.

Engines: 3.0 **CO2:** 266g/km **MPG**: 24.6 **Gearbox:** Auto **Boot:** 530-900 **Seats**: 5 **Doors:** 3

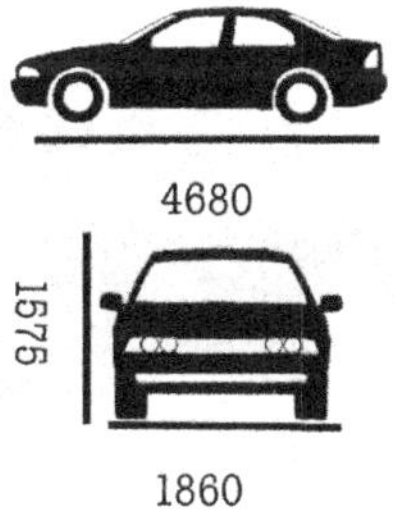

Renault Vel Satis 2002 - 2004

Sensible Avantime, with useful extra doors. It has fancy comfortable seats which are big and adjustable. There is a large 3.5 petrol and 3.0 diesel, but really it is the 2.2 diesel which the best lump to have. The ride is exceptionally French as can be the reliability. Oil leaks from diesels, failing turbo chargers, all the electrics though can be part time.

Engines: 2.0 • 3.5 • 2.2D • 3.0D **CO2:** 188-275g/km **MPG**: 24.6-39.8 **Gearbox:** Auto **Boot:** 459 **Seats**: 5 **Doors:** 5

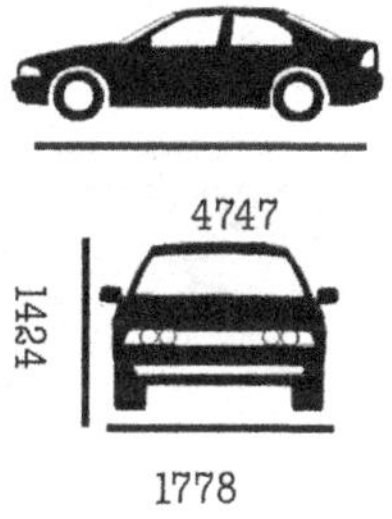

Rover 75 1999 - 2005

A BMW retro-reimagining of what a retired colonel drove in 1955. Three petrols and diesel and carefully crafted trim levels. Avoid head gasket K-Series issues and uninspiring 2.0 V6 engines. The diesel is probably the pick of the bunch. Rotten sills and wheelarches, blown head gasket, diesel pump failure is major grief. Corroded suspension.

Engines: 1.8 • 2.5 • 2.0D **CO2:** 163-225g/km **MPG**: 29.2-48.8 **Gearbox:** Man/Auto **Boot:** 432 **Seats**: 5 **Doors:** 4

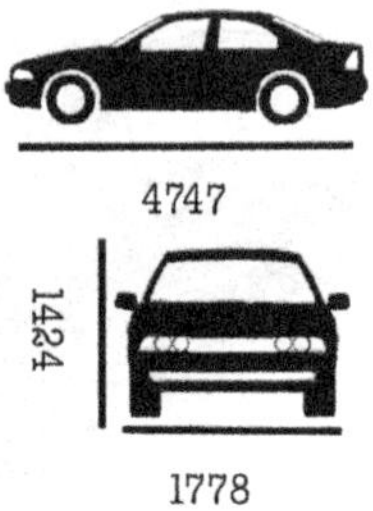

Rover 75 2004 - 2005

Before it left us the 75 got a questionable nose job. Long wheelbase version MG rover expected local mayors to buy. Not sure who was in the market for a Mustang V8. The last ones not necessarily as well built as the worthy originals. Rotten sills and wheelarches, blown head gasket, diesel pump failure is major grief. Corroded suspension.

Engines: 1.8 • 2.5 • 4.6 • 2.0D **CO2:** 163-225g/km **MPG**: 29.2-48.8 **Gearbox:** Man/Auto **Boot:** 432 **Seats**: 5 **Doors:** 4

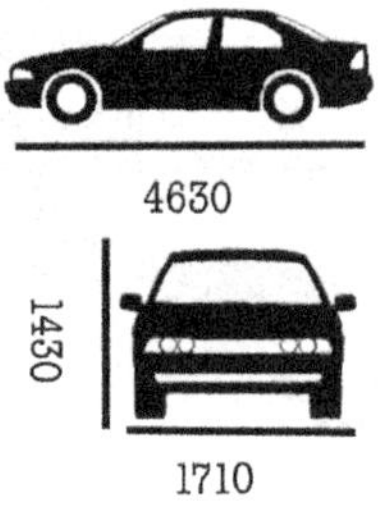

Saab 9-3 1998 - 2002

Swedish hatchbacked 3-Series according to nomenclature. Bemusing engine and model combinations. HOT engine was just that but the 2.0 turbo petrol and diesels are the ones to own. Specifications were comprehensive, worth aiming for an SE. On the whole pretty tough and reliable. Diesel fuel and oil leaks, worn timing chains, coil packs don't last long.

Engines: 2.0 • 2.2D **CO2:** 164-216g/km **MPG**: 31.4-45.6 **Gearbox:** Man/Auto **Boot:** 494/1314 **Seats**: 5 **Doors:** 5

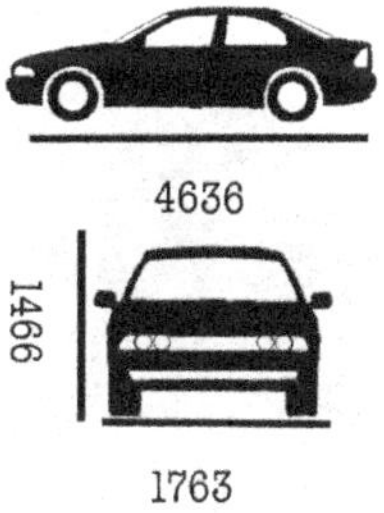

Saab 9-3 2002 - 2011

Last gasp small Saab in saloon forma. Lots of engines and specification combinations. Go for the 2.0 litre petrol or 2.2 diesel. There are multiple power outputs and even a bioethanol on offer which is now irrelevant. Safe and reasonably sound buy. Electrics, windows and locking not great and engine ECU means rough running. Check aircon.

Engines: 1.8 • 1.8B • 2.0 • 2.0B • 2.8 • 1.9D **CO2:** 157-216g/km

MPG: 35.8-48.7 **Gearbox:** Man/Auto **Boot:**425 **Seats**: 5

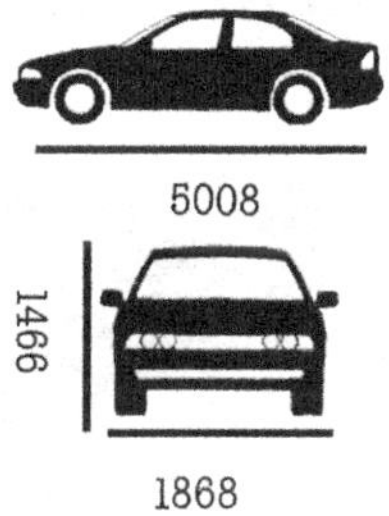

Saab 9-5 1997 - 2010

Swedish 5-Series according to the nomenclature. Huge inside and very comfortable, just not special. 2.0 and 2.3 and top of the range V6. 2001 revisions all round with diesel engines. 2005 facelift amounted to a headlamps with goth level mascara, still lovable. Coil pack failure, air con issues, automatic 'box + suspension bushes.

Engines: 2.0 • 2.3 • 1.9D • 2.0B • 2.3B **CO2:** 184-212g/km

MPG: 31.7-41.5. **Gearbox:** Man/Auto **Boot:** 500 **Seats**: 5

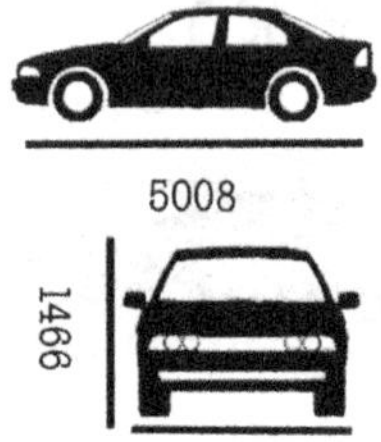

Saab 9-5 2010 - 2011

High Tech finale. Head up display telling you important stuff via the windscreen plus clever adaptive headlights. Still looks the future executive car part, remains safe, practical and sophisticated. Not many around and although this isn't strictly relevant oddly resistant to depreciation. Suspension wear, brake discs need replacing, slipping clutch.

Engines: 2.0 • 2.8 • 2.0D **CO2:** 139-244g/km **MPG**: 26.6-47.1

Gearbox: Man/Auto **Boot:** 515 **Seats**: 5 **Doors:** 4

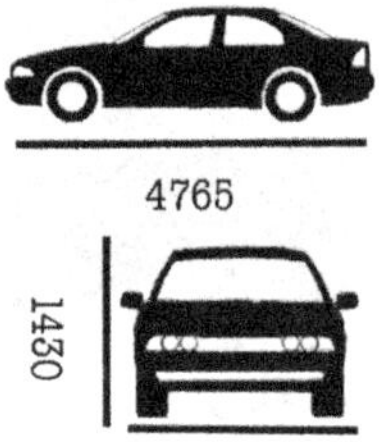

Toyota Camry 1996 - 2001

A simple enough line up with just a four and V6 cylinder engine in Standard and Sport flavours. 1999 update meant new lights and bumpers, but nothing else remotely interesting. The Camry is and was all about not breaking down. Wear and tear are the only issues and that will be brakes, tyres and suspension parts. Expensive to fix.

Engines: 2.2 • 3.0 **CO2:** 214-276g/km **MPG**: 24.4-32.1

Gearbox: Auto **Boot:** 524 **Seats**: 5 **Doors:** 4

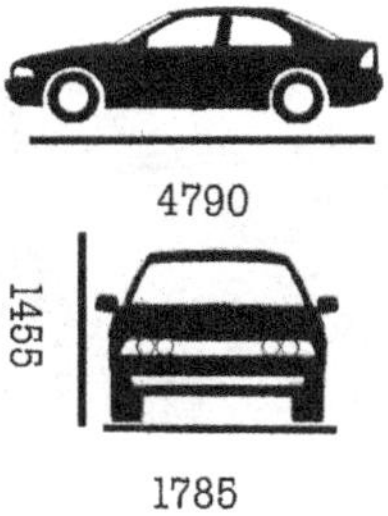

Vauxhall Omega 1994 - 2003

Made in Germany so they were well screwed together Better looking from 1999 and the more powerful DTi diesels are the ones to have, unless you can run a V6 petrol for fun. A motorway missile with plenty of equipment, especially top end Elite and CDX. Worn suspension bushes, oil leaks, fuel injector seals, air conditioning and heating system.

Engines: 2.0 • 2.2 • 2.6 • 3.2 • 2.2D **CO2:** 192-284g/km **MPG**: 23.9-39.8 **Gearbox:** Man/Auto **Boot:** 540 **Seats**: 5 **Doors:** 4

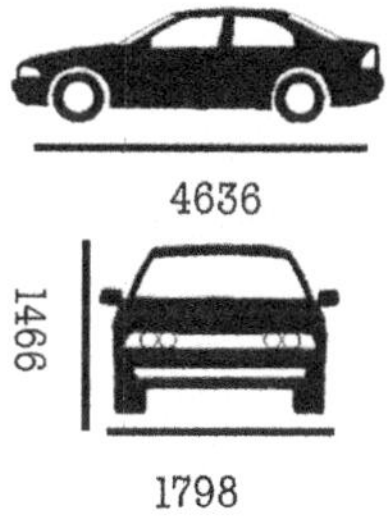

Vauxhall Signum 2003 - 2008

Vectra determined to put the CEO in the big back seat. Nobody got it, or bought it. All about the FlexSpace system so you could have five or ideally four seats with sliding and flipping seats. Revamped for 2005 with better engines and yes there is a V6. Alternative Vectra estate then. Worn diesel engine, smoke, all electric gadgets need checking.

Engines: 1.8 • 2.0 • 2.2 • 2.8 • 1.9D • 3.0D **CO2:** 159-214g/km **MPG**: 31.8-47.9 **Gearbox:** Man/Auto **Boot:** 365/1410

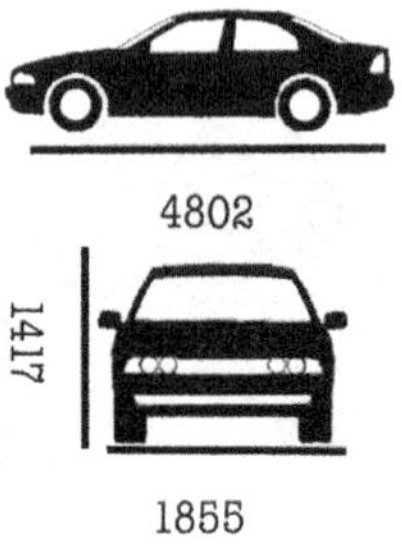

Volkswagen CC 2008 - 2012

It's a 'Comfort Coupe' which is longer, lower and wider than the deputy area manager's saloon, but with one less seat and a console splitting the interior down the middle and dividing the rear. Mind you in 2010 they retro fitted a proper three-seat rear. A couple of specs CC and GT so basically everything is standard.

Engines: 1.8 • 2.0 • 3.6 • 2.0D **CO2:** 125-182g/km **MPG**: 30-61.4 **0-60**: 6.8-9.8 **Gearbox:** 6spd 7spd A **Boot:** 452

SportiBangers - Pauper Performers

There is no rule which says that a Banger ought to be a sad little MOT borderline hatchback with all sorts of corrosion issues. Well actually you could get all of those wonderful qualities all wrapped inside a low down, two door sports car. This is reasons to be careful when buying a Banger, part 33.

Going for one of these is great if you do want to turn heads and change the sports car owning demographic. You know, from old and retired to someone a bit younger who actually appreciates the performance/handling of the chosen coupe and doesn't need hair dye/restorer.

Easy wins for us are getting a next generation 6 Series which be storing up a who;e heap of problems, whereas a Z4 is easier option that looks like a villains coupe. Much less would be a Chrysler Crossfire, whereas a Jaguar XK would be the best way to reclaim the biggest and bestest old cats.

What we have here is the perfect example of how the world of used cars actually works. Sometimes you may be offered a motor at a budget that you can afford and it turns out to be a sporty number. Obviously you are not going to squeeze a scout cub pack of nippers in the back, it's just you and a significant other. The insurance is OK and it has a years MOT. There's your Banger. A coupe. That's how it happens.

If cash is tight the most dull models qualify like an old Xsara, or more up to date C4, Citroens with issues. Better to go for Honda Civics and Accords for long term reliability. For real character the legendary Puma and Cougar are still around in a time before they were renamed and madeover into SUVs.

All you have to do is make sure you can pay the bills and there is your next set of stylish wheels. Job done. Ideally not previously owned by an idiot whose slammed it (taken away the suspension) and boosted the bhp to infinity with one of those fancy chips. Unless it actually was a Fast and Quite Furious film extra, better to buy from an OAP with a fully stamped service book. Anyway, let's have some fun.

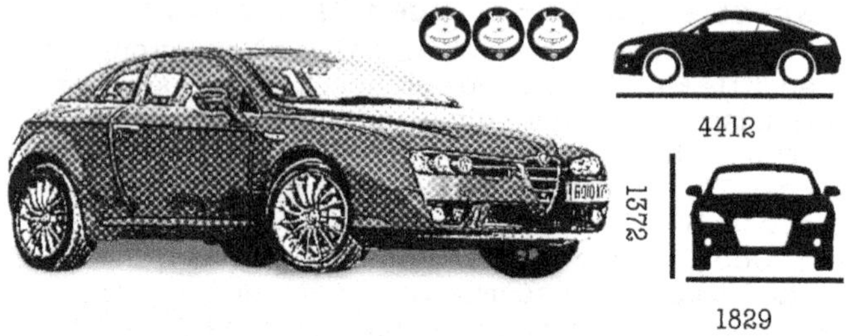

Alfa Romeo Brera 2006 - 2012

The Brera is now generally accepted as the Alfa that you don't need to make any allowances for. It's reliable, great value and very, very pretty. Not as sporty to drive as some other coupes, despite some sharp steering. Entry level 2.2 litre four-cylinder develops 185bhp. It sounds good too thanks to a trick exhaust. It is an Alfa, so electrics and ECUs.

Engines: 2.2 • 3.2 • 2.4D **CO2:** 179-273g/km **MPG**: 24.6-41.5
Top Speed: 138-142 **0-60:** 6.8-8.6 **Boot:** 300/610 **Seats**: 4

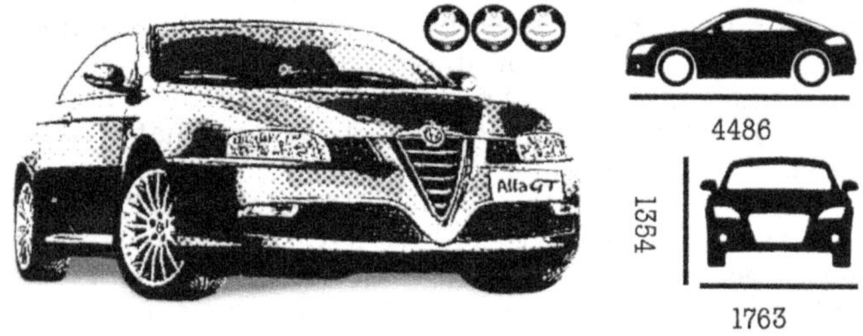

Alfa Romeo GT 2004 - 2010

Overlooked and underrated, here is a four-seater coupe that looks goods and is tidy when pushed. As with all Alfas these days the specification list is reason enough to consider a GT, as one trim level fits all. V6 for the enthusiast, but a diesel turbo for everyone else. Electrics, locks and windows. Water pump failure, worn brakes, low level rust.

Engines: 1.7 • 2.0 • 1.9D **CO2:** 176-207g/km **MPG**: 32.5-42.2
Top Speed: 130-135 **0-60:** 8.9-9.6 **Boot:** 320/905 **Seats**: 4

Alfa Romeo GTV 1996 - 2004

Head turning body by Pininfarina helped by a twin spark 2.0 litre under the bonnet from 1996, followed by a characterful 3.0 V6 in 1998. Standard specification was twin airbags and an alarm. Lusso pack though added air, leather and 16-inch alloys. Either looked after or falling to bits. Proper risky Alfa buy. MOT failure level rust, smokey engine.

Engines: 2.0 • 3.0 **CO2:** 220-278g/km **MPG**: 24.1-30.7 **Top Speed:** 137-154 **0-60:** 8.9-9.6 **Boot:** 155 **Seats**: 4

Audi A5 2007 - 2011

A TT for grown ups, well at least drivers with a small family, really. Plenty of petrol and diesel engines as is the Audi way. Reasonable standard kits includes 17 inch alloys, leather, MP3, rain sensing wipers, parking sensors and automatic boot opener. Mostly electrical issues, central locking and in car entertainment, ECUs and brakes.

Engines: 2.0 • 3.2 • 4.2 • 2.0D • 2.7D • 3.0D **CO2:** 140-191g/km **MPG**:39.2-53.3 **Top Speed:** 140 **0-60:** 5.9-8.3 **Boot:** 455/829

Audi TT 1999 - 2006

Automotive art. Thanks Audi for making traffic jams more stylish. Classic in waiting. Let gushingly it is just a Golf. The choice is what state of tune for your 1.8T and whether or not you want to Quattro or not. From 2005 there was the option of a 3.2 V6 and S-Line spec option packs. Coil packs, ECUs, brakes, body damage.

Engines: 1.8 • 3.2 **CO2:** 194-238g/km **MPG**: 26.6-35.0

Top Speed: 141-155 **0-60:** 6.5-7.8 **Boot:** 220/490 **Seats**: 4

Audi TT 2006- 2014

Seemingly pointless update, except that it is better to drive. Still a Golf and rather like that shopper it came in a diesel flavour too. Entry point 1.8 and 2.0FSi models as well as the TTRS and TT-S models. Lots of aluminum in the construction making suspension parts pricey if pranged. Check electric windows, Xenon headlamps, brakes and suspension bushes.

Engines: 2.0 • 3.2 • 2.0D **CO2:** 139-247g/km **MPG**: 27.4-53.3

Top Speed: 140-155 **0-60:** 5.9-7.5 **Boot:** 220 **Seats**: 4

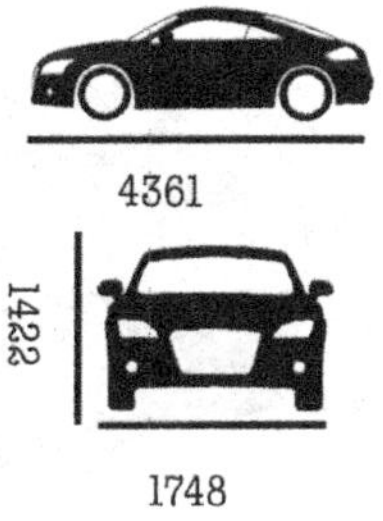

BMW 1 Series 2007 - 2011

Return of the 2002 as a handy sized (though bigger than the hatchback)point and squirt device. There were the hot ones 135i and also the others for shopping 120d, 120i and 118d. Got the M Series treatment as well. The right models will become collectible. 2.0 diesels need caution. Dual mass flywheels. Turbo smoke. ECU failures.

Engines: 1.6 • 2.0 • 3.0 • 2.0D • 3.0D **CO2:** 128-220g/km **MPG**: 30.7-58.9 **Top Speed:** 142-155 **0-60:** 5.3-7.6 **Boot:** 370

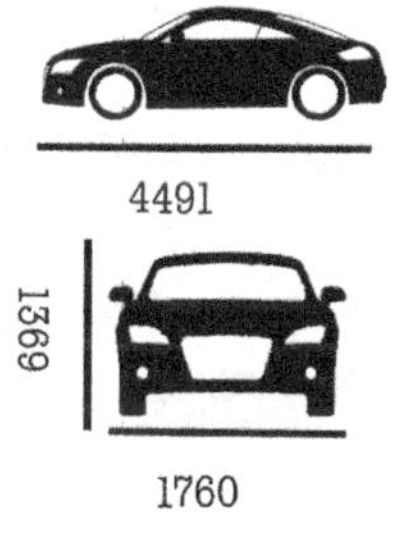

BMW 3 Ci Series 1999- 2006

It needs to be repeated, repeatedly, these were the very last of the remotely handsome BMWs and the coupe is even better. Buy because of the looks, but also the sheer value for money of the bigger engined ones in particular. Motorsports will never Bangers of course, but there are some lovely 6 cylinder alternatives for less. Bore wear, suspension bushes.

Engines: 2.0 • 2.2 • 2.5 • 3.0 • 2.0D • 2.9D **CO2:** 213-218g/km **MPG**: 28.5-31.7 **Top Spd:** 140-155 **0-60:** 6.5-8.2 **Boot:** 410

BMW 3 Ci Series 2007 - 2013

Still good to look at and if you are counting we are at the E92 model designation now. Not as many engine options as the standard model. Efficient Dynamics models, designed to save the planet and drivers a fortune at the pumps. Got a re-style in 2010, so those will cost a bit more, not least because the specification was upgraded, such as leather all round. ECU failures, electrics, windows and locks, plus brakes.

Eng: 1.6 • 2.0 • 2.5 • 3.0 • 4.0 • 2.0D • 3.0D **CO2:** 174-218g/km
MPG: 31.0-46.3 **Top Spd:** 153-155 **0-60:** 5.5-6.9 **Boot:** 440

BMW 6 Series 2004 - 2010

Designed for overweight, middle aged buyer whose eyesight was probably going and three passengers. V8 645 was perfect for them and if they were poorer, a 630i six. M Series for the fan boys, but every 6 costs a fortune to run and fettle and best not to be tempted by entry level pricing. ECU failures, electrics, oil leaks and suspension bushes.

Engines: 3.0 • 4.8 • 3.0D **CO2:** 183-279g/km **MPG**: 24.1-40.9
Top Speed: 155 **0-60:** 5.1-6.3 **Gearbox:** Auto **Boot:** 450

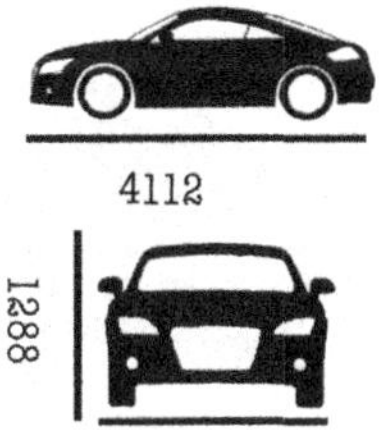

BMW Z4 2006 - 2009

Elegantly funky looking thing whereas the Z3 was a piggy compromise. Offered with the good six cylinder engines in SE and Sport trim. Lots of competition and unfashionable so they can be spectacular value. Seller profile is mostly well behaved people which is excellent news. Steering racks, minor rust, electrics, broken door handles.

Engines: 3.0 • 3.2 **CO2:** 213-292g/km **MPG**: 23.3-31.7

Top Speed: 155 **0-60:** 5.5-5.7 **Gearbox:** Man **Boot:** 243/300

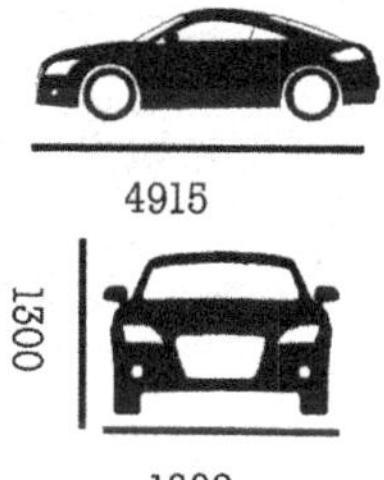

Chevrolet Camaro 1998 - 2004

Not something British coupe buyers ever asked for. Worth bearing in mind because a scrappy one could tempt you. Optional pop out targa panels mean that you can appreciate the 5.7 V8 soundtrack better. Otherwise pushed along by a more workaday V6. Parts are cheap, freight may not be. Unstressed, might have seized brakes and minor rust.

Engines: 3.8 • 5.7 **CO2:** 275-294g/km **MPG**: 20.9-22.8 **Top Speed:** 125-157 **0-60:** 5.5-8.5 **Gearbox:** Auto **Boot:** 187

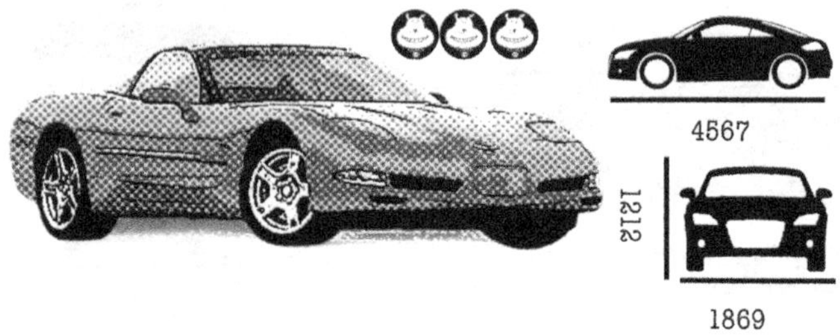

Chevrolet Corvette 1998 - 2002

The C5. Unlikely ever to be a Banger, but worth a mention, just in case there is a damaged repairable or partly immolated one to consider. Targa roof was an option, right hand drive less so. Left hook only, otherwise most things are standard. Sorting a scruffy one out may not cost a fortune in cheap mechanical parts. We'll forget about the C6.

Engines: 5.7 **CO2:** 299g/km **MPG**: 21.6 **Top Speed:** 170
0-60: 5.1 **Gearbox:** Auto **Boot:** 187 **Seats**: 2

Chrysler Crossfire 2003 - 2009

Looks like a cut price Batmobile it's an out of date SLK. It shares Mercedes underpinnings and just the one 3.2 V6 petrol engine. It can handle the power, rather underrated in that respect, so best to buy from someone who has driven it to the shops rather than chipped it to Mars. Electrics, ECU failure, but mostly not much to worry about.

Engines: 3.2 **CO2:** 256g/km **MPG**: 22.7 **Top Speed:** 144
0-60: 7.8 **Gearbox:** Man/Auto **Boot:** 215 **Seats**: 2

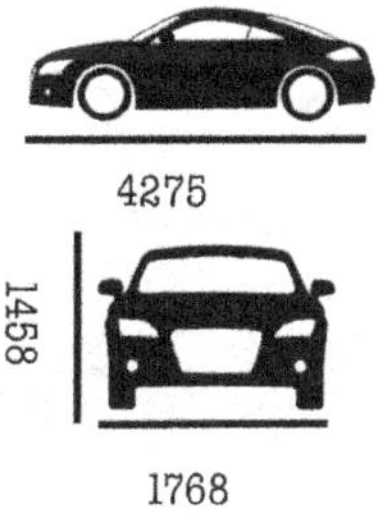

Citroen C4 Coupe 2004 - 2010

All the clever dick electrics that you find in luxury barges squeezed into a cubist coupe/hatch. Our favourite gadget is a toss up between the automatic fragrance distribution from the dash and translucent instruments. Lots of things to go wrong which is a warning. Dual mass flywheel failure. Air con system, electrics, dashboard and electric windows.

Engines: 1.4 • 1.6 • 2.0 • 1.6D • 2.0D **CO2:** 125-200g/km

MPG: 39.8-60.1 **Top Spd:** 113-116 **0-60:** 8.3-14.2 **Boot:** 320

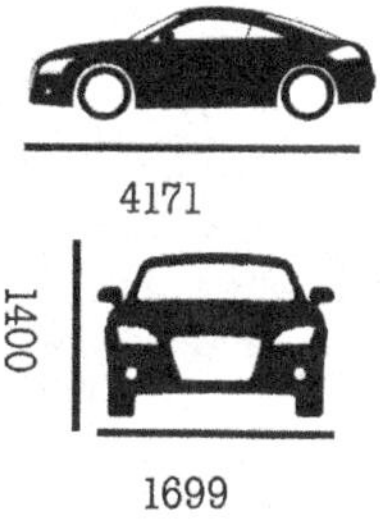

Citroen Xsara Coupe 1998 - 2004

Dull but worthy range of vehicles that will last a lifetime, proper Banger. The HDi 110 does have a bit of attitude which delivers thumpingly economical entertainment. There is a traditional hot hatch in the shape of the 167bhp VTS. No one really wants them. But they should. Still fixable. ECU issues, some late '90s diesel engine. Diesel clutches weak.

Engines: 1.6 • 2.0 • 2.0D **CO2:** 138-215g/km **MPG**: 30.4-54.3

Top Speed: 121-137 **0-60:** 30.4-54.3 **Boot:** 480/595

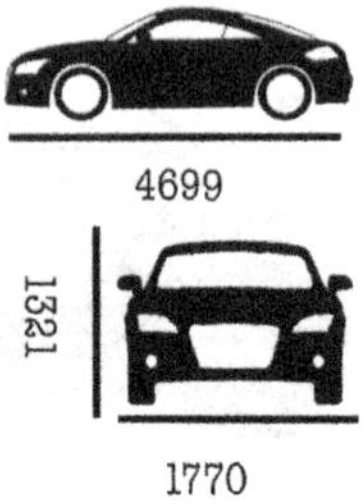

Ford Cougar 1998 - 2002

Those who wanted it to be the new Capri were either retired or dead. Simply a collection of sharp shapes and ordinary engines. Unloved and effectively a less practical Mondeo. Good standard spec though for safety and comfort, whilst a V6 adds leather. Overlooked gem. Electrics and smokey engine, suspension and brake wear.

Engines: 2.0 • 2.5 **CO2:** 202-228g/km **MPG**: 29.7-34.4 **Top Speed:** 130-140 **0-62:** 7.5-9.6 **Boot:** 351 **Seats**: 4

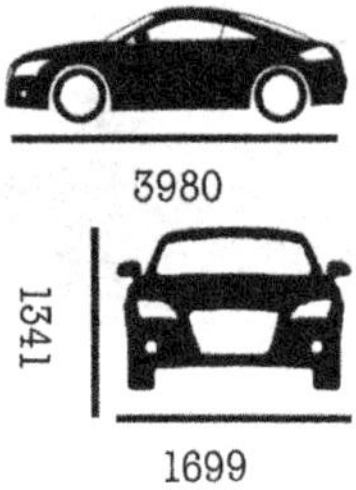

Ford Puma 1997 - 2002

The Fiesta was great to drive and popping a fat feline body on top was greatly appreciated. The 1.7 was powerful enough although the 1.4 rather underwhelming. There was a limited edition Racing but the 1.6 was better. Fun to drive, reliable and cheap to run, easy to get parts for, but sadly rampant rust is the real killer.

Engines: 1.4 • 1.6 • 1.7 **CO2:** 171-178g/km **MPG**: 38.2-38.7 **Top Spd:** 117-122 **0-60:** 8.6-10.8 **G'box:** Man **Boot:** 241/725

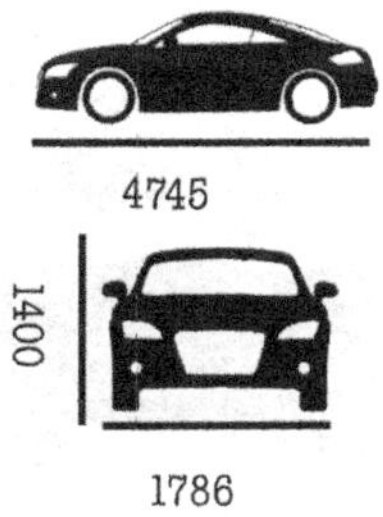

Honda Accord Coupe 1998 - 2001

Large comfy, anonymous and inoffensive two door barge. That's a shame as it is brilliantly engineered and totally reliable. 2.0 ES uses less petrol than the 3.0 V6. Both have got big boots for the owners sand wedge things, but also luggage. Great for long distance, two up travel and long term ownership. Simply brakes and suspension.

Engines: 2.0 • 3.0 **CO2:** 212-253g/km **MPG**: 26.4-31.4 **Top Speed:** 129-139 **0-60:** 8.8-10.0 **Boot:**467 **Seats**: 5

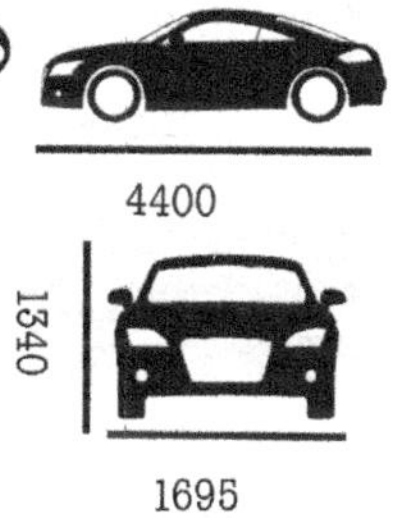

Honda Civic Coupe 1994 - 2003

Not much to look at, but the engines are universally excellent and obviously unlikely to break down. Reasonable amounts of room inside and at least four can get on board without long term physical damage. VTEC engines for the enthusiast. Ideally, buy from someone who does not obviously identify as a Fast and Furious fan.

Engines: 1.7 **CO2:** 159g/km **MPG**: 42.2- **Top Speed:** 125 **0-62:** 9.6 **Gearbox:** Man/Auto **Boot:** 325 **Seats**: 5

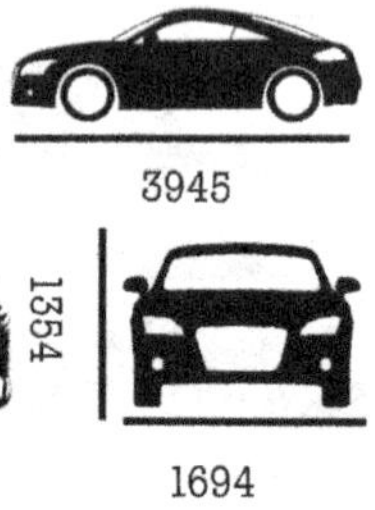

Honda Insight 2000 - 2005

Included because we love the Dan Dare Pilot of the Future looks. Will eventually need a battery pack which kills its Banger credentials. Honda build quality is a plus point, maybe converting it to a V6 would be inappropriately amusing. There is the US Market Civic IMA saloon and Civic based Insights if you want to do Hybrids properly.

Engines: 1.0H **CO2:** 80g/km **MPG**:83.1 **Top Speed:** 106 **0-62:** 12.9 Gearbox: Auto **Boot:** 139 **Seats**: 2

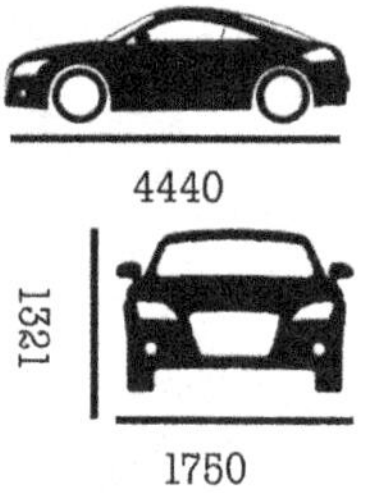

Honda Prelude 1997 - 2001

The go to company for various sizes and formats of coupe, not least because buyers know that it will be well built and never break down. The Prelude is reasonably interesting with four wheel steering and those VTEC engines. As ever buy from an owner that cared, or at least had it over serviced, whatever the mileage. Getting rare. Just wear and tear, ideally not a noisy and rough VTEC.

Engines: 2.0 • 2.2 **CO2:** 220-235g/km **MPG**: 29.5-30.7 **Top Speed:** 126-140 **0-60:** 7.1-9.2 **Boot:** 284 **Seats**: 2+2

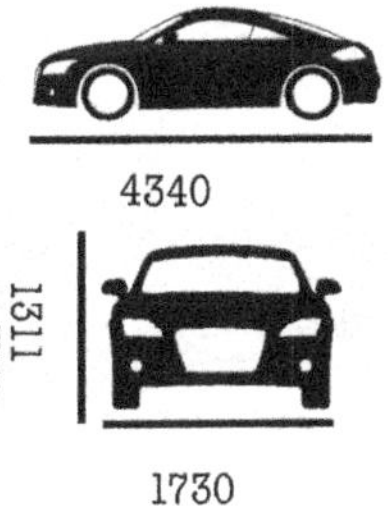

Hyundai Coupe 1996 - 2002

Hasn't got the badge or the looks, but it will at least fail to break down then be easy and cheap to look after. So not much going for this on the face of it, but as a set of stop gap wheels it will be better than sexier, originally more expensive alternatives. Rust otherwise wear and tear to brakes, exhaust, tyres.

Engines: 1.6 • 2.0 **CO2:** 184-212g/km **MPG**: 31.7-36.7 **Top Speed:** 120-125 **0-60:** 8.6-11.2 **Boot:** 362 **Seats**: 2+2

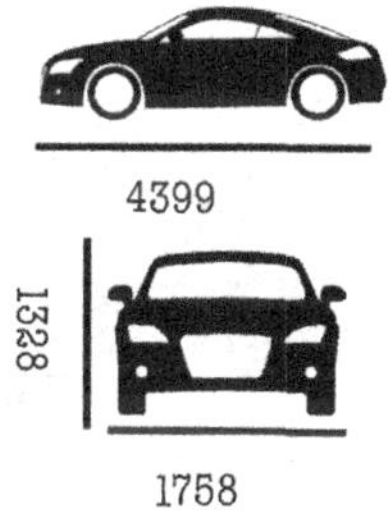

Hyundai Coupe 2002 - 2009

Sort of miniaturised Jaguar XK, but that didn't make it remotely desirable, because it is still a Hyundai. Even with a 2.7 V6 it really isn't any more desirable. For the Banger buyer though it will always be cheap and cheerful. There's the budget 1.6 and 2.0 options if the V6 is too complicated and costly. Rust, brakes, suspension and cosmetic damage.

Engines: 1.6 • 2.0 • 2.7 **CO2:** 182-236g/km **MPG**: 28.5-37.2 **Top Speed:** 115-136 **0-60:** 8.2-11.6 **Boot:** 418 **Seats**: 4

Infiniti G37 Coupe 2009 - 2015

Marque that nobody loved despite Nissan underpinnings, it was mostly indifference. Came to Britain with a 3.7 litre petrol engine making this is just a higher quality finish Hyundai. Actually rather impressive package and the rampant depreciation has to be seen as a positive. Expensive to insure/run, easy to buy if looked after. Must be perfect.

Engines: 3.7 **CO2:** 248g/km **MPG**: 26.7 **Top Speed:** 155
0-60: 5.8 **Gearbox:** Auto **Boot:** 274 **Seats**: 4

Jaguar XK8 1996 - 2005

Lots of ancient XJ-S underneath and various parts can cost a fortune and take time to fit. So Banger money, for what is a classic, but hardly something to smoke around in without fix it skills or a thick wallet. Biggest later update was in 2002 with bigger engines, Xenon headlamps and fresh alloys, then a final facelift in 2004. Electrics, ECUs, rust, suspension.

Engines: 4.0 • 4.2 **CO2:** 284-305g/km **MPG**: 22.2-24.1
Top Speed: 155 **0-60:** 4.4-7.1 **Gearbox:** Auto **Boot:** 326

Jaguar XK 2006 - 2014

Clever update of the previous model. Lighter and more focused than before with quite a lot of aluminum in its make up. 2+2 but more room inside and keyless starting complication. Restyled a bit in 2009, don't get better with age as a used buy. Being a Jag it will eventually be cheap and problematic. Electrics, ECUs, gearbox, locks and suspension.

Engines: 4.2 • 5.0 **CO2:** 269-292g/km **MPG**: 23-25.0 **Top Speed:** 155 **0-60:** 4.6-5.9 **Gearbox:** Auto **Boot:** 300

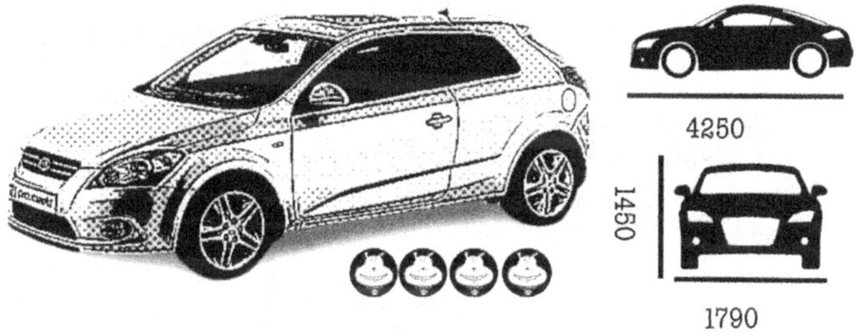

Kia Pro_'ceed 2008 - 2012

A three door hatch despite the weird name, but it's here amongst the pretty coupes as this is a prime Banger. You can get bogged down in the specification quagmire, basically it is small petrols 1.4 and 1.6, and a 2.0 in a load of trims. Entry level spec is generous and being a Kia means it will be solidly reliable. Tyres, brakes, exhaust and high wear parts.

Engines: 1.4 • 1.6 • 1.6D **CO2:** 119-166-g/km **MPG**: 44.1-62.8 **Top Speed:** 116-128 **0-60:** 9.5-12.4 **Boot:** 385/1210

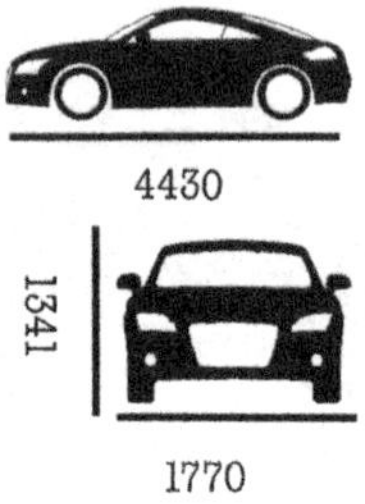

Mazda RX-8 2003 - 2010

Buyers are inevitably attracted to the oddest of coupes, but should of course run the other way. Despite being well built, the Rotary engine is a harbinger of oil leaks and expensive doom. Despite the cleverness of it all and the micro rear doors on no level does it makes sense, even with a generous specification. You have been warned.

Engines: 1.3 **CO2:** 299g/km **MPG**: 24.6 **Top Speed:** 146
0-60: 6.4 **Gearbox:** Auto **Boot:** 290 **Seats**: 4

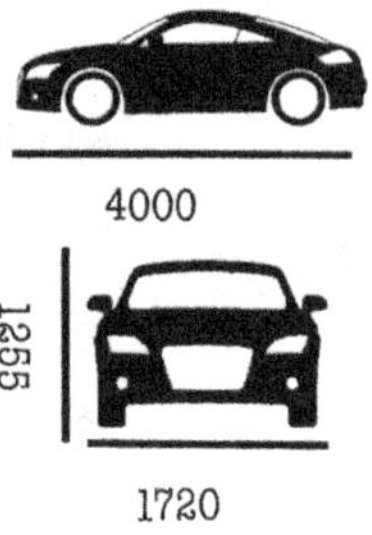

Mazda MX-5 RC 2005 - 2015

The point at which Mazda joins the folding tin roof club, with much success. The roadster-coupe has a three part job which takes just 12 seconds to operate, so all very well engineered. Old school buyers could stick with a fabric roof if they wished. Still has 1.8 and 2.0 engines. So complicated folder, or simple, easy roadster life canvas. Dock a Slog for that.

Engines: 1.8 • 2.0 **CO2:** 174-193g/km **MPG**: 36.7-38.7
Top Speed: 122-130 **0-60**: 7.9-9.4 **Boot:** 150 **Seats**: 2

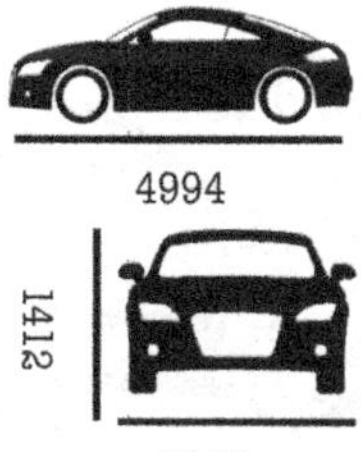

Mercedes CL 2000-2007

A Banger purely on the basis of its massively depreciated used price. Only a mad fool would take the bait. Effectively a two door S-Class and usually a little bit more complicated. 12 way adjustable electric seats and insanely complicated door hinges, never mind the 600 V12 engine are all designed to come back and bite the ill-prepared buyer on the bum.

Engines: 5.0 • 6.0 **CO2:** 270--353g/km **MPG**: 19.2-24.8 **Top Speed:** 155 **0-60:** 4.8-6.3 **Gearbox:** Auto **Boot:** 450 **Seats**: 4

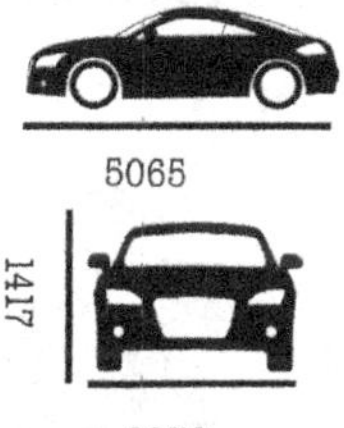

Mercedes CL 2007 - 2014

More of the same, but this time wrapped up in a rather less attractive package. Still has the massaging, air conditioned seats, which tells you all you need to know. Updated in 2010 with even more gadgets and oddly pointless BlueEFFICIENCY engine magic. Buyers should be confused rather than impressed and just buy a more expensive CLK.

Engines: 5.0 • 6.0 **CO2:** 270-353g/km **MPG**: 19.1-23.3 **Top Speed:** 155 **0-60:** 4.4-5.4 **Gearbox:** Auto **Boot:** 490 **Seats**: 4

Mercedes C Class Sports Coupe 2001 - 2008

They call it a coupe, but this is a stumpy little hatch surely? Quite a nice package actually, good to drive, parts will cost more. Diesel and petrol options of course, the 2.0 petrol is fine if mpg not a priority. Not to be confused with a sports car. Just a posh hatchback and there's nothing wrong with that. Electrics and ECUs, oil and coolant leaks.

Engines: 1.8 • 2.0 • 2.5 • 3.5 • 2.1D **CO2:** 172-192g/km **MPG**: 35.3-43.5 **Top Spd:** 127-145 **0-60:** 9.1-11.7 **Boot:** 310/1100

Mercedes CLC 2008 - 2012

On the face of it a made-over Sports Coupe with a new model designation. SE and Sport levels of trim and a similar bunch of diesel and petrol engines. Panoramic roof is a good looking option. Decent boot space. Low spec and no leather have traditionally been the cheap ones. It is electrics mainly and blocked particulate filters if locally used.

Engines: 1.8 • 2.0 • 3.5 • 2.0D • 2.2D **CO2:** 170-191g/km **MPG**: 39.2-48.7 **Top Spd:** 128-137 **0-60:** 9.7-11.2 **Boot:** 310/1100

Mercedes CLK 1998 - 2003

Arguably the pick of Merc's huge range of stylish two doors, although the earlier models not built in the best era for quality control. So there may be rust and other issues. Massive range of engines from the ordinary 2.0 petrol to the V8s and CDI diesels. Beware rampant rust and electrical issues just about everywhere.

Engines: 2.0 • 2.3 • 3.2 • 4.3 **CO2:** 230-281g/km **MPG**: 24.2-28.8 **Top Speed:** 129-155 **0-60:** 5.4-9.7 **Boot:** 419

Mercedes CLK 2002 - 2009

Their middle order model is upped for quality and spec all bundled into Elegance, Avantgarde and Sport to cover all the bases. Big and small engines. Steer clear of top level AMGs and choose your diesel carefully. Better to drive and more expensive to fix if not as sharp looking as before. Electrics, windows and locks, minor corrosion still, diesel oil leaks.

Eng: 1.8 • 2.6 • 3.2 • 3.5 • 5.0 • 2.2D • 3.2D **CO2:** 195-220g/km **MPG**: 29.7-36.7 **Spd:** 143-155 **0-60:** 7.2-9.3 **Boot:** 381/435

Mercedes E-Class 2009 - 2017

Just under the Bangerpedia wire and possibly the finest example of a two door car for the middle to old age target market. Not the finest example of a future Banger though as these are not cheap to fix or fettle at all. Interesting that AMG option dropped, but all the usual petrols and diesels. Depreciates fast. Electrics, ECUs +baffling complication.

Engines: 1.8 • 5.4 • 2.2D • 3.0D **CO2:** 129-179g/km **MPG**: 36.2-41.5 **Top Speed:** 155 **0-60:** 6.7-7.2 **Boot:** 450 **Seats**: 4

Nissan 200SX 1994 - 2001

Drifter, originally bought by old fellas with chest hair then modified later by teenagers. Just makes it into the Bangerpedia and the high spec Tourings seem to make up most of the survivors. The secret is to dodge the Fast & Ferocious ones that have been chipped and over boosted to infinity and very much beyond. Rust, mods, crashes etc...

Engines: 2.0 **CO2:** 267g/km **MPG**: 21.6 **Top Speed:** 146 **0-60:** 7.1 **Gearbox:** Man/Auto **Boot:** 240/397 **Seats**: 4

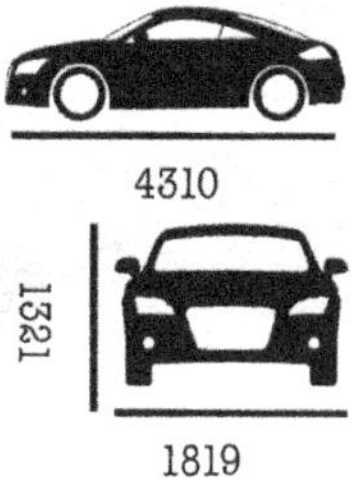

Nissan 350Z 2003 - 2009

Return of the Brutish sports car, not a spelling mistake. Big Healey for the new century but reliable and considerably safer. Restyle in 2006 inside and out. Has climate control, heated and folding door mirrors, GT Pack at heated leather seats, cruise and a banging stereo as the kids say. Ideally buy from someone who was a kid in 1963 and you'll be fine.

Engines: 3.5 **CO_2:** 273g/km **MPG**: 24.8 **Top Speed:** 155
0-60: 6.9 **Gearbox:** Man **Boot:** 235 **Seats**: 2

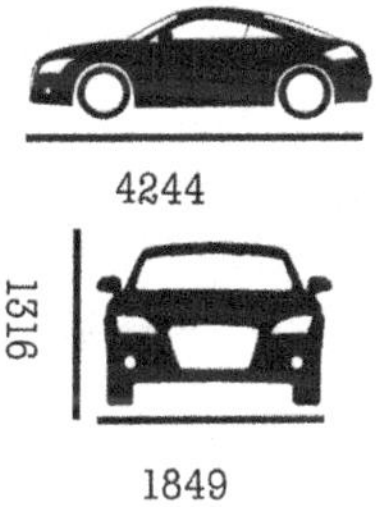

Nissan 370Z 2009 - 2020

Bigger engine, shorter wheelbase, weight saving bodywork and a seven speed auto/rev control for the manual. The British sportster gets even hairier. Inside there is a flipping brace bar where the parcel shelf should be, so this is the full on road racing experience. Everything comfy is standard especially with GT Ultimate trim. Avoid write offs.

Engines: 3.7 **CO_2:** 220g/km **MPG**: 26.9 **Top Speed:** 155
0-60: 5.3 **Gearbox:** Man **Boot:** 235 **Seats**: 2

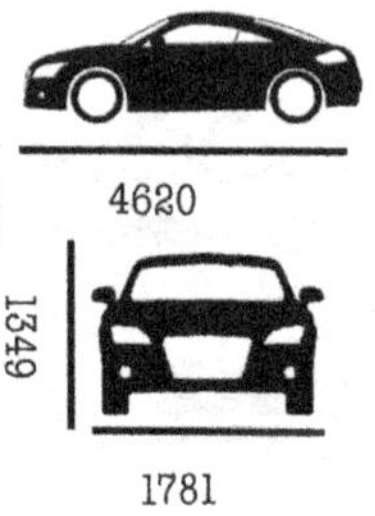

Peugeot 406 1997 - 2003

Elegant Pug that depreciation made even more attractive. 2.0 and 3.0 V6 petrols, but the excellent HDi diesels are the practical picks plus they all have split rear seats. Good specification levels even pre SE. Those drop down windows cost a bit to fix if playing up. Corroded radiators, rusty brake discs, engine ECU issues, suspension bushes.

Engines: 2.0 • 3.0 • 2.2D **CO2:** 168-260g/km **MPG**: 25.9-44.1 **Top Speed:** 146 **0-60:** 7.9-10.9 **Boot:** 391 **Seats**: 4

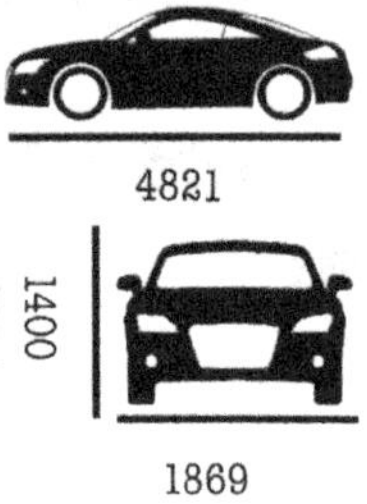

Peugeot 407 2005 - 2011

Here is a huge two door, non posh motor and that makes it instantly a prime used buy. A couple of petrols, 2.0i and 3.0 V6 and a raft of Hdi engines available for a time, becoming all diesel from 2008. Proper overhaul in 2009 with brand new engines and great big wing mirrors. Dashboard failure, clutch failure, ABS sensors, diesel throttle issues.

Engines: 2.2 • 2.0D • 2.7D **CO2:** 219-226g/km **MPG**: 30.7-33.2 **Top Speed:** 138-143 **0-60:** 8.5-9.2 **Boot:** 400

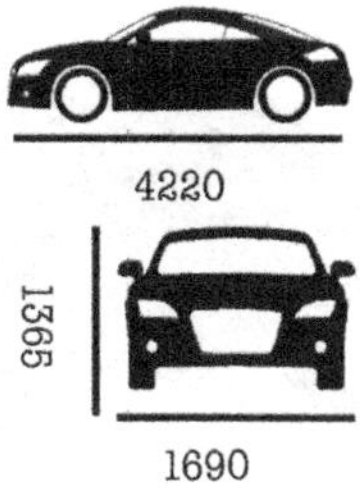

Proton Coupe 1997 - 2001

Badge does not help, then neither does the styling, real bonus here is back to basics Mitsubishi Lancer engineering and zero desirability. Also has boot spoiler and Recaro seats. Decent basic spec for the time and a split rear seat. Quite likely to last until the end of time. Reliable, maybe check for head gasket, oil leaks and just worn out suspension parts.

Engines: 1.8 **CO2:** 208g/km **MPG**: 32.7 **Top Speed:** 126
0-60: 9 **Gearbox:** Man **Boot:** 400 **Seats**: 4

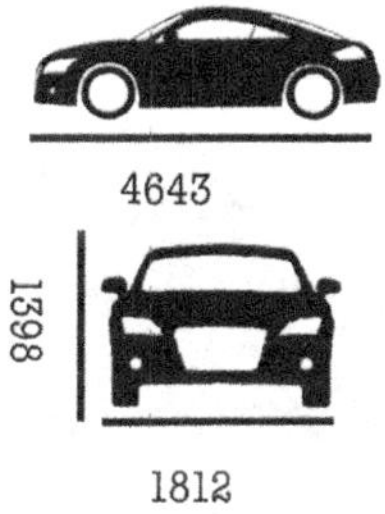

Renault Laguna 2008 - 2012

Inevitably unsuccessful coupe with limited appeal. Only marginally smaller than the five door hatch, the boot access is a bit narrow, but there are split rear seats. Lots of things to go wrong such as climate, rain sensing wipers and automatic lights. Petrol engines deleted 2010. Check electrics, EGR valves and gearbox operation.

Engines: 2.0 • 3.5 • 2.0D • 3.0D **CO2:** 157-195g/km **MPG**: 28.0-47.1 **Top Speed:** 128-146 **0-60:** 7.6-11.7 **Boot:** 423

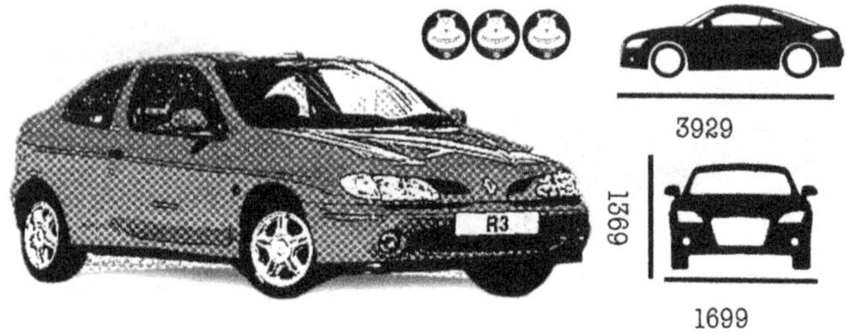

Renault Megane 1996 - 2003

Popular compact sized coupe with a smallish boot and quite honestly only room for kiddies in the back. Properly sporty when driven hard it had a major restyle in 1999. Came with a wide range of petrol engines. Easy to sort models and no diesels keeps it simple. Gearbox and electrics. Clutch repair difficult, coil packs and flywheel sensors.

Engines: 1.4 • 1.6 • 2.0 **CO2:** 153-177g/km **MPG**: 37.2-43.5

Top Speed: 116-132 **0-60:** 8.6-11.4 **Boot:** 289 **Seats**: 5

Toyota Celica 1999 - 2006

Seventh generation is a smart looking thing. Lightweight and sharp it came with cleverly engineered 1.8 engine in two different states of tune. Decent standard kit includes electric windows and allows, whilst the Premium added climate, leather and a CD. Compliant adults can get in the back. Many examples, not passengers, have been hammered.

Engines: 1.8 **CO2:** 185-200g/km **MPG**: 33.6-36.7

Top Speed: 127-140 **0-60:** 7.2-8.7 **Boot:** 345 **Seats**: 4

Vauxhall Astra Coupe 2000- 2004

Stumpy yet stylish and this had some proper poke as well, especially the Turbo. Being a Vauxhall is good because parts are not a problem and relatively cheap. Being so closely related to the hatch means a big old boot and the split fold rears good take fully formed adults. Standard equipment is alloys, air and electric windows. Avoid the trashed ones.

Engines: 1.8 • 2.0 • 2.2 **CO2:** 187-214g/km **MPG**: 31.7-36.2

Top Speed: 127-152 **0-60:** 7.0-9.5 **Boot:** 460/1150 **Seats**: 4

Vauxhall Astra Coupe 2005 - 2010

There are some good engines and diesels now as the 1.7 CDTi ecoFLex is supposed to deliver over 60mpg, plus the performance focused and equipped VXR models. Still has a high boot lid, but shopping friendliness is not why you buy. Shop carefully though. Worn suspension bushes and rusty turrets. Niggly electrical, lights, wipers, locks.

Engines: 1.6 • 1.8 • 2.0 • 2.2 **CO2:** 165-214g/km **MPG**: 27.0-42.0 **Top Speed:** 121-149 **0-60:** 11.0-7.2

Gearbox: Man/Auto **Boot:** 370/1210 **Seats**: 4

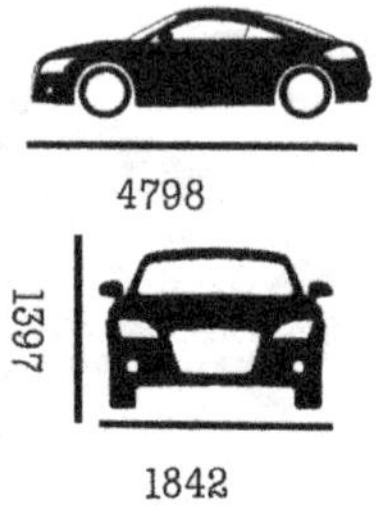

Vauxhall Monaro 2004 - 2006

Not a Calibra resurrection but an Australian importation courtesy of Holden with 5.7 litre V8 from a Corvette. Proper working class supercar. Revised VXR model even hairier and the engine went up to 6.0 litres. Easier and cheaper to look after than an Aston DB9 and likely to be rather more rare. Working blokes Banger supercar.

Engines: 5.7 • 6.0 **CO2:** 360-365g/km **MPG**: 18.5 **Top Speed:** 160-170 **0-60:** 5.4-6.0 **Gearbox:** Man/Auto **Boot:** 370

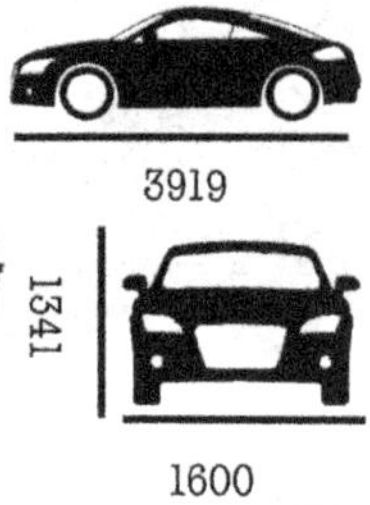

Vauxhall Tigra 1994 - 2001

Strange little 2+2 thing that appealed to the young, female buyer. Or hairdresser. Adequate 1.6 and less than adequate 1.4 petrols, but incredibly Lotus sharpened up the handling in 1997. Never cherished so tend to end up in a bit of state unless a truly one family owner example. Cheap to run. Rust, smokey engines, mostly worn out.

Engines: 1.4 • 1.6 **CO2:** 173-183g/km **MPG**: 39.0 **Speed:** 118-126 **0-62:** 9.8-10.5 **G'box:** Man **Boot:** 215/425

Volkswagen Scirocco 2008 - 2014

Return of the legendary windy name badge and yet again underneath the body work it was still Golf Mark 5. Petrol 2.0 TSI then a 2.0 TDI and in 2010 a 1.4 petrol. Otherwise this was a fancy hatchback with good boot access and reasonable rear seats. Caring private sellers are the best. Coil pack failures on petrols, suspension bushes, ECU failures.

Engines: 1.4 • 2.0 • 2.0D **CO2:** 124-179g/km **MPG**: 37.2-57.6

Top Speed: 136-146 **0-60:** 7.2-8.1 **Boot:** 292/755

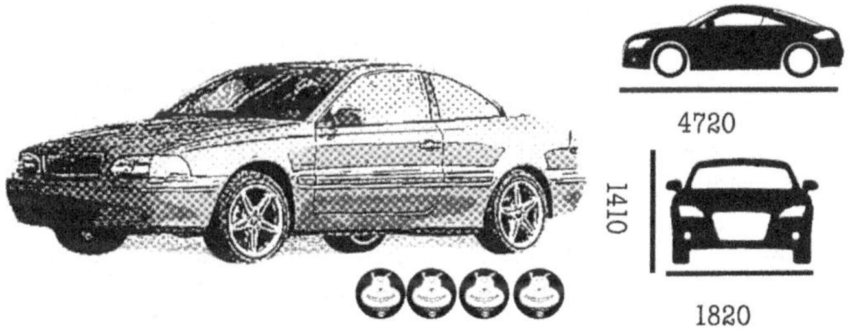

Volvo C70 1997 - 2005

These look smart, not P1800 smart. Full on four seater with luggage space for all. Five cylinder engines some with turbos were good and eventually a cheaper to own 2.0 arrived. The value for money specification included air conditioning, electric most things and a decent stereo. Niggly problems with electrical relays sometimes. Kenny loves them.

Engines: 2.0 • 2.4 • 2.5 **CO2:** 226-236g/km **MPG**: 28.5-29.7

Top Speed: 124-155 **0-60:** 6.9-9.9 **Boot:** 371 **Seats**: 4

Bangvertibles - Fresh Air Funmobiles

The question is, can an open top car be Banger? Absolutely. Same reason as any other motor you can actually afford falls instantly into the Banger category without really trying.

Convertibles are not simply about showing off, although that is one of the many upsides. Another benefit is that you may be able to get some 4' x 2' planks of wood on board. Provided of course that you tie a handkerchief, ideally red, on any overhang.

That's just being silly, when it is possibly time to go and treat yourself to an Alfa Romeo Spider before they all go away. Meanwhile the raft of Audi A Series are the default open air models that you can usually rely on which become amazing volume after a decade or so. The Audi TT is probably the best real world Golf related roadster and there are always the raft 3 Series contenders. For grown ups the Mercedes range is there for your consideration.

Apart from the serious and sometimes seriously expensive to own and run quality open tops, the Chrysler PT Cruiser, Daihatsu Copen and Mitsubishi CZC are always there for your mild amusement and entertainment.

Unless it is raining, or you are parked up in a town centre multi-story, there is no reason to ever put the roof up. That's the whole point of buying a car with half the metal missing. Just put a hat on. Mind you, those folding roofs are somewhat worrying. They really are fabulously complicated if they go wrong, or have been mistreated, make sure it is either stuck open, or stuck closed. Then pay an appropriate amount.

The simple Bangervertible fact is that car owners really don't need them, but thank goodness they do. It means that there can be a lot of low mileage, well maintained perfect condition convertibles out there. Caution is of course required because when they fall into the wrong hands you can end up with what is a skip on wheels.

Anyway, good luck shopping for one and probably the best thing to do is give in and get yourself a Mazda MX-5.

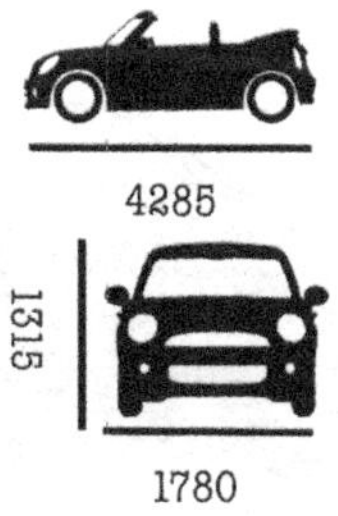

Alfa Romeo Spider 1996 - 2004

Latecomer to GB market with twin spark 2.0, then V6 arrived. Quality roof lining shrouds mechanism and whole system neatly folds away beneath the metal roof cover. In 2003 there was a mild facelift and improved 2.0 JTS and 3.2 V6 engines. Caution: used Alfa. It's a proper risky Alfa buy. MOT failure level rust, smokey engine damaged roof.

Engines: 2.0 • 3.0 **CO2:** 220-278g/km **MPG**: 24.1-30.7

Top Speed: 137-154 **0-60:** 8.9-9.6 **Boot:** 147 **Seats**: 2

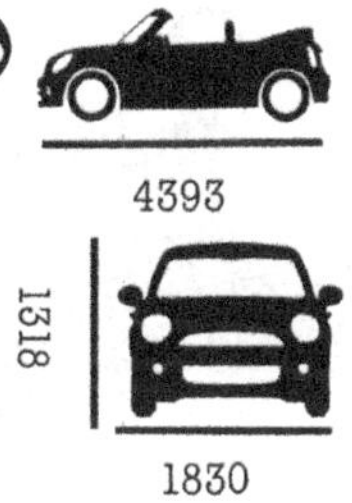

Alfa Romeo Spider 2007 - 2010

Brera with the roof missing, so it looks pretty good. Mostly front wheel drive 2.2 petrol and 2.4 diesel, but also a 3.2 V6 with Q4 four wheel drive. The folding roof takes a longish 25 seconds to pack away. Behind the seats are a couple of useful lockable storage bins that is in addition to the boot. It is an Alfa, so electrics and ECUs, wear and tear.

Engines: 2.2 • 3.2 • 2.4D **CO2:** 179-273g/km **MPG**: 24.6-41.5

Top Speed: 138-142 **0-60:** 6.8-8.6 **Boot:** 200 **Seats**: 2

Audi A3 Cabriolet 2008 - 2013

Diesel and petrol engines and the 2.0 Tdi seems like the best option. A bit bigger than the hatch the roof can be semi or fully automatic. The fully auto has extra insulation and erects in a rapid 9 seconds. Proper heated rear window. Electrics can let it down. Clutch wear, steering racks, radiators, ECU and related sensor problems.

Engines: 1.4 • 2.0 • 1.6D • 2.0D **CO2:** 143-154g/km **MPG**: 43.5 -52.3 **Top Speed:** 127-144 **0-60:** 9.7-7.4 **Boot:** 320/275

Audi A4 Cabriolet 2001 - 2005

Quattro all wheel drive an option, goes from 1.8 Turbo petrol to a 2.5 TDi diesel. High quality electric three layer roof packs out of sight beneath the aluminum cover and takes 24 seconds without any human involvement. Proper glass rear window. Optional hard top. Early model ignition coils, water pumps, auto gearbox unreliable.

Engines: 1.8 • 2.4 • 3.0 • 1.9D • 2.5D **CO2:** 149-252g/km
Top Speed: 130-149 **MPG**: 26.9-38.0 **Boot:** 315

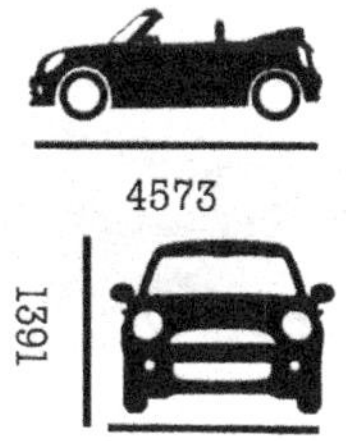

Audi A4 Cabriolet 2006 - 2009

Loads of engines from the 1.8 Turbo, to a 2.0 TDI and onwards to a 4.2 V8 in the madly quick S4. The electric hood was upgraded so that buyers could opt for an acoustic model with dense fabrics that soaked up exterior road noise. Uprated suspension and steering. Blocked Particulate Filters, worn brakes and suspension.

Engines: 1.8 • 2.0 • 3.2 • 4.2 • 2.0D • 2.5D **CO2:** 120-252g/km

MPG: 34.0-44.1 **Top Speed:** 131-144 **Boot:** 315 **Seats**: 4

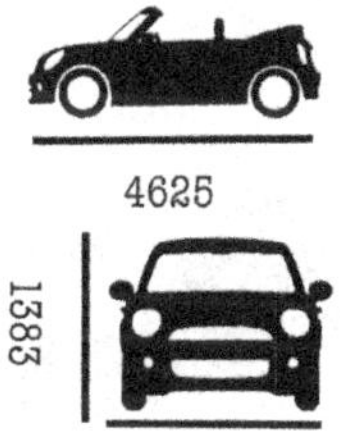

Audi A5 Cabriolet 2009 - 2017

Based on the A5 Coupe which delivers a decent amount of luggage room for the four it can pack on board. Hood packs away in just 15 seconds and takes just 2 seconds more to go up. It is packed with some sound deadening foam to make it one of quietest hood up environments ever. Mostly electrical issues, central locking and in car entertainment.

Eng: 2.0 • 3.0 • 3.2 • 2.0D • 2.7D • 3.0D **CO2:** 140-191g/km

MPG: 39.2-49.6 **Top Spd:** 140-155 **0-62:** 5.9-8.3 **Boot:** 360

Audi TT Roadster 1999 - 2006

The choice is what state of tune for the 1.8T engine and whether or not Quattro four wheel drive is required. There was the option of a 3.2 V6 and S-Line spec option packs. The simple manual hood is easy to use and there is an electronic glass wind deflector which is useful. Ignition packs were not great. Also ECUs, brakes, bodywork and hood damage.

Engines: 1.8 • 3.2 **CO2:** 194-238g/km **MPG**: 29.4-35.0

Top Speed: 133-147 **0-60:** 6.5-7.8 **Boot:** 220 **Seats**: 2

Audi TT Roadster 2007 - 2014

Sharper to look at and drive with lots of engine updates. The fabric hood is fairly nifty and takes 12 seconds to do its stuff. Plenty of engine options from the familiar 1.8T and on to diesels and a V6. Incidentally, lots of costly aluminum in the construction not just the pretty details. Electric windows, Xenon headlamps, brakes and suspension bushes.

Engines: 2.0 • 3.2 • 2.0D • 2.5D **CO2:** 139-247g/km **MPG**: 27.4-53.3 **Top Speed:** 140-155 **0-60:** 6.1-7.5 **Boot:** 250

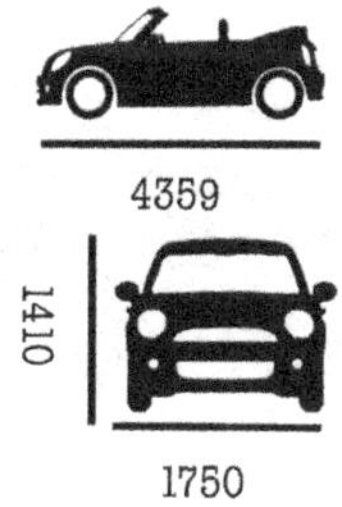

BMW 1 Convertible 2008 - 2013

Handy sized drop top with a number of different engine options from the hot 135i and the cheaper to fuel 120d, 120i and 118d. A four seater but something of a squeeze in the back. Quality roof takes 22 seconds and that leaves a reasonable amount of space in the boot, at just over 200 litres. Dual mass flywheels. Turbo smoke. ECU failures.

Engines: 2.0 • 3.0 • 2.0D • 3.0D **CO2:** 128-220g/km **MPG**: 30.7-58.9 **Top Speed:** 142-155 **0-60:** 5.3-7.6 **Boot:** 260/305

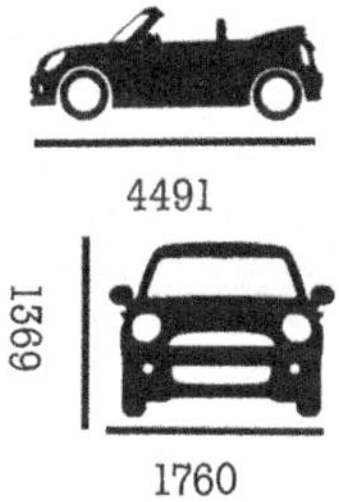

BMW 3 Series Convertible 2000 - 2006

Last of the remotely handsome BMWs. Got all the decent six cylinder engines from the off with the budget fours and diesels added later plus the mighty M3 version. Triple layer hood is amazingly snug, plus there is a proper rear window and defrost element. Clever luggage boosting gadget is a removable hood cartridge. Bore wear, suspension bushes.

Engines: 2.0 • 2.2 • 2.5 • 3.0 • 2.0D • 2.9D **CO2:** 213-218g/km **MPG**: 28.5-31.7 **Speed:** 140-155 **0-60:** 6.5-8.2 **Boot:** 260

BMW 3 Series Ci 2007 - 2013

Joining the folding metal roof club with their three part system which takes an OK 22 seconds to do its work. That roof does compromise the boot space but this is about looking good and so it does. Usual range of outstanding six cylinder petrol engines and selected diesels. ECU failures, electrics, windows and locks, brakes.

Engines: 2.0 • 2.5 • 3.0 • 4.0 • 2.0D • 3.0D **CO2:** 174-218g/km **MPG**: 31.0-46.3 **Speed:** 153-155 **0-60:** 5.5-6.9 **Boot:** 440

BMW 6 Series Ci 2004 - 2010

Fabric hood was an engineering work of art. It is a big car, but the rear seats are for children and the SunReflective seats to keep everyone cool. Has iDrive, but never mind. Came with a V8 645, a 630i six was added and the 630d to save on the running costs. Not repair ones though. ECU failures, electrics, oil leaks and suspension bushes.

Engines: 3.0 • 4.8 • 3.0D **CO2:** 183-279g/km **MPG**: 24.1-39.2 **Top Speed:** 155 **0-60:** 5.1-6.3 **Gearbox:** Auto **Boot:** 300/350

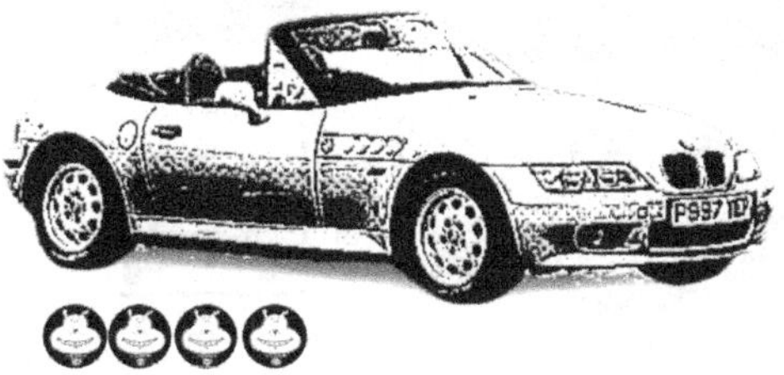

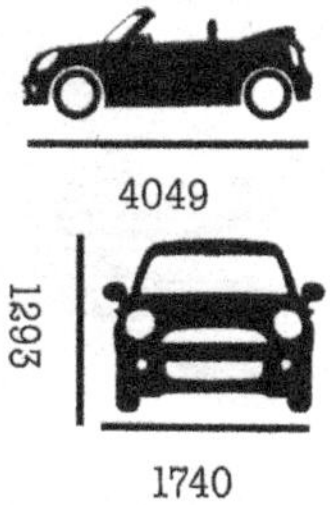

BMW Z3 1996 - 2002

Underneath it is a very old 3 Series effectively from the '80s so easy to live with and fix. Basic 1.9 petrol engine from the off followed by the 2.8 6 cylinder. Slightly face lifted in 1999 to look a bit more like the full on M Coupe. Certain models more 'classic' than others, proper Banger Bimmer with the roof missing. Bore wear, suspension bushes.

Engines: 1.9•2.2•3.0•3.2 **CO2:** 189-268g/km **MPG:** 25.4-36.2 **Top Speed:** 122-155 **0-60:** 5.5-10.4 **Boot:** 165

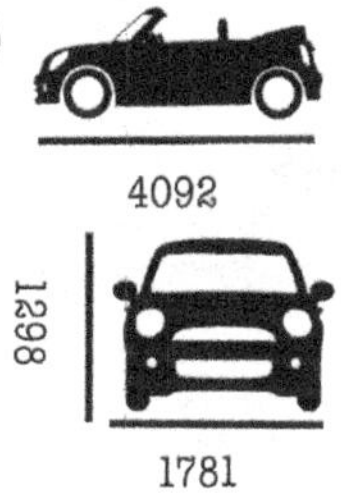

BMW Z4 2003 - 2008

Offered with the good six cylinder engines before the four cylinders arrived. That hood is fully automatic, press a button and it does everything in just 10 seconds. Full on M cars won't ever be Bangers, whereas the relatively slow ones will. Buy low previous owner examples with confidence. Steering racks, minor rust, electrics, broken door handles.

Engines: 2.0•2.2•2.5•3.0 **CO2:** 213-292g/km **MPG:** 23.3-31.7 **Top Speed:** 140-155 **0-60:** 6.2-7.7 **Boot:** 240/260

Chrysler Crossfire 2004 - 2008

Mercedes underpinnings and comes with just the one 3.2 V6 petrol engine, although a higher performance SRT was available in 2005. Deep but narrow boot. Electric folding roof takes 22 seconds and hides neatly away under a metal cover. Plasticky inside, tacky outside. Cheap Fauxcedes. Electrics, ECU failure, but mostly not much to worry about.

Engines: 3.2 **CO2:** 240g/km **MPG**: 28.0 **Top Speed:** 150
Gearbox: Auto **Boot:** 104/190 **Seats**: 2

Chrysler PT Cruiser Cabriolet 2005 - 2008

Retro, MPV, cabrio is one way of looking at it. Doesn't do anything right (initially left hand drive) except look odd. Boot is basically a low down slot, getting in the back awkward, but the electric hood operates in 20 seconds. Just the one 2.4 petrol engine. Cheap open air oddity. Leaking air con, head gasket failure, power steering fluid leak, clutch.

Engines: 2.4 **CO2:** 231g/km **MPG**: 29.1 **Top Speed**: 121
0-60: 11.3 **Gearbox:** Man/Auto **Boot:** 210/377 **Seats**: 4

Chrysler Sebring Cabrio 2001 - 2002

There are anywhere between 0 and 50 of these left hand drive, four seat 2.7 V6 petrol powered Yanks in circulation. Good boot, electric roof with a glass rear window. LX specification gives you almost everything from cruise to air conditioning. Good luck finding one. Unstressed, might have seized brakes and minor rust.

Engines: 2.7 **CO2:** 255g/km **MPG**:26.0 **Top seed:** 131

0-60: 9.9 **Gearbox:** Auto **Boot:** 317 **Seats**: 4

Citroen Pluriel 1994 - 2010

Lego kit with lots of plastic parts that can be deconstructed into a convertible. Eventually. There is nowhere to store the bits as you move from sunroof to targa so hopefully it won't rain too much. A humble C3 with a decent diesel option. A cheap but hands on way of getting a convertible. Fragile. Electricals fail such as locks and windows.

Engines: 1.4 • 1.6 • 1.4D **CO2:** 120-163g/km **MPG**: 41.5-62.8

Top Speed: 98-117 **Boot:** 137/266 **Seats**: 4

Daihatsu Copen 2004 - 2010

Shrunk in the wash TT Roadster, with folding tin top which takes 25 seconds. Also beats the Germans with its Cuore based mechanicals, but sourcing parts could be fun. 660 cc original engine replaced by a 1.3 in 2007 which delivered better economy and less fraught performance. Lovable. Rust, brakes and calipers. Roof damaged by idiot owners.

Engines: 700cc **CO2:** 151g/km **MPG**: 44.1 **Top Speed:** 106
0-60: 11.8 **Gearbox:** Man **Boot:** 210 **Seats**: 2

Fiat Barchetta 1995 - 2005

Left hand drive although there were aftermarket right hook conversions. Front wheel drive puts some purists off, but they are twits. Manual hood is far easier to operate than an old rubbish MGB plus it packs away under a metal tonneau cover. Slight revamp in 2003. Overall charming, but rust and Fiat related fragility can kill it. Check everything.

Engines: 1.8 **CO2:** 198g/km **MPG**: 33.2 **Top Speed:** 124
0-60: 8.9 **Gearbox:** Man **Boot:** 165 **Seats**: 2

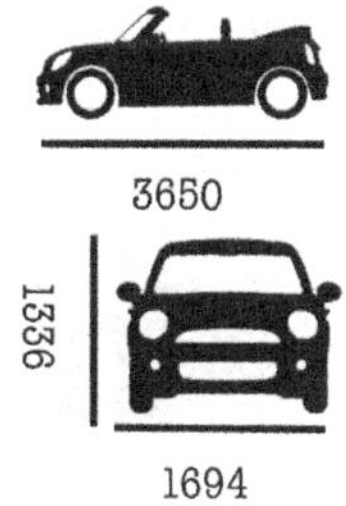

Ford StreetKa 2003 - 2007

Being Ka based it is fun to chuck around and powered by a perky 1.6 Duratec engine. Topped off with a fabric sunroof which takes around 30 seconds to deal with and folds away behind the metal cover. Lots of basic equipment, the Luxury trim delivers leather and a heated screen. Rust, neglect, rusty brake lines, smoky engine.

Engines: 1.6 **CO2:** 189g/km **MPG**: 35.8 **Top Speed:** 107
0-60: 11.8 **Gearbox:** Man **Boot:** 214 **Seats**: 2

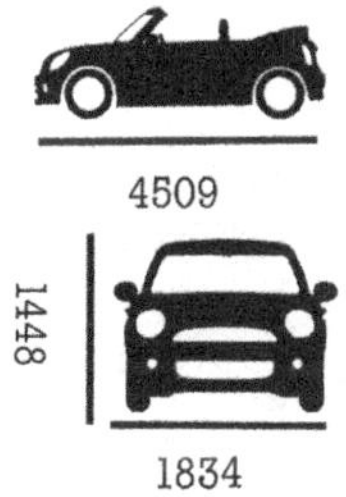

Ford Focus CC 2006 - 2010

Brilliant peoples hatch which made more sense than a Golf sometimes. This was a bit of a stretch. Powered with 1.6 and 2.0 litre petrols and 2.0 TDCi, but not the most elegant of executions and restyled in 2007. One button operation for the folder, it did though have issues with regards to leaks and the rear windows not working.

Engines: 1.6 • 2.0 • 2.0D **CO2:** 139-164g/km **MPG**: 37.7-47.9
Top Speed: 114-130 **0-60**: 10.3-13.6 **Boot:** 248/534 **Seats**: 4

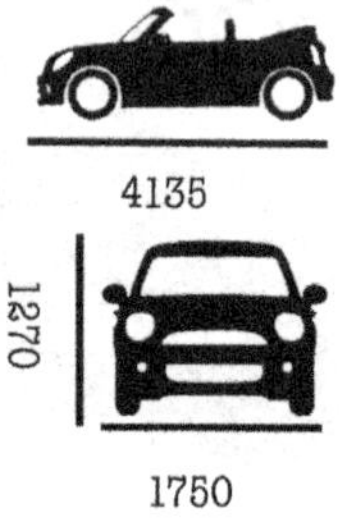

Honda S2000 1999- 2009

Incredibly well built and a rear wheel drive model in its own right, rather than being an Accord underneath. Engine is a work of engineering genius, a hugely powerful four cylinder 2.0 litre. Almost as quick is the electric roof which takes just 10 seconds to do its stuff. It's a genuine classic. Some rust, not cheap to fix if engine shot.

Engines: 2.0 **CO2:** 236g/km **MPG**: 28.2 **Top Speed:** 150

0-60: 6.2 **Gearbox:** Man **Boot:** 143 **Seats**: 2

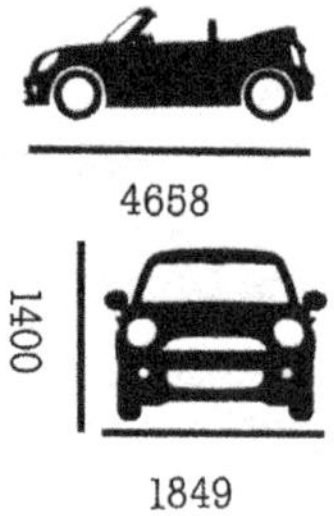

Infiniti G Convertible 2009 - 2015

Amodel no one knows, it is all about value for money, which helps owners put petrol into the 3.7 litre engine. Has a neat three piece tin top which takes its time to do any folding, at 31 seconds. Little room in the boot when packed away. GT and GT Premium covers all specification bases. Full service history will give you a couple of trouble free years.

Engines: 3.7 **CO2:** 275g/km **MPG**: 23.7 **Top Speed:** 155

0-60: 6.2 **Gearbox:** Auto **Boot:** 366 **Seats**: 4

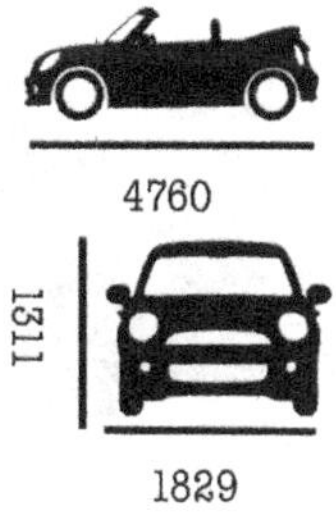

Jaguar XK8 1996 - 2005

Still related to the XJS, 2+2 tight in the back, the classic reason to buy is that it is a sheer value for money Jag. Comes with an electric roof which still needs to be manually packed away beneath the tonneau cover and that robs the boot of space. The V8 is thirsty but good. Needs proper, costly, looking after. Electrics, ECUs, rust, suspension.

Engines: 4.0 • 4.2 **CO2:** 284-305g/km **MPG**: 22.2-24.1

Top Speed: 155 **0-60:** 4.4-7.1 **Gearbox:** Auto **Boot:** 306

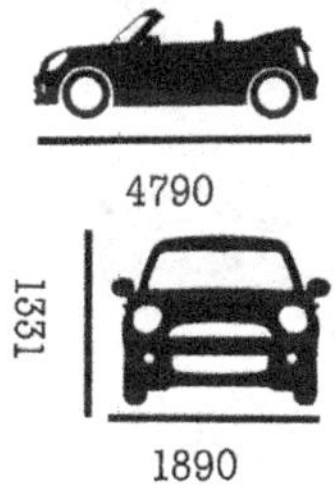

Jaguar XK 2006 - 2014

2+2 body based on Kate Winslet. An actress. She is probably cheaper to run and lighter. Now with a folding triple layer roof, which takes a reasonable 18 seconds from being up to going under the tonneau cover. Yet it does not mean more luggage space than metal roofed rivals. Bigger 5.0 V8 from 2007. Electrics, ECUs, gearbox, locks and suspension.

Engines: 4.2 • 5.0 **CO2:** 269-292g/km **MPG**: 23-25.0

Top Speed: 155 **0-60:** 4.6-5.9 **Gearbox:** Auto **Boot:** 313

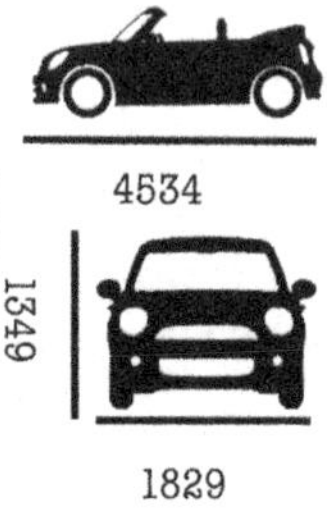

Lexus SC430 2001 - 2009

Looks like a bar of soap with alloy wheels. Desirability is pretty much zero, so a value for money buy. What buyers get is one of the most involved folding roofs ever invented which takes 25 seconds to disappear/reappear. Suspension progressively stiffened and blink and you miss it restyle in 2005. Has a V8. Rust, brakes, neglect.

Engines: 4.3 **CO2:** 287g/km **MPG**: 23.5 **Top Speed:** 155
0-60: 6.2 **Gearbox:** Auto **Boot:** 135/368 **Seats**: 4

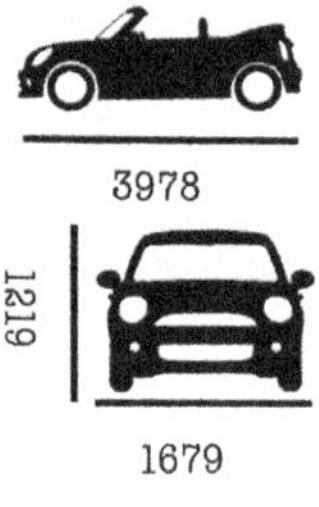

Mazda MX-5 1998 - 2005

All new MX-5 without pop up lights but with a (non contact) 1.6, or 1.8 engine. Easy to live with and a simple manual hood. There are still Japanese imports with extra kit like air con and less underseal. It is a cliché but this really is the best convertible in history. More special editions than Talbot (ask your Grandad). Rust, brakes, hood, nothing unfixable.

Engines:1.6 • 1.8 **CO2:** 188-215g/km **MPG**: 31.7-36.2
Top Speed: 119-120 **0-60**:8.9-9.7 **Boot:** 127 **Seats**: 2

Mazda MX-5 2005 - 2015

Wider, longer, heavier, plusher with more powerful engines a 124bhp 1.8 and 148bhp. Fans reckoned it was not as sharp anymore, but the facelifted models from 2009 onwards sorted out the chassis and brought extra zip to the 2.0-litre engine. Pick of the modern MX-5s with the canvas top and can do everything fun attitude.

Engines: 1.8 • 2.0 **CO2:** 174-193g/km **MPG**: 36.7-38.7
Top Speed: 122-130 **0-60**: 7.9-9.4 **Boot:** 150 **Seats**: 2

Smart Cabriolet 2001 - 2007

Pram hood cabriolet based on the Passion with same 600 cc petrol power. Crossblade version for show offs in 2002 without the pram ingredient. 2004 the City Cabrio was rebadged as the Fortwo Cabrio. Showoffy version had Brabus ingredients boosted the output (74 bhp). As ever, ECU failure kills the car and also suspension, gearbox and hood.

Engines: 700cc **CO2:** 113-127g/km **MPG**: 53.3-60.1
Top Speed: 113-127 **Gearbox:** Semi Auto **Boot:** 150-260

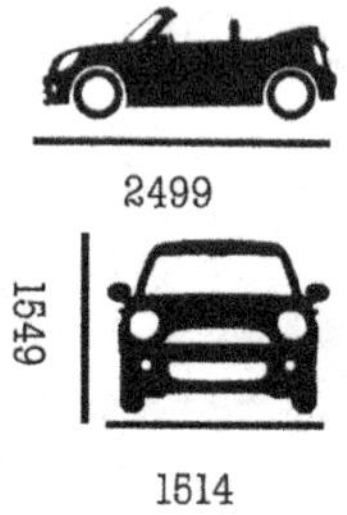

Smart Cabriolet 2007 - 2014

Nothing else so titchy on the market with a fully electric sliding roof that has a glass rear window. Also has removable side bars that need to be removed for the full open experience. Engine available in two states of tune plus a turbo Brabus. Standard air, electric and Bluetooth things. ECU failure it, also check suspension, gearbox and hood.

Engines: 1.0 • 1.0D **CO2:** 87-116g/km **MPG**: 57.6-85.6- **Top Speed:** 105-116 **Gearbox:** Semi-Auto **Boot:** 220 **Seats**: 2

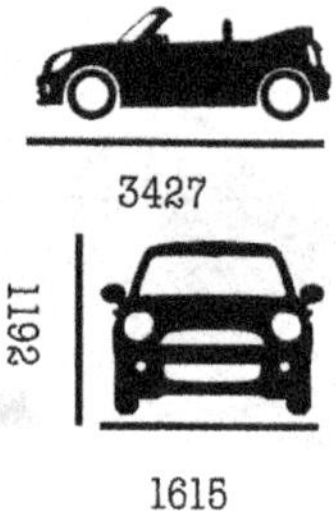

Smart Roadster/Coupe 2003 - 2007

MG Midget without leaks/rust. Not quick, light and purposeful, 101bhp is more than enough shove.. Also left hand drive imports. Coupe version has removable hardtop panels and wider tyres. Not much storage once those are panels are popped out. So Roadster the mildly more practical one. Brakes, suspension, ECUs.

Engines: 700cc **CO2:** 121-126g/km **MPG**: 53.5-55.4

Top Speed: 126 **0-60:** 9.8-10.9**Gearbox:** Semi Auto **Boot:** 86

Mercedes CLK 1998 - 2003

Proper 4-seat family sized cab. Three layered hood is electrically operated and manually released from the driver's seat. Although it gets complicated there are the supercharged 230K and 320 and 430 V6 are Sport, Elegance and Avantgarde trim. MOT failure level rust, suspension and brakes worn or need replacing.

Engines: 2.0 • 2.3 • 3.2 • 4.3 **CO2:** 230-281g/km **MPG**: 24.2-28.8 **Top Speed:** 129-155 **0-60:** 5.4-9.7 **Boot:** 419

Mercedes CLK 2003 - 2009

State of the art fully automatic hood which is well lagged so no one gets cold or bothered by noise and can be operated from the key fob. Standard kit: climate control, rain sensing wipers and cruise control, plus Elegance and Avantgarde trim. Fuel pump failures, tired fuel injectors, catalytic converter issues, broken suspension springs.

Eng: 1.8 • 2.6 • 3.2 • 3.5 • 5.0 • 2.2D • 3.2D **CO2:** 195-241g/km **MPG**: 29.7-36.7 **Top Spd:** 140-155 **0-60:** 6.7-9.3 **Boot:** 381

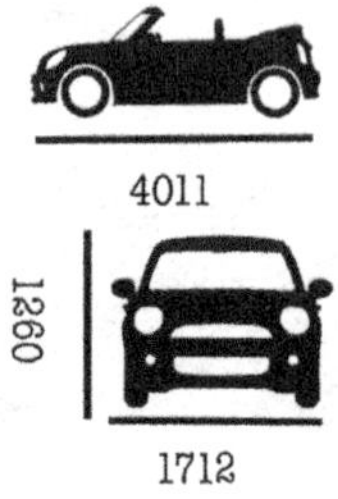

Mercedes SLK 1996 - 2004

Trick folding roof something of a novelty, an SL for the masses. Little 2.0/2.3 supercharged engines, then a 3.2 V6 and eventually AMGs. Just the two metal panels in that roof which takes around 25 seconds to fold away. Comes with leather and most electric things. Long in the tooth and roof. Caution needed, some serious rust, roof can be knackered.

Engines: 2.0 • 2.3 • 3.2 **CO2:** 237-274g/km **MPG**: 24.8-28.5
Top Speed: 138-142 **0-60**: 5.2-8.2 **Boot:** 345/144 **Seats**: 2

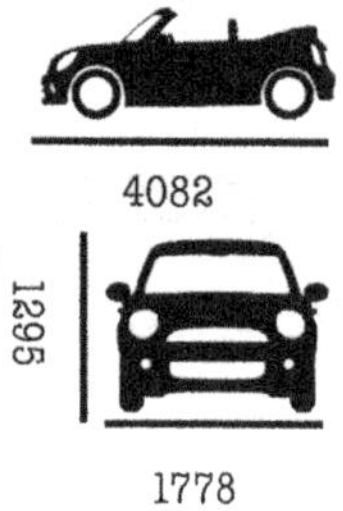

Mercedes SLK 2004 - 2011

Beefed up in most departments and consequently a bit bigger. Folding roof took 22 seconds and the rear window swiveled so that there was more room in the boot. Novel system in the headrests to keep neck and above areas warm when the top is down. Better put together than before and it remains expensive to fix or just tidy up.

Engines: 2.0 • 2.8 • 3.5 **CO2:** 182-288g/km **MPG**: 23.5-36.7
Top Speed: 147-155 **Gearbox:** Auto **Boot:** 208/300 **Seats**: 2

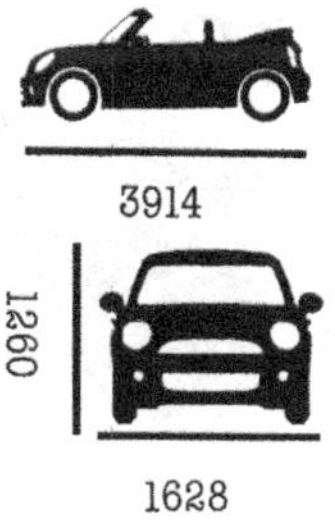

MGF 1995 - 2002

Unexpectedly rather good when it rolled up with two versions of the 1.8 engine, which was positioned behind the seats. Revisions in 1999 meant a mild restyle, a Steptronic gearbox and electronic steering. Five models from 2001 with 1.6 models. Beware blown head gaskets and that Hydragas suspension could give owners sleepless nights.

Engines: 1.6 • 1.8 **CO2:** 177-190g/km **MPG**: 36.3-38.4

Top Speed: 112-130 **0-60:** 7.0-9.3 **Boot:** 210 **Seats**: 2

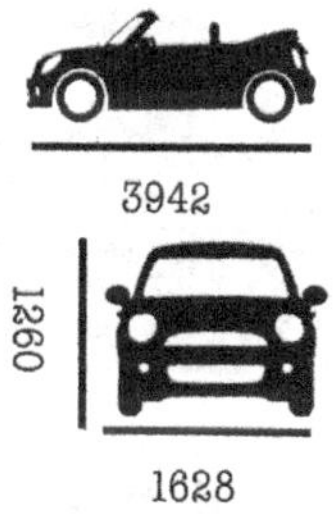

MG TF 2002 - 2005

Hydragas replaced by conventional coil springs. Still with the 1.6 and 1.8 engines in various states of output. Last gasp mods in '05 saw a glass heated rear screen, suspension was fiddled with and trim changed. Remains surprisingly excellent to drive. Flat cap not required, just a few bob for parts. Head gaskets, brakes and sundry parts.

Engines: 1.6 • 1.8 **CO2:** 169-194g/km **MPG**: 33.9-39.8 **Top Speed:** 118-127 **0-60**: 6.9-9.7 **Gearbox:** Man **Boot:** 210

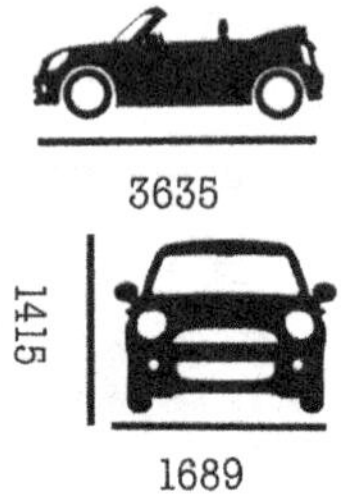

MINI Convertible 2004 - 2010

Clever fabric roof. Press button and it slides back to make a sunroof. Press again and it folds fully back behind the seats in 15 seconds, arranging itself in a pram like fashion. That leaves a lot of boot room. The lid itself is clever in that it can be used as platform to load stuff. Cockpit electrical failures, ABS pump, suspension wear, gearboxes, electric steering.

Engines: 1.6 • 1.5D **CO2:** 109-221g/km **MPG**: 32.1-49.6

Max Speed: 123-134 **0-60**: 7.2-11.4 **Boot:** 125/660 **Seats**: 4

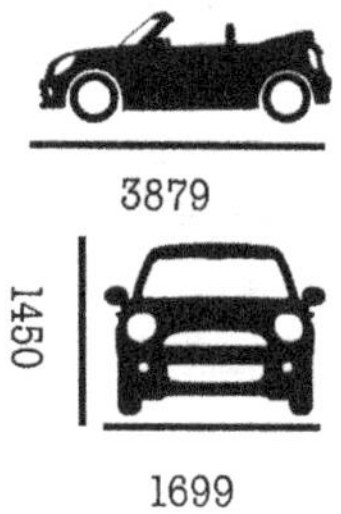

Mitsubishi CZC 2006 - 2009

Effective two piece folding metal roof. Incredible that they bothered with rear seats which are quite pointless. 1.5 petrol engine is not going to let anyone down, plus there is also a turbo option which did not last long. Reliable and well finished. Just wear and tear brakes and suspension plus making sure that the complicated roof does its stuff.

Engines: 1.5 **CO2:** 157g/km **MPG**: 42.8 **Top Speed:** 114

0-60:10.5 **Gearbox:** Man **Boot:** 155/460

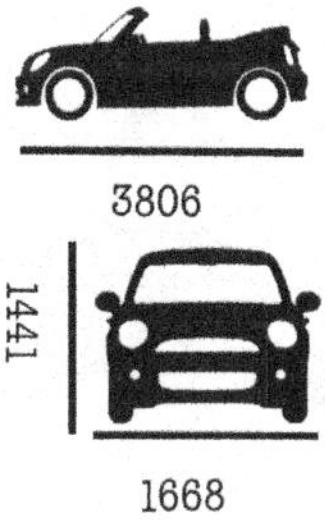

Nissan Micra C+C 2003 - 2010

Just a Micra with a trick metal roof that you can rely on because it's a Datsun. Standard equipment is pretty good. Has a decent sized boot, so a practical buy with optional sunshine. Not for a big burly bloke or girl as they won't fit. Better than being stylish its actually useful. Roof needs to work and there may be ECU, brake and suspension issues.

Engines: 1.4 • 1.6 **CO2:** 158-160g/km **MPG**: 42.2-42.8

Top Speed: 109-119 **0-60**:10.6-12.1 **Boot:** 255/457

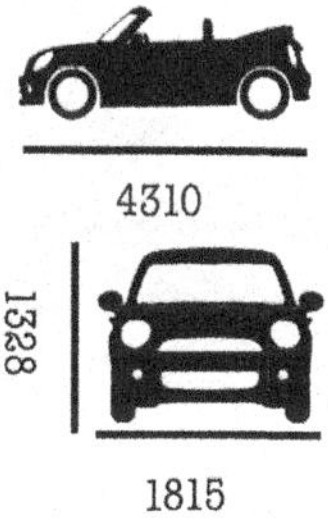

Nissan Roadster 350Z 2004 - 2009

After generations away the proper Z car was back and not only that they chopped off the roof to make a Roadster. Had the same incredible engine and performance whilst after turning a crank the electronic fabric roof folded away in 20 seconds. Upgraded engine with bonnet bulge in 2007. Wonderfully noisy. Sticky gearbox, rusty brake pipes.

Engines: 3.5 **CO2:** 273g/km **MPG**: 24.8 **Top Speed:** 155

0-60: 5.9 **Gearbox:** Man **Boot:** 102 **Seats**: 2

Nissan 370Z Roadster 2010 - 2014

Now with a 3.7 V6 with a shorter wheelbase and an emphasis on saving some weight. Came with a larger rear window, but the hood operated as before. Considerable standard specification puts air conditioning, 18 inch alloys, automatic xenon headlamps, otherwise there was a GT and a Premium Pack for more gear. Often hammered, so be careful.

Engines: 3.7 **CO2:** 220g/km **MPG**: 26.9 **Top Speed:** 155
0-60: 5.3 **Gearbox:** Man **Boot:** 235 **Seats**: 2

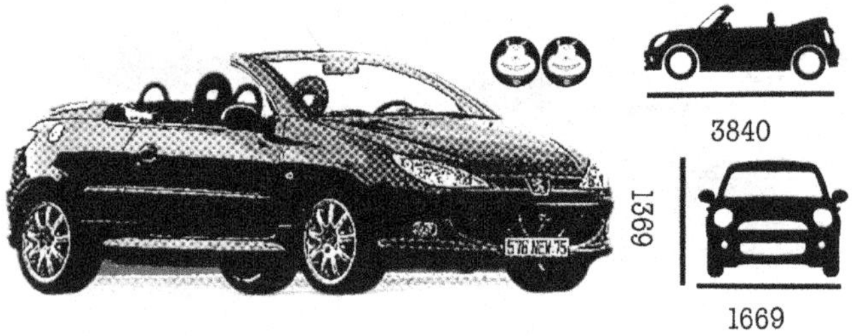

Peugeot 206 CC 2001 - 2007

Undo the two catches behind the sun visors then press the button beside the handbrake and 20 seconds later it is packed into the boot. Rear seats for nippers only. All that folding metal was engineered on the cheap side, so it could be seized and need sorting. Worn suspension, broken engine mounts, ECU problems with auto gearbox.

Engines: 1.6•2.0 **CO2:** 166-191g/km **MPG**: 35.3-40.9 **Top Speed:** 114-127 **Gearbox:** Man/Auto **Boot:**175/410 **Seats**: 4

Peugeot 207CC 2007 - 2010

Stylish enough, good to drive and decently equipped. Mostly petrol options which are the right buys although the diesel works well enough. Easy to own and drive which is all most buyers are going want plus a roomy boot for shopping. Loadsa safety kit, airbags and stuff. ABS ECU failure. Dashboard goes dark. Gearbox needs to be checked.

Engines: 1.6 • 1.6D **CO2:** 136-171g/km **MPG**: 39.2-54.3 **Top Speed:** 120-129 **Gearbox:** Man/Auto **Boot:**187/449 **Seats**: 4

Peugeot 306 Cabriolet 1994 - 2002

Pininfarina's cabriolet of beauty. Hood was well insulated for the time and electric powered. Simply unclip and push the button and it disappears into the raised cover in front of the boot. The optional hardtop is worth having as it is so stylish. Facelifted in 1997 and it got ABS brakes. Most are worn out or scrapped sadly. Everything broken or rusty.

Engines: 1.8 **CO2:**198g/km **MPG**: 30.7 **Top Speed:** 117 **0-60:** 12.2 **Gearbox:** Man/Auto **Boot:** 274 **Seats**: 2+2

Peugeot 307 CC 2003 - 2009

Scaled up 206 with a 2.0 litre engine in two states of tune and a very nicely integrated roof. Decent specification included climate and cruise control. Tidied up in 2005 with the important addition of a diesel Hdi engine. This is a very complicated model and there are electrical issues, never mind an ill fitting roof due to misuse.

Engines: 2.0 **CO2:** 194-210g/km **MPG**: 32.1-34.5 **Top Speed:** 128-140 **0-60:** 10-10.9 **Boot:** 204/350 **Seats**: 4

Porsche Boxster 1996 - 2004

Sort of cut price 911 cabriolet, but with rather more luggage space in the front boot. Had the same sort of flat air cooled engines, so sounds like a Porsche, in 2.7 and 3.2 litre sizes. Major engine issues from cracked blocks to crankshafts and the poor engine access only part of the issue. Porsche running costs, but otherwise worth a go if you are brave.

Engines: 2.7•3.2 **CO2:** 248g/km **MPG**: 26.9-29.1 **Top Speed:** 157-164 **0-60**: 5.7-6.4 **Gearbox:** Man/Auto **Boot:** 260

Renault Megane 1997 - 2003

A strong seller, with 1.6 or 2.0 litre power. That fabric roof is though easy to operate by unclipping and twisting a handle on the screen then pressing a button followed by a gentle 25 second folding procedure. New nose for '99 and restyle, plus some airbags and the potent 2.0 IDE engine. Gearbox and electrics. Clutch repair difficult, coil packs +flywheel sensors.

Engines: 1.6 • 2.0 **CO2:** 165-177g/km **MPG**: 37.2-40.4

Top Speed: 124-131 **0-60:** 9.0-10.1 **Boot:** 280 **Seats**: 4

Renault Megane CC 2003 - 2010

Sharpened up safer, this time with a folding metal and glass roof. 2.0 petrol and 1.9 diesel remain the best options. Dynamique is the middle order spec which delivers most of what owners need, including air conditioning. Facelifted a bit in 2006 with detail changes. Arguably best of the cheapy folders. Roofs can leak. Electrics sulk, blocked EGR valve.

Engines: 1.6 • 2.0 • 1.9D **CO2:** 146-196g/km **MPG**: 34.4-51.4

Top Speed: 121-137 **0-60**: 9.9-11.8 **Boot:** 190/490 **Seats**: 4

Saab 9-3 Cabriolet 1998 - 2003

Four seater, Vauxhall related and no bad thing. Fantastic seats and always sensational value. Lots of criticism about the hard ride in town and the floppiness when being driven enthusiastically. Nobody cares, it is a classy four seater. Best not bought as any sort of project. Parts getting expensive. Oil leaks, worn timing chains, coil packs don't last long.

Engines: 2.0 • 2.3 **CO2:** 210-216g/km **MPG**: 31.4-32.1

Top Speed: 134-146 **0-60**:6.9-8.0 **Boot:** 294

Saab 9-3 Cabriolet 2003 - 2011

Middle class four seater stiffer than previously. Triple skinned hood, obviously is quiet and snug, but also has a sewn internal gutter to prevent drips. Also, the boot expands to fit the load, so this is a very clever vehicle. Last chance to buy. Electrics, windows and locking not great and engine ECU means rough running. Check aircon.

Engines: 1.8 • 2.0 • 1.9D **CO2:** 154-200g/km **MPG**: 31.0-45.6

Top Speed: 125-143 **0-60**:8.0 **Boot:** 235 **Seats**: 4

Toyota MR2 2000 - 2006

It is like they copied the MGF and made it better, apart from the luggage space, because there isn't any, or very much. More importantly putting the hood down is a flick of the wrist from the driver's seat is a doddle, plus it has a proper heated glass window. Even has air con, great if you have the optional hardtop. Reliable and will take a ton of abuse.

Engines: 1.8 **CO2:** 178g/km **MPG**: 38.2 **Top Speed:** 130 **0-60:** 7.2 **Gearbox:** Man **Boot:** 90 **Seats**: 2

Vauxhall Astra Convertible 2001 - 2005

Blunt, basic and effective convertible for the people. Four people.They can be protected from the elements by the three layer hood which can even be operated by the key fob (high tech at the time) and takes 30 seconds to pack away. Can be had with the 2.0 Turbo. Corroded ECUs, cockpit leaks, failing Air conditioning and water pumps.

Engines: 1.6 • 1.8 • 2.0 • 2.2 **CO2:** 161-214g/km **MPG**:34.4-42.2 **Top Speed:** 118-149 **0-60:** 7.5-12.0 **Boot:** 330 **Seats**: 4

Vauxhall Astra Twintop 2006 - 2010

The roof does its tricks at the press of a button in the cabin, or on the key fob and that takes 26 seconds. Good to drive and the choice some decent engines including a 2.0 Turbo and a 1.9 CDTI. Probably one of the better current near moderns. Worn suspension bushes and rusty turrets. Niggly electrical, lights, wipers, locks. **Engines:** 1.6 • 1.8 • 2.0 • 1.9D **CO2:** 165-214g/km **MPG**:34.4-42.2 **Top Speed:** 118-149 **0-60:** 7.5-12.0 **Boot:** 205/440

Vauxhall Tigra 2004 - 2009

Pocket sized Coupe Cabriolet. It takes 20 seconds to fold away the roof. Those who pay attention may notice the circa 2000 Corsa interior. No rear seats so has better boot than you might expect plus a rear compartment. Small engines and diesel sip fuel. Steering racks, wheel bearings, water pumps and timing belt servicing.

Engines: 1.4 • 1.8 • 1.3D **CO2:** 124-185g/km **MPG**:36.7-61.4 **Top Speed:** 98-127 **0-60:** 9.4-16.8 **Boot:** 250/440 **Seats**: 2

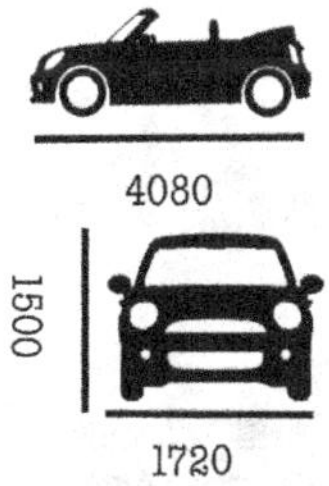

Volkswagen Beetle 2003 - 2010

Certainly better than one of those terrible rusty originals from the '60s. Built for cruising hence the no larger than 2.0 petrol and a 1.9 Tdi. Electric hood on all but the basic models. Good room for four and only claustrophobic hood up. Solidly built and reassuringly old fashioned. Minor electrical failures, windows, locks etc. Engine ECU. Now bubbling rust.

Engines: 1.6 • 1.8 • 2.0 • 1.9D **CO2:** 151-199g/km **MPG**: 34.0-50.4 **Top Speed:** 110-125 **0-60:** 9.3-12.4 **Boot:** 201

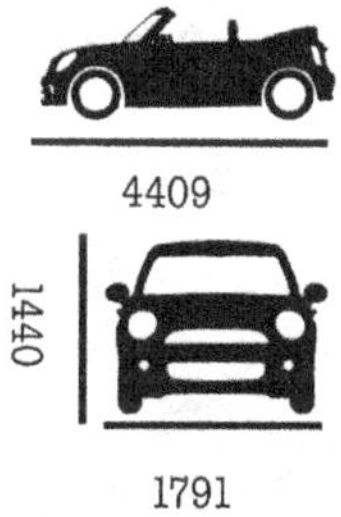

Volkswagen EOS 2006 - 2014

Name is not even an anagram of Golf, but that's what it is, with a coupe/cabriolet topping. Great petrol FSI engines including a V6, plus the diesels to cover all bases. Usual folding roof system, but the added complication is a sliding glass roof. Something for everyone then including a ton of electrics like memory seats. They can go wrong, also ECUs.

Engines: 1.4 • 2.0 • 3.2 • 2.0D **CO2:** 125-182g/km **MPG**:39.8-58.9 **Top Speed:** 118-144 **0-60:** 7.8-12.0 **Boot:** 205/380

Volkswagen Golf Cabriolet 1994 - 2003

Mark 3 Golf was never the best, but this was better made. Folding rear seats boosted the boot area and apart from the basic model, the hood was electrically operated. Just the petrols, a 1.8, 1.6 and 2.0 in the noughties are all just adequate, but certainly reliable and should last forever. 1998 Mark 4 front end, but not interior. Rust, bore wear.

Engines: 2.0 **CO2:** 194g/km **MPG**: 34.9 **Top Speed:** 119 **0-60:** 10.7 **Gearbox:** Man/Auto **Boot:** 269/447 **Seats**: 4

Volvo C70 1997- 2006

Advanced electric roofage which saw just one press of the button to make it all work, plus had a proper glass window. Incredibly four roof colours available to complement the bodywork. A ton of interesting engines from a basic 2.0 right up to the superconvertible T5. Bear in mind these are decades old now. Rust, electrics, brakes and suspension.

Engines: 2.0•2.3•2.4 **CO2:** 226-236g/km **MPG**:28.5-29.7 **Top Speed:** 130-146 **0-60:** 7.5-9.9 **Boot:** 201

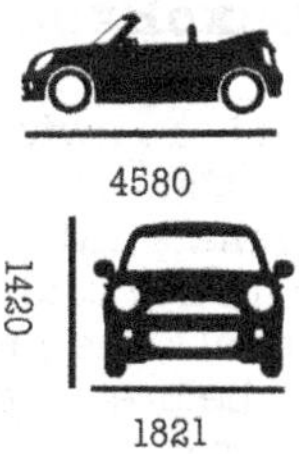

Volvo C70 2006 - 2009

The best looking four seat convertible turns into a part time coupe as well with its folding roof. 2.0 diesel entry models and still the T5. The thing is that it is based on the Focus CC which had its problems, but the C70 seems much better in that build quality respect. Got a facelift and trim upgrade in 2009. Try and plug it into a diagnostic tool.

Engines: 2.4 • 2.5 • 2.0D • 2.4D **CO2:** 193-217g/km **MPG:** 31.0-38.7 **Top Spd:** 137-150 **0-60:** 31.0-38.7 **Boot:** 200/400

LuxoBanger Barge - Blingtastic Bangers

These aren't Bangers surely? Well what you are getting is an awful lot of comfy car for not very much money. That's the point. When I bought a cheap, but utterly immaculate XJ Jaguar the specialist dealer letting it go at trade money said to me, 'don't bother getting it serviced, just run it into the ground.' He was right. It lasted a couple of years before expiring with perfect timing in a Jaguar breakers yard. Apparently it never started again.

This isn't about me, because what you have are the standard issue XJ8s, but the last gasp model run out XJ is worth buying for old times sake, although it will cost an absolute fortune to keep going. A Honda Legend is even more rare, more reliable, but no less pricey when parts are needed. The default big barge choice remains the Lexus LS.

Pretend luxury cars remains the Volvo S80s which last forever, the Volkswagen Phaeton is an impressive effort, but has the wrong badge and the same set of expensive car problems. There is always the Mercedes S waiting to waft you away, the BMW is more sporty and Audi A8 very subtle but all need careful selection.

Everything mentioned so far in the buying rules about any class of Banger applies here in absolute spades. In theory these are big unstressed lumps of luxury which should be fine with a less intensive Banger related lifestyle. However, they will be KO'd by the cost of replacing a 5p washer which necessitates a whole engine out scenario, if you are lucky, but that's many modern cars these days. What can be the real killer is the running costs.

If you don't do many miles then the mpg is not a big issue, however... a £100K car when new means that the servicing and parts are six figure related. Unless you can do the work or you have a creative garage, who are prepared to actually fix things, a Luxobarge may not be for you. Then again, finding parts via a salvage yard and speaking to other owners may result in a cheaper repair. After all what have you got to lose except money and your sanity?

Let's go Barging for a Banger.

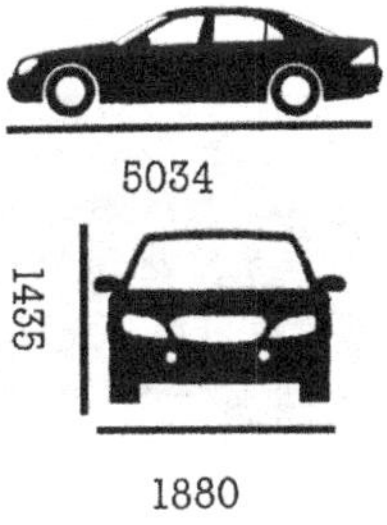

Audi A8 1994 - 2002

Xerox'd up A6 in an age when these barges mostly did not have diesel engines. The aluminum body once dented, costs a bomb to fix, so you probably wouldn't bother once it's a Banger. V8 engines are great, Quattro makes it a super snowplough. Suspension, brakes may need sorting, automatic gearbox , air con and general electrics/ECUs.

Engines: 2.8 • 3.7 • 4.2 **CO2:** 286-334g/km **MPG**: 20.2-26.4
BHP: 193-360 **0-60**: 6.4-9.8 **Top Speed:** 155 **Boot:** 524

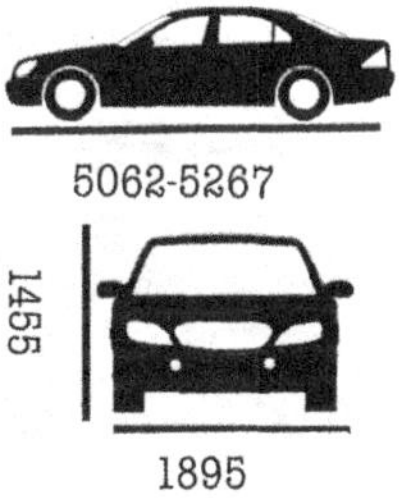

Audi A8 2003 - 2011

Makes some sort of complicated sense now, Diesels, including a full fat V8 will appeal to some. Longer Limo ones part of the line-up and it got a new face in 2005. FSI engines the best long term option. Lashings of luxury and a great big boot. ECU fails, tired diesels and suspension/brake overhauls, any bodywork issues costly to fix.

Engines: 3.2 • 3.7 • 4.2 • 6.0 • 3.0D • 4.2D **CO2:** 231-350g/km
MPG: 23.7-32.8 **BHP:** 207-443 **0-60**: 5.7-7.6 **Boot:** 500

Bentley Arnage 1998 - 2009

This is the Bentley T, Roller Shadow of the modern era. It is dead smart and destined for wedding work. RL is 5640 long. originally Green and Red Labels replaced by R and T models. Everything costs a bomb, but a lot of BMW bits in the early ones, so you need to be clever. Brakes, suspension and the complicated electrics. Cosmetics kills asking price.

Engines: 6.75 **CO2:** 495g/km **MPG**: 13.7 **BHP:** 450-500 **0-60**: 5.2 - 6.2 **Top Speed**: 168-180 **Boot:** 374

BMW 7 Series 1994 - 2002

E38 is one of the very best barges ever built. As fun to drive as a little 3, but with lots more kit and comfort. The engine should go on forever, it is the suspension and brakes where the big expenses will be. V12 possibly an engine too far the 6 and V8 cylinders are the pick of the units. They don't come any bigger, better or harder. Buy one. LWB is 5029.**Engines:** 2.8 • 3.5 • 4.4 • 5.4 **CO2:** 267-332g/km **MPG**: 20.5-25.4 **BHP**: 193-326 **0-60**: 7-9 **Top Speed:** 141-155 **Boot:** 501

BMW 7 Series 2002 - 2008

A level of complexity that renders it useless if a few parts go pop. There was a blandifying restyle in 2005. Longer limos are 5169. 3.0 diesels around in the most numbers. The 735i petrol is the best compromise and candidate for an LPG conversion. Electrics the big issue must be plugged into a proper fault code reader which could be really frightening.

Engines: 3.0 • 4.0 • 4.8 • 6.0 • 3.0D **CO2:** 216-325g/km **BHP:** 214-438 **0-60:** 5.3-8.0 **MPG**: 21.1-34.4 **Boot:** 500

BMW 7 Series 2008 - 2015

Will depreciate to become temptingly cheap, but should you be tempted? No bits carried over from the last one so used parts are pricey. Lighter than before, but not on your wallet. Mostly diesel so stick to petrol. Sealed automatic gearbox may need overhaul. Software and ECUs may need upgrading or replacement. Electric everything a nightmare. LWB 5212

Engines: 3.0 • 4.4 • 6.0 • 3.0D **CO2:** 192-266g/km **MPG**: 24.8-39.2 **BHP:** 241-536 **0-60**: 4.5-7.1 **Boot:** 500/750

Cadillac Seville 1998 - 2002

Right hand drive Caddy 4.6 V8 was available at selected Vauxhall dealers. Front wheel drive as well. Parts not cheap and it may be worth taking a transatlantic holiday to get bits, Not that many around, so it is rather exclusive. Just not as classy as the Germans. Proper Yank Tank. Oil leaks, coolant leaks, rusty bodywork and suspension. Electrics.

Engines: 4.6 **CO2:** 330g/km **MPG**: 19.9 **BHP**: 275-300

0-60: 6.9 - 7.7 **Boot:** 445 Weight: 1805- 1839

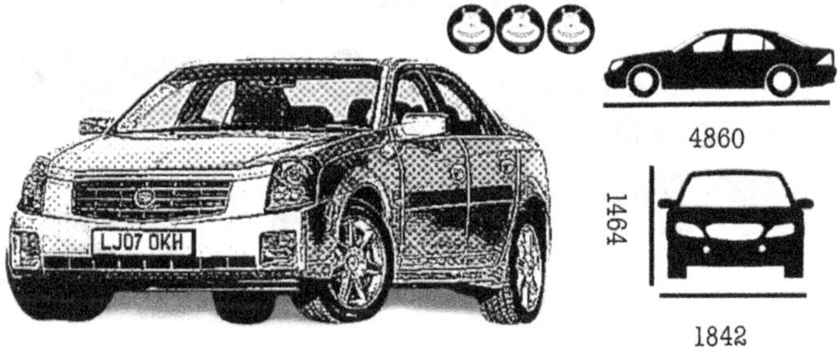

Cadillac CTS 2005 - 2010

Some might say this is just a Vauxhall Omega with an American accent. Good value new and even better used, a couple of petrol V6s will go on forever and there is a ton of electrical gadgets. Can have oil and coolant leaks and random electrical failures. Not hugely special inside, but it is a big and comfy enough cruiser.

Engines: 2.8 • 3.6 **CO2:** 263-264g/km **MPG**: 25.4-25.7

BHP: 208-306 **0-60**: 6.1-8.7 **Boot:** 373 **Weight**:1590-1880

Citroen C6 2006 - 2012

Le Luxobarge Extreme, great, so long as you don't have to own it. Comfy with loads of equipment. Unless you are the President of France, this isn't special, just different and there are no prizes for that. Will break down and be expensive to fix. Suspension ball joints, wheel bearings, ABS sensors, ECU failures, there is almost too much to list.

Engines: 3.0 • 2.2D • 2.7D **CO2:** 230-266g/km **MPG**: 25.2-32.5 **BHP:** 173-240 **0-60**: 8.6-9.2 **Boot:** 488

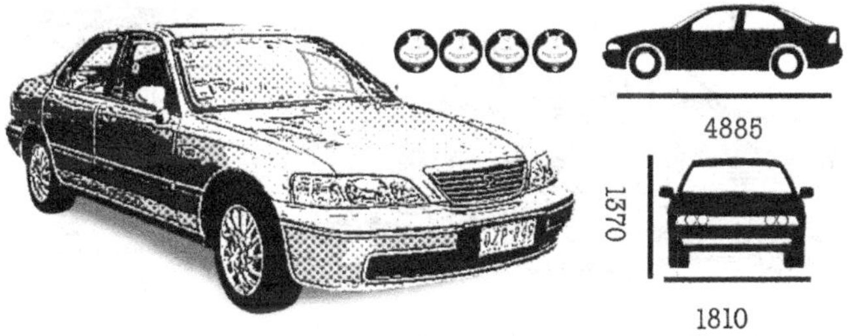

Honda Legend 1991 - 2004

It isn't really a Legend but it is a Honda. A great big comfy saloon that won't break down provided you look after it. This motor delivered ABS brakes and driver, then passenger airbags before they were widely available. A scaled up Accord and just as dull to look at and drive. Heavy use items like brakes and suspension take a bashing, also electrics.

Engines: 3.5 **CO2:** 291g/km **MPG**: 23.2 **Gearbox:** Auto **BHP:** 203 **0-60**: 8.8 **Boot:** 450 **Weight**: 1600 **Doors:** 4

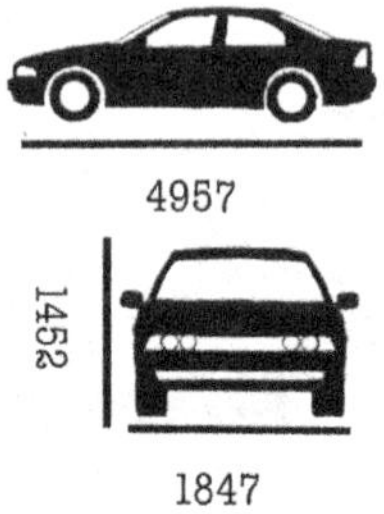

Honda Legend 2006 - 2007

Returned after a break when Honda wondered what the point was of bothering to finance a Lexus LS rival. This mark 4 comes with a ton of very clever tech, but existed for just a year although some sources say it was technically available to 2010. Sort of cleverer Citroen C6 but reliable. Simply wear and tear issues, brakes, suspension and some electrics.

Engines: 3.5 **CO2:** 282g/km **MPG**: 23.7 **Gearbox:** Auto

BHP: 290 **0-60:** 7.1 **Boot:** 452 **Weight**: 1920 **Doors:** 4

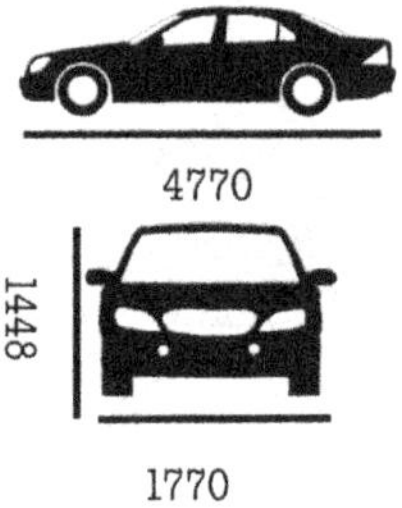

Infiniti G37 2009 - 2013

Not entirely sure this is a luxury car because no one knows it exists. This is a Lexus without the charm. Anyway, because it is engineered by Nissan it will last as long as a driving school Micra, which is forever. All the kit, but not that special. Needs a full on service history to make some sense as a run into the ground zero option.

Engines: 3.7 **CO2:** 248g/km **MPG**: 26.7 **BHP:** 315-328

0-60: 5.3 **Top Speed:** 155 **Boot:** 450 **Weight:**1652:**Doors:** 4

Jaguar XJ8 1994 - 2003

Lovable old school Jag which got even better with a V8 from '97. Fairly snug inside and there's a long wheelbase for work which is 5149. Running costs can be containable. faults tend to be minor, niggly and irritating. Worn brakes and suspension, cracked windscreens, rattly timing chains. Blown head gaskets. Rust MOT failure.

Engines: 3.2 • 4.0 **CO2:** 287-305g/km **MPG**: 22.4-23.7

BHP: 208-370 **0-60**:6.9-8.1 **Boot:** 408 **Weight**: 1710-1975

Jaguar XJ8 2003 - 2009

XJ styling with 21st century high tech underpinnings. Aluminum underneath means major work is pricey. not for the bodger. Diesels from '05. Revamp in 2007 so bigger specs and funkier front end. Still has a pokey boot. Still manages to be proper bounder mobile. Timing chain tensioner failure, complicated suspension failures.

Engines: 3.0 • 4.2 • 2.7D **CO2:** 214-249g/km **MPG**: 27.0-35.0

BHP: 262-400 **0-60:** 5-7.3 **Boot:** 470 **Weight**: 1755-1892

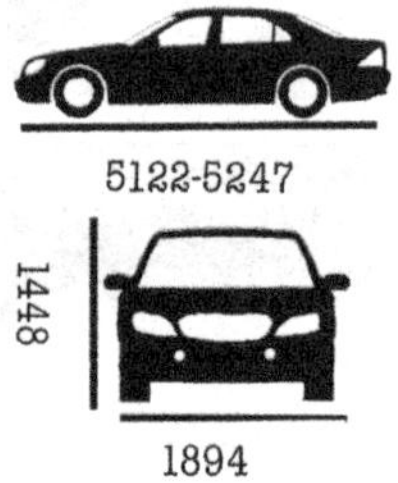

Jaguar XJ 2010 -2019

The last proper XJ, a model which has always been the quintessential big Banger that costs buttons to buy. This model has struggled in the new market. What you get is a huge comfy car which may have some electrical and gearbox niggles. If that 'glass' cockpit display goes on the blink you are staring at bankruptcy.

Engines: 3.0 • 5.0 • 3.0D **CO2:** 184-268g/km **MPG**:25.5-40.1
BHP: 203-275 **0-60**: 4.7 - 6.4 **Boot:** 520 **Weight**: 1769-1914

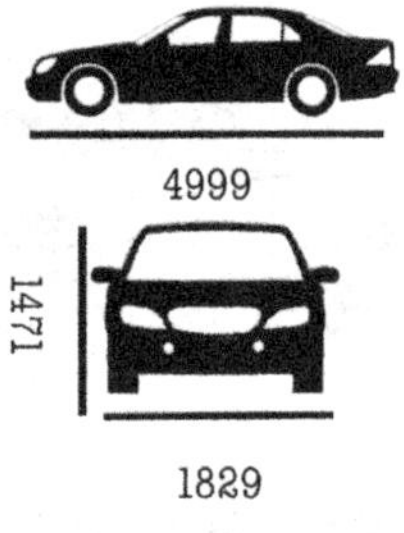

Lexus LS430 2000 - 2006

If you want to tick all the luxury boxes then this is roomy, comfy and over-equipped. The engineering is exemplary to make it a Stepfordluxosaloon. Very reliable, parts pricey. There was a Premium Pack, but really everything is included. No flaws if serviced and looked after. Just wear and tear, brakes and suspension. Any problems don't bother.

Engines: 4.3 **CO2:** 270g/km **MPG**:24.8 **BHP:** 278
0-60: 6.1-6.5 **Top Speed**: 155 **Boot:** 573 **Weight**: 1795

Lexus LS430 2007 - 2017

Giant Camry air of ordinariness. the 600h is the hybrid flavour, being a Toyota make it dependable. Premier model means colour special spec options and an F-Sport for fun. If you were going to take a risk on a high tech Banger the LS is it. Caution required when hybrid powerplants as they must have finite lives, and a cost benefit analysis will be needed.

Engines: 4.6 • 5.0H **CO2:** 219-261g/km **MPG**: 25.4-30.4 **BHP**: 381-445 **0-60**: 6.3 **Boot:** 330/505 **Turning**: 10.4

Mercedes S-Class 1999 - 2005

Difficult time for Stuttgart, but the S-Class was better than a pretty ordinary bunch. Needs to be looked after otherwise deteriorates rapidly. Most examples are the 320 diesel. Longer models are available for private hire. V8 petrols and good standard kit. Oil and coolant leaks, bodywork and suspension corrosion. ECU and electrical failure.

Engines: 2.8 • 3.7 • 4.3 • 5.0 • 5.5 • 3.2D **CO2:** 204-355g/km

MPG: 19.1-36.7 **BHP:** 197-612 **0-60:** 4.3-9.5 **Boot:** 560

Mercedes S-Class 2006 - 2013

Properly back on form with this one. Big in all the right areas, including the boot. 320 diesel is there of course, but this is yet another example of the frighteningly complex Banger purchase. Seems to have had its unfair share of recalls so you really do need a service history before buying. Just heavy use items suspension, brakes tyres etc.

Engines: 2.8 • 3.5 • 5.4 • 3.0D **CO2:** 220-279g/km **MPG**: 24.1-34.0 **BHP**: 227-603 **0-60**: 6.9-7.6 **Boot:** 560

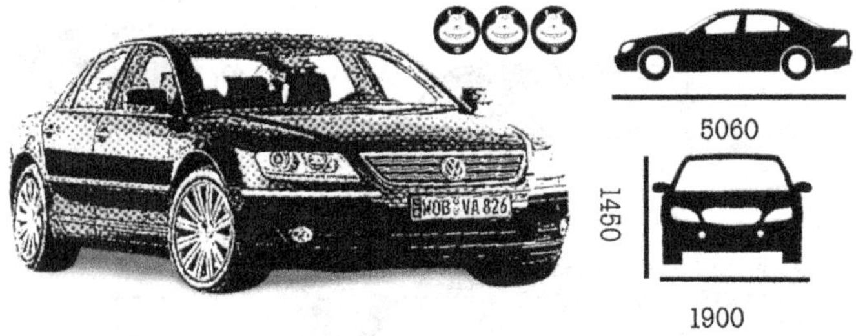

Volkswagen Phaeton 2003 - 2015

An inexplicably luxurious Passat, or a working class Bentley. Buttons to buy, comes with big car running costs. Everyone loves the diesels and 4-Motion on everything except the petrol V6. It is an odd choice, but likely to be a value and price point purchase. Niggly electrics sometimes, feedback from second third, fourth owners is positive.

Engines: 3.2 • 4.2 • 6.0 • 3.0D • 5.0D **CO2:** 239-374g/km **MPG**: 21.6-29.4 **BHP:** 221-443 **0-60:**6.1-8.3 **Boot:** 500

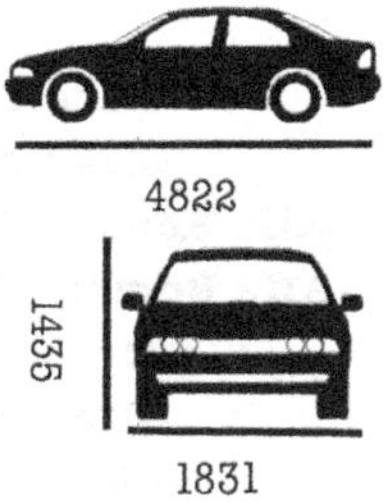

Volvo S80 1998 - 2005

Size matters and they don't come much bigger than the S80. For those who want lots of equipment and a comfortable way to cover long distances this is the perfect vehicle. Underrated and ignored for not being posh enough it is a proper Swedish tank that will rack up massive mileages. Suspension wear, diesel ECUs and general electrics.

Engines: 2.0 • 2.9 • 2.4D **CO2:** 172-232g/km **MPG**: 29.1-43.5 **BHP:** 140-272 **0-60:** 7.1 - 9.9 **Boot:** 460 **Weight**: 1670-1708

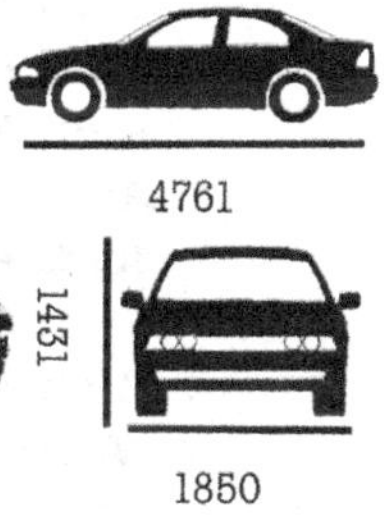

Volvo S80 2006 - 2016

Front wheel drive. It continued being comfortable. Surprisingly the boot is not massive, at least the rear seats fold. More technology to go wrong from 2013 with cyclist/ pedestrian detection systems. Slightly restyled in 2009 and later on in 2013, but no one noticed or cared. Blocked Diesel filters, ECU rough engine running, complicated electrics.

Engines: 2.5 • 4.4 • 1.6D • 2.0D • 2.4D **CO2:** 151-206g/km **MPG**:30.7-49.6 **BHP:** 107- 310 **0-60**:6.0-12.4 **Boot:** 422

Bangindex by Car Builder

James Ruppert still regrets ever starting this Bangerpedia and committing to a whole series of these volumes. Always interested in motoring history (grade 1 CSE), he bought all sorts of MOT borderline Bangers that were prehistorically awful. Once sold cars then switched to writing about used ones. Spent many years driving around the country looking at car lots and talking to blokes who sold them. Wrote for newspapers including The Independent, The Sunday Times and Evening Standard. Best known as a writer for Car magazine, Autocar, Performance Car, Supercar Classics and many others. Dreamt up the word Bangernomics in 1990. Founded Free Car Mag in 2015 the only pro motorist publication. Believes we are being Demotorized. Currently banned from being in the mainstream motoring media. Lives in The Rabbit Hole Underground Car Park. Banging on about marginal motoring only makes a marginal living. Please consider the following titles for inclusion on your bookshelves to help pay for food, fuel and automotive tat.

Order Books from www.bangernomics.com which takes you to flyscreenqueen.co.uk

Stories on Substack as Bangernomics

www.freecarmag.com

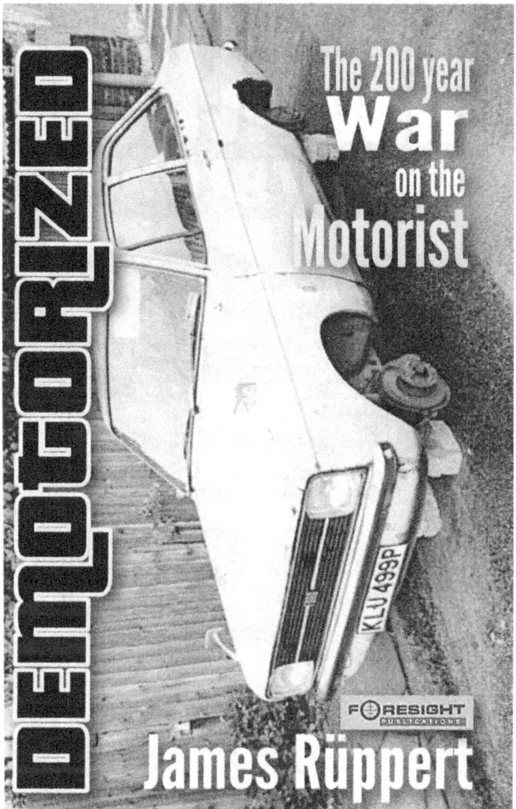

Demotorized. This is the not quite definitive history of how parking became an industry, governments overtaxed motorists, carmakers lied about how safe and environmentally friendly they were and tested their emissions on monkeys and humans. Also why robot cars are so dangerous and electric cars are not the answer. Also, who is behind the ongoing plan to Demotorize the world. Printed in 2020 it predicted the future with terrifying accuracy.

Buy it before it gets banned.

Rabbit Hole Underground Car Park Join the Conspiracy Rabbit as he goes deep, deep undercover to find out why fuels are not fossil, cars that run on water, how those Germans survived WW2, a Jeep that went to the moon, how evil the World Ecommunist Forum is, how to win a war with a pick up truck, why Commies can't build cars, but the Chinese can copy them, all that big trouble in Japan, them electricity cars being a gateway drug, why Castro could not ban cars, but then again who really owns your motor, while at the end of the day: Yuri Bezmenov was right.

The Award Winning and so good the BBC based a documentary (Das Auto) on it **The German Car Industry, My Part in its Victory.** Winner of the Mercedes-Benz Montagu of Beaulieu Trophy for the best motoring book of year. Contains Yuppies, BMW 3, 5, 6 & 7-Series (yes that's all there was back then) and references to contemporary 1980s music.

The Critically acclaimed and best selling **The British Car Industry Our Part in Its Downfall**, which has been described as "Beautifully written with considerable wit" by the Daily Telegraph and The Daily Express said, "It's informative, well-written and extremely funny" which was really nice of them. The story of the motor industry from 1945 to 2005, explaining just why you can't buy a British built family hatchback from a British owned company. Plus all the cars, James's Dad bought.

Channeling the absurdist spirit of Spike Milligan" said Classic Car, others just called it "Witty", that's **My Mini Cooper Its Part in My Breakdown** which has the full history of the world's favourite small car, plus the history of what came before but not after. Also details the on-off restoration of a Mini Cooper that cost £200 in 1979. The expense of restoring the bodywork and mechanicals ought to put off most sane people. Except that original minis are always worth far more than you could imagine.

Ruppert's Modern Bangerpedia seems to prove that these cars are indeed rubbish. This may contain hundreds of models, over 600, but it is far from complete as any electric ones, not hybrids, have been edited out in the interests of commonsense as no Battery car will ever become a Banger. Slog The Bangernomic Dog still provides a ridiculous rating system. There is an almost definitive indication as to whether it will fit into a modern garage, or not.

Ruppert's Bangerpedia Modern Classic covers the epic '90s decade which was a golden period for Bangers. The Bangerpedia does this with a fairly ridiculous rating system involving Slog the Bangernomic Dog. The briefest of guides to each of the 528+ models includes some fairly marginal buying advice, a sarcastic appraisal and an almost definitive indication as to whether it will fit into your garage, or not.

Ruppert's Bangerpedia Practical Classic proves that old cars don't die they just become Classics. The 1980s is whne everyday motors came good. Euroboxes were everywhere, the VW Golf was the best car in the world, BMW 3 Series set the executive car standard, Datsuns turned into Nissans and the British Car industry turned Japanese. As usual there's a fairly ridiculous rating system involving Slog the Bangernomic Dog.

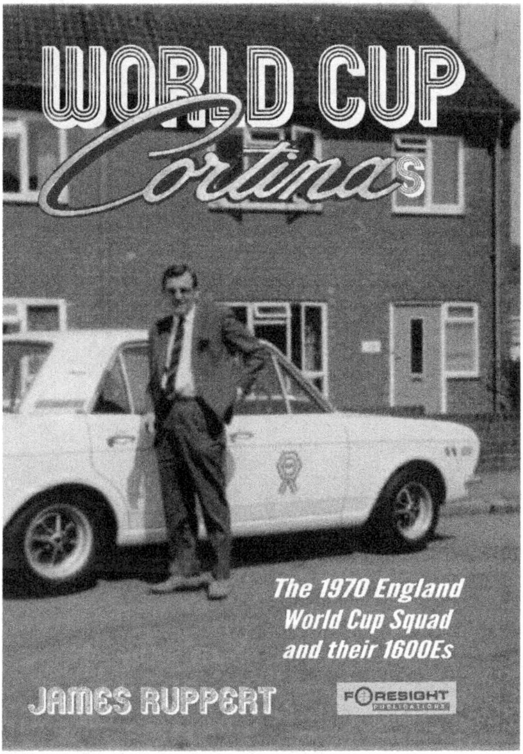

World Cup Cortinas In 1970 something very remarkable happened. Not only had England qualified for the World Cup in Mexico because they were defending champions, but most important of all Ford had loaned every squad member the greatest saloon car in the world. A book about what happened to those Ford Cortinas and how footballers bought better cars when they earned more money and no longer owned the same motors that your mum and dad drove.

Free Car Mag - 0-100 (issues) in 6.9 Years A very brief appreciation of the art of printing car magazines in the UK and a 'how to' guide of how not to do it. In many ways just an excuse to reprint all of our amazing covers in one place purely for reference purposes. It also chronicles how the Mag went from commercial to controversial.

Autofutroplis What if BMC had not been so terrible, that Elvis ran Ford, levitating motors were a reality along with steam powered gentleman's carriages and an electricity car designed by and for gentleladies?...Well, wonder no more. Here is a fully illustrated book which rewrites and redraws automotive history.

Bangernomics The Board Game Here is a book that thinks it is a board game. Well, you have to take it to bits, but the cover is the actual playing service. All you need to do is pinch a few Matchbox scale models from your toy box and a dice from a proper game like Monopoly. Then set about destroying the book which contains playing cards with Bangers on them. Simply chase your mates around the board)cover of the book and follow the daft instructions. The idea is to get the most Bangers and not to go bankrupt in the process. A bit like real marginal motoring. Fun for all the family. Probably. Contains a fake currency which is not exactly legal tender. You can download all the parts to play, apart from the model car and dice, from Bangernomics.com

Bangernomics Diet is a faux academic course designed to get you a Doctorate on the cheap. It is dead easy, just pay your money and fast forward to the end where you can tear out and frame the Proficiency Certificate, or print it off if you have the digital one. You can then call yourself a Doctor of Bangernomics. Lots of scanned articles from Bangernomics Mag and Free Car Mag, so buy the cheap digital one.

In the beginning, 1993, there was Bangernomics and now there is **Bangernomics Rebooted** which contrasts the absurd expense of buying a new car with the supreme good sense of buying well used. At a stroke depreciation no longer became an issue, running costs were slashed and there were no finance charges to be endured. Here is the original book, republished and effectively rebooted. So it is absolutely useless to any car buyer unless they have access to a time machine.

The world of motoring is in a desperate crisis. Demonised, despised and attacked by every wrong thinking government, local authority, pressure group and even car manufacturers themselves. Bangernomics is here to bring motoring back to the masses, saving time, money and automotive headaches. Squashing together the original Bangernomics, Bangernomics Bible and Bangernomics Diet into one easily digestible guide to buying and running an older cost efficient motor. Only **Bangernomics can save the World.**

There is a new version of the **Bangernomics Bible**. Most of it is now in the Bangernomics Will Save the World so this has become the top 200 Bangernomic tips. Which may change from time to time.

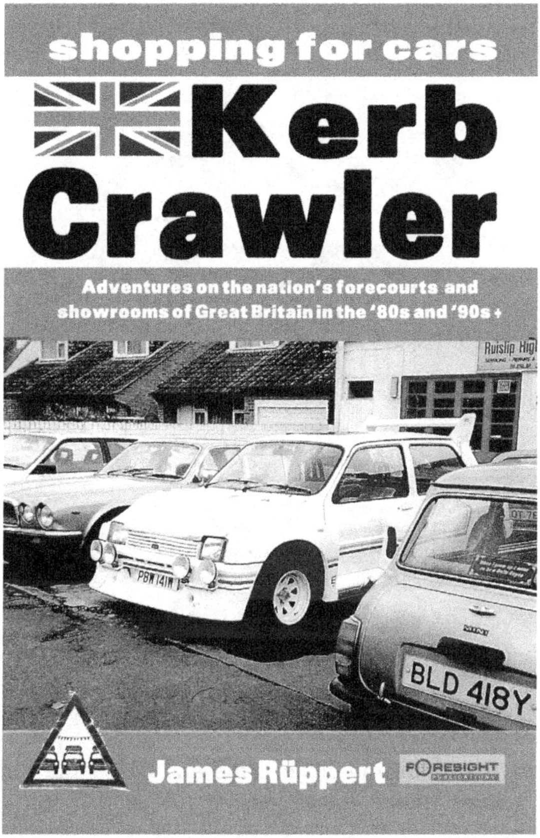

Shopping for cars can be a full time job, well it was for me so here's **Kerbcrawler** In the pre Interweb age it meant actually going to look at cars for sale. It meant talking to the blokes selling the cars face to face. Find out how I got sued by one of the 'stars' of the Italian Job, was told off by the Deputy Prime Minister, find out what the worst car Sir Ian Botham ever owned was, and that time I lent my car to Bjork. A book to dip in and out of...

He was a man running out of time and that time was 1973. **Spencer Haze** is a comic superhero who first hit the 21st century in the year 2000. First on the 4Car website, then later in national newspapers and car magazines. Some say he was even on the television and might well have been real. Here is just about every frame he ever squeezed into, plus a TV script and random pictures that have never been seen before. Here is everything possibly no one ever wanted to know about a fictional bloke called Spencer who loved cars, people and the absolute truth. Here's his **Comicarpocalypse**

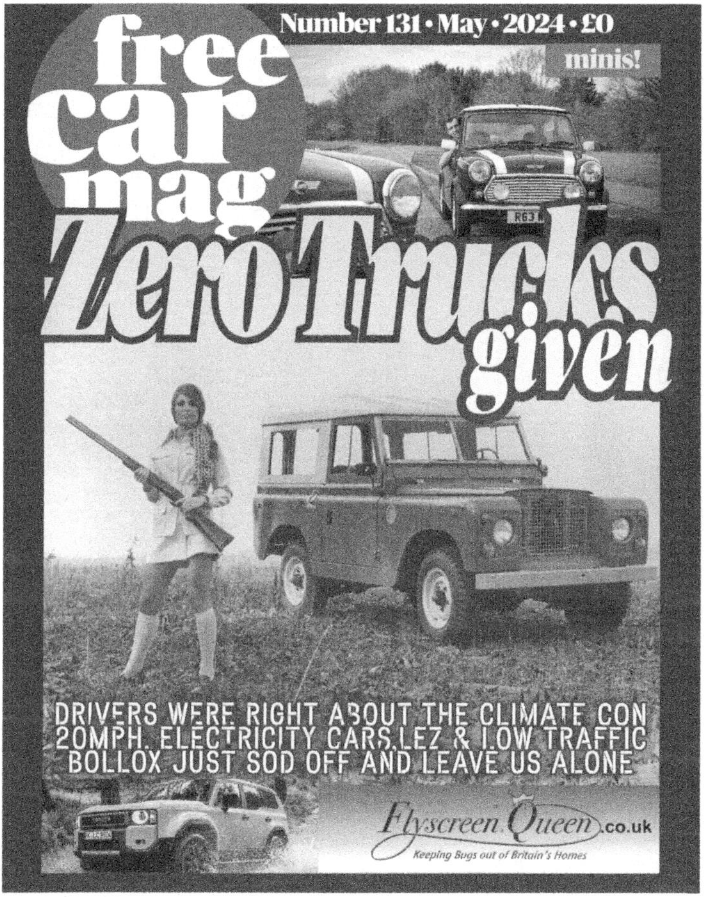

The only pro-motorist car mag not in captivity*It costs nothing, but contains everything you need to be a more anarchistic, bolshie and less co-operative car driver.* **www.freecarmag.com**

Made in the USA
Monee, IL
04 November 2026

67377560R00233